The Growth of AI Data Centers and the Impact on Global Water Resources

Hitesh Mohapatra
https://orcid.org/0000-0001-8100-4860
School of Computer Engineering, Kalinga Institute of Industrial Technology (Deemed to be) University, Bhubaneswar, India

Amiya Kumar Rath
Biju Patnaik University of Technology (BPUT), Chhend Colony, Rourkela, India & Veer Surendra Sai University of Technology (VSSUT), Burla, Sambalpur

Fadi Al-Turjman
https://orcid.org/0000-0001-6375-4123
AI and Software Engineering Dept., AI and IoT Research Center, AI and Informatics Faculty, Near East University, Mersin, Turkey

Manjur Kolhar
Department of Health Information Management and Technology, College of Applied Medical Sciences, King Faisal University, Al-Ahsa, Saudi Arabia

IGI Global
Scientific Publishing
Publishing Tomorrow's Research Today

Vice President of Editorial — Melissa Wagner
Director of Acquisitions — Mikaela Felty
Director of Book Development — Jocelynn Hessler
Production Manager — Mike Brehm
Cover Design — Jose Rosado

Published in the United States of America by
IGI Global Scientific Publishing
701 East Chocolate Avenue
Hershey, PA, 17033, USA
Tel: 717-533-8845 | Fax: 717-533-7115
Website: https://www.igi-global.com E-mail: cust@igi-global.com

Library of Congress Cataloging-in-Publication Data

Names: Mohapatra, Hitesh, 1985- editor | Rath, Amiya Kumar, 1968- editor | Al-Turjman, Fadi, 2003- editor | Kolhar, Manjur editor
Title: The growth of AI data centers and the impact on global water resources / edited by: Hitesh Mohapatra, Amiya Kumar Rath, Fadi Al-Turjman, Manjur Kolhar.
Description: Hershey PA : IGI Global Scientific Publishing, [2027] | Includes bibliographical references. | Summary: "This book aims to raise awareness and encourage informed decision-making about the environmental impacts of AI data center operations"-- Provided by publisher.
Identifiers: LCCN 2026005947 (print) | LCCN 2026005948 (ebook) | ISBN 9798337384344 hardcover | ISBN 9798337384351 paperback | ISBN 9798337384368 ebook
Subjects: LCSH: Data centers--Environmental aspects | Water-supply
Classification: LCC TJ163.5.D38 G76 2026 (print) | LCC TJ163.5.D38 (ebook)
LC record available at https://lccn.loc.gov/2026005947
LC ebook record available at https://lccn.loc.gov/2026005948

British Cataloguing in Publication Data
A Cataloguing in Publication record for this book is available from the British Library.

This book is dedicated to the researchers, innovators, policymakers, and environmental stewards around the world who strive to advance technological progress while safeguarding our planet's most precious resources.

We dedicate this volume to present and future generations whose well-being depends upon the responsible management of water resources and the sustainable development of emerging technologies. As artificial intelligence continues to reshape society, may our collective pursuit of innovation remain guided by principles of environmental responsibility, social equity, and global sustainability.

We also dedicate this work to all scholars and practitioners committed to building a future where technological advancement and ecological preservation coexist in harmony for the benefit of humanity and the Earth.

The Editors

Table of Contents

Preface xi

Acknowledgment xiii

Introduction xv

Chapter 1
AI Infrastructure Expansion and Water-Energy Sustainability Challenges 1
Rampelli Manojkumar, BVRIT HYDERABAD College of Engineering for Women, India

Chapter 2
AI Data Centers in Cooling Mechanisms and Freshwater Consumption: Challenges, Technologies, and Sustainable Pathways 33
Oiendri Mukhopadhyay, Kalinga Institute of Industrial Technology, India
Banasree Maji, Kalinga Institute of Industrial Technology, India
Hitesh Mohapatra, School of Computer Engineering, Kalinga Institute of Industrial Technology (Deemed to be University), Bhubaneswar, India

Chapter 3
Cooling Technologies and Comparative Sustainability Studies: AI Workloads, High-Performance Computing, and Energy-Water Relationships 61
Riadh Hanachi, Laboratory Desalination and Natural Water Valorization, Center of Water Researches and Technologies (CERTE), Tunisia
Emna Melliti, Laboratory Desalination and Natural Water Valorization, Center of Water Researches and Technologies (CERTE), Tunisia
Hamza Elfil, Laboratory Desalination and Natural Water Valorization, Center of Water Researches and Technologies (CERTE), Tunisia

Chapter 4
Harnessing Predictive Analytics for Efficient Water Resource Management in Data Center Cooling Systems 113
Robinson Joel, KCG College of Technology, India

Chapter 5
Predictive Analytics for Water Management in Data Center Operations 145
Shanthini Sampath, SRM Institute of Science and Technology, India
D. Rajeswari, SRM Institute of Science and Technology, India
J. Dhanalakshmi, SRM Institute of Science and Technology, India
M. G. Sumithra, Sri Krishna College of Technology, India

Chapter 6
Water Scarcity Challenges in Regions Hosting Large Cloud Infrastructures ... 181
Emna Melliti, Center of Water Researches and Technologies, Tunisia
Riadh Hannachi, Center of Water Researches and Technologies, Tunisia
Nesrine Kamoun, Center of Water Researches and Technologies, Tunisia
Hamza Elfil, Center of Water Researches and Technologies, Tunisia

Chapter 7
The Hidden Cost of Intelligence: Water Sustainability in AI-Driven Data Ecosystems 219
Sonali Srivastava, Jaypee Institute of Information Technology, Noida, India
Manisha Singh, VIT Business School, VIT Bhopal University, Bhopal, India

Chapter 8
Water as a Structural Requirement for AI Data Centers: Thermodynamics, Workload Intensity, and the Limits of Air Cooling 247
Krishna Chaitanya Sunkara, Oracle, USA

Conclusion 283

Compilation of References 287

About the Contributors 309

Index 315

Detailed Table of Contents

Preface xi

Acknowledgment xiii

Introduction xv

Chapter 1
AI Infrastructure Expansion and Water-Energy Sustainability Challenges 1
Rampelli Manojkumar, BVRIT HYDERABAD College of Engineering for Women, India

The rapid expansion of AI-driven data centers has intensified concerns beyond energy consumption, particularly regarding water use for cooling. This chapter examines how high-density AI workloads increase thermal loads, driving demand for water-intensive cooling systems and expanding both direct and indirect water footprints. It analyzes the water–energy nexus, compares cooling technologies, and highlights impacts in water-stressed regions. Regulatory gaps and sustainability metrics such as WUE are discussed alongside emerging solutions, including advanced cooling, water reuse, and AI-based optimization. The chapter provides an integrated perspective to support sustainable and responsible AI infrastructure development.

Chapter 2
AI Data Centers in Cooling Mechanisms and Freshwater Consumption: Challenges, Technologies, and Sustainable Pathways 33
Oiendri Mukhopadhyay, Kalinga Institute of Industrial Technology, India
Banasree Maji, Kalinga Institute of Industrial Technology, India
Hitesh Mohapatra, School of Computer Engineering, Kalinga Institute of Industrial Technology (Deemed to be University), Bhubaneswar, India

The rapid growth of AI infrastructure has increased freshwater demand in global data centers due to the cooling requirements of high-density GPU clusters. While carbon emissions from AI systems receive major regulatory attention, water consumption remains largely overlooked despite its impact on water-stressed communities and ecosystems. This chapter examines the hidden water footprint of AI data centers across three scopes: direct cooling water use (Scope 1), indirect water use from electricity generation (Scope 2), and embodied water in hardware manufacturing (Scope 3).

It reviews cooling technologies such as air cooling, liquid cooling, and immersion cooling. It also analyzes intelligent optimization methods, including reinforcement learning controllers, digital twins, and IoT-based predictive maintenance. Regional disparities are explored with emphasis on water-stressed regions and Africa. Policy measures and future research directions are discussed for sustainable AI infrastructure.

Chapter 3
Cooling Technologies and Comparative Sustainability Studies: AI Workloads, High-Performance Computing, and Energy-Water Relationships 61
Riadh Hanachi, Laboratory Desalination and Natural Water Valorization, Center of Water Researches and Technologies (CERTE), Tunisia
Emna Melliti, Laboratory Desalination and Natural Water Valorization, Center of Water Researches and Technologies (CERTE), Tunisia
Hamza Elfil, Laboratory Desalination and Natural Water Valorization, Center of Water Researches and Technologies (CERTE), Tunisia

This chapter examines the sustainability challenges of AI-intensive and high-performance computing (HPC) data centers, focusing on cooling technologies and their impacts on energy and water use. It situates these infrastructures within the broader energy–water nexus and compares major cooling approaches—air, water-based, and liquid—using key metrics such as PUE, WUE, and CUE. The chapter highlights how the rapid growth of AI workloads, marked by high power densities and dynamic thermal constraints, increases resource demand and complicates thermal management. It also explores regional variations, showing how climate, water availability, and electricity mix affect sustainability outcomes. By integrating technical, environmental, and policy perspectives, it identifies trade-offs between energy efficiency and water consumption and emphasizes context-aware design. Ultimately, it argues that combining thermal modeling, energy optimization, hydrological analysis, and carbon assessment supports more informed infrastructure and location decisions.

Chapter 4
Harnessing Predictive Analytics for Efficient Water Resource Management in Data Center Cooling Systems 113
Robinson Joel, KCG College of Technology, India

Efficient water resource management has become a critical challenge in modern data centers due to increasing computational demands and environmental concerns. This study explores the application of predictive analytics to optimize water usage in data center cooling systems. By leveraging historical sensor data, machine learning models, and real-time monitoring, the proposed approach forecasts cooling demand and dynamically adjusts water consumption. Techniques such as time-series analysis and regression models are employed to predict temperature fluctuations, workload variations, and cooling requirements. The system enhances decision-making by identifying inefficiencies, detecting anomalies, and enabling proactive maintenance. Experimental results indicate significant reductions in water wastage while maintaining optimal thermal performance and system reliability. Furthermore, the integration of predictive analytics supports sustainable operations by minimizing environmental impact and operational costs. This research demonstrates that data-driven water management strategies.

Chapter 5
Predictive Analytics for Water Management in Data Center Operations 145
Shanthini Sampath, SRM Institute of Science and Technology, India
D. Rajeswari, SRM Institute of Science and Technology, India
J. Dhanalakshmi, SRM Institute of Science and Technology, India
M. G. Sumithra, Sri Krishna College of Technology, India

The fast development of AI has grown the size and the concentration of data centers and raised the alarm about their water footprint. The cooling of high-performance computing infrastructure uses a lot of water and new facilities are usually placed in areas that are already experiencing water pressure. This chapter discusses predictive analytics as a solution to proactive and adaptive water management in such settings. Predictive systems optimize future cooling demands, water use, and supply risk by using historical operational data streams, real-time sensor feed, weather predictions, and machine learning classifiers. This reactive control to anticipatory planning helps the operations to be more stable, has less avoidable water use, and is more resilient to climate variability and changing workloads. It has been discussed how short-term forecasting, long-term capacity planning, and scenario-based risk assessment is related to governance and regulatory reporting and sustainability goals and has covered the issue of data quality, uncertainty, and challenges of integration.

Chapter 6
Water Scarcity Challenges in Regions Hosting Large Cloud Infrastructures ... 181
Emna Melliti, Center of Water Researches and Technologies, Tunisia
Riadh Hannachi, Center of Water Researches and Technologies, Tunisia
Nesrine Kamoun, Center of Water Researches and Technologies, Tunisia
Hamza Elfil, Center of Water Researches and Technologies, Tunisia

Cloud computing Cloud computing has transformed the global digital economy, driven by the rapid expansion of hyper scale data centers. While these facilities enable scalable storage, real-time analytics, and AI deployment, their environmental footprint, especially water consumption, has become a critical issue. In fact, Data centers require permanent cooling systems, often relying on fresh water-intensive resources that consume millions of liters annually, placing tension on local water resources, especially in water-stressed regions. This chapter examines the connection of data centers growth and water scarcity, analyzing world distribution patterns, conventional and non-conventional cooling technologies, and sectoral water competition. This study also underlines sustainable strategies, which include water-efficient cooling, reuse systems, alternative sources, and policy interventions. It argues that integrated planning, technological innovation, and governance are essential to guarantee resilient and sustainable cloud development.

Chapter 7
The Hidden Cost of Intelligence: Water Sustainability in AI-Driven Data Ecosystems 219
Sonali Srivastava, Jaypee Institute of Information Technology, Noida, India
Manisha Singh, VIT Business School, VIT Bhopal University, Bhopal, India

The fast development of artificial intelligence (AI) has changed the world industries and has intensified healthcare, financial, manufacturing, education, and government services innovations. However, the hidden side of this mighty digital revolution is a very simple environmental fact: it takes a ton of water to cool AI data centers. This chapter is a critical analysis of the invisible water footprint of data ecosystems in the digital age powered by AI, and why the digital age is in dire need of sustainable water management strategies. The chapter suggests that to have sustainable AI the world needs a holistic approach that balances the growth of technologies and the ecological concern. To ensure that the water-smart data ecosystems will have minimal ecological footprint without reducing the pace of AI innovation, cooperation between technology firms, policymakers, environmental scientists, and the local communities is necessary.

Chapter 8
Water as a Structural Requirement for AI Data Centers: Thermodynamics, Workload Intensity, and the Limits of Air Cooling 247
Krishna Chaitanya Sunkara, Oracle, USA

AI data centers do not use water because they are inefficient; they use it because physics demands it. GPU-heavy workloads turn most electrical power into heat at the silicon layer, and that heat has to be removed continuously. Once rack densities cross roughly 40–50 kW, air cooling struggles to keep systems stable, so liquid cooling becomes the practical path. A large share of facility energy already goes into cooling, and it keeps rising as AI shifts to long training runs and steady inference. Water here is not a consumable, it acts as a heat transport medium, carrying energy away from the rack. That energy is not lost and can often be reused. This chapter looks at water through thermodynamics and real rack power, using measures like liters per inference and temperature rise to show that water demand follows workload intensity and time, not just facility size. Unlike prior work that treats cooling as a secondary optimization problem, this analysis frames it as a first order constraint coupled directly to workload duration and compute intensity

Conclusion 283

Compilation of References 287

About the Contributors 309

Index 315

Preface

ABOUT THIS BOOK

The rapid advancement of artificial intelligence (AI) has ushered in a new era of computational demand, fundamentally transforming the global digital infrastructure landscape. At the heart of this transformation lies the unprecedented growth of AI data centers massive computational hubs that power machine learning models, generative AI systems, and intelligent applications across industries. While these data centers enable technological progress, innovation, and economic development, they also introduce critical environmental challenges that demand urgent attention.

One of the most pressing yet underexplored concerns is the substantial water footprint associated with AI-driven data centers. Water is extensively used for cooling high-performance computing systems, maintaining operational efficiency, and preventing thermal degradation. As the scale and density of AI workloads increase, so does the demand for water resources, often in regions already facing water scarcity. This raises important questions about sustainability, resource allocation, and the long-term ecological implications of AI infrastructure expansion.

This book aims to provide a comprehensive examination of the growth of AI data centers and their impact on global water resources. It brings together interdisciplinary perspectives, combining insights from computer engineering, environmental science, and policy studies. The discussion extends beyond technical architecture to include water usage metrics, geographic disparities, regulatory frameworks, and emerging sustainable practices such as liquid cooling innovations, renewable energy integration, and water-efficient system design.

In addition, the book highlights the role of intelligent optimization techniques including artificial intelligence itself in mitigating the environmental footprint of data centers. Concepts such as energy-aware scheduling, reinforcement learning-based resource allocation, and edge computing paradigms are explored as potential solutions to balance performance with sustainability. Through case studies, empirical analyses, and forward-looking discussions, the book seeks to bridge the gap between technological progress and environmental responsibility.

This work is intended for researchers, policymakers, industry practitioners, and students who are interested in understanding the complex interplay between AI infrastructure and natural resource sustainability. By shedding light on the hidden environmental costs of digital intelligence, the book aspires to encourage more responsible innovation and informed decision-making in the design and deployment of future AI systems.

Ultimately, the growth of AI must be aligned with the principles of sustainability and stewardship. As we stand at the intersection of technological evolution and environmental constraint, it is imperative to rethink how we build, operate, and regulate the digital ecosystems that increasingly shape our world.

Acknowledgment

The successful completion of this edited volume, *The Growth of AI Data Centers and the Impact on Global Water Resources*, would not have been possible without the support, expertise, and dedication of numerous individuals and organizations.

First and foremost, we express our sincere gratitude to all contributing authors whose scholarly efforts, innovative research, and commitment to excellence have enriched this volume. Their valuable contributions have enabled a comprehensive exploration of the complex relationship between artificial intelligence infrastructure and global water sustainability.

We are deeply indebted to the reviewers who generously devoted their time and expertise to the rigorous evaluation of submitted chapters. Their constructive feedback and critical insights played a vital role in maintaining the academic quality, relevance, and integrity of this publication.

We would also like to extend our appreciation to the editorial and publishing team at IGI Global for their continuous guidance, professionalism, and support throughout the development of this project. Their assistance has been instrumental in bringing this volume to fruition.

Special thanks are due to our fellow editors for their collaborative spirit, dedication, and unwavering commitment during the chapter review, selection, and editorial processes. The successful completion of this volume reflects the collective efforts of a team united by a shared vision of advancing knowledge at the intersection of artificial intelligence, environmental sustainability, and resource management.

We also acknowledge the researchers, practitioners, policymakers, and institutions worldwide whose ongoing work continues to advance understanding of sustainable AI development and responsible water governance. Their contributions have inspired many of the discussions and perspectives presented in this book.

Finally, we express our gratitude to our families, colleagues, and professional networks for their encouragement, patience, and support throughout this endeavor.

It is our hope that this volume will contribute meaningfully to academic research, industry practice, and policy discussions concerning the sustainable growth of AI infrastructure and the responsible stewardship of global water resources.

The Editors

Introduction

Artificial Intelligence (AI) has emerged as one of the most transformative technologies of the twenty-first century, driving innovation across industries, economies, and societies. From generative AI and machine learning applications to autonomous systems and predictive analytics, AI technologies are increasingly shaping how organizations process information, make decisions, and deliver services. Behind these technological advancements lies a vast and expanding digital infrastructure comprising data centers, cloud computing platforms, and high-performance computing systems that power the AI revolution.

As the demand for AI capabilities continues to grow exponentially, so too does the scale and complexity of the data centers required to support them. Modern AI workloads require significant computational power, energy consumption, and sophisticated cooling systems to maintain operational efficiency and reliability. While much attention has focused on the carbon footprint and energy requirements of AI infrastructure, another critical resource has received comparatively less scrutiny: water.

Water plays a fundamental role in the operation of data centers. Cooling systems used to dissipate heat generated by servers and computing equipment often rely on substantial quantities of freshwater. As AI applications become more computationally intensive, the demand for water-intensive cooling solutions is increasing, raising important questions about resource sustainability, environmental stewardship, and equitable access to water resources. In many regions, particularly those already experiencing water stress, the expansion of AI data centers may intensify competition for limited freshwater supplies among industrial, agricultural, ecological, and domestic users.

The intersection of AI development and water resource management represents an emerging area of global concern. Governments, industry leaders, researchers, and environmental organizations are increasingly recognizing that the sustainability of digital infrastructure must be evaluated not only in terms of energy efficiency and greenhouse gas emissions but also through the lens of water consumption and resource governance. As climate change continues to affect water availability across many parts of the world, understanding the water implications of AI infrastructure has become both a scientific necessity and a policy imperative.

This edited volume, *The Growth of AI Data Centers and the Impact on Global Water Resources*, seeks to address these pressing challenges by bringing together diverse perspectives from researchers, practitioners, policymakers, and experts across multiple disciplines. The volume explores the environmental, technological, economic, legal, and governance dimensions of AI-driven data center expansion and its implications for global water sustainability.

The chapters included in this volume collectively examine the evolving relationship between AI technologies and water resources through a variety of analytical lenses. Contributors investigate emerging cooling technologies, predictive analytics applications, sustainable resource management strategies, environmental impacts, regulatory frameworks, and governance mechanisms aimed at balancing technological advancement with responsible resource stewardship.

A central theme throughout the book is the recognition of the water-energy nexus that underpins AI infrastructure. Data centers require substantial energy to perform computational tasks and significant water resources to maintain operational temperatures. Understanding these interdependencies is essential for developing sustainable solutions that minimize environmental impacts while supporting continued innovation. Several chapters explore how advances in cooling technologies, resource optimization techniques, and predictive analytics can contribute to more sustainable data center operations.

Another important focus of this volume is the application of artificial intelligence itself as a tool for improving water management. AI-powered systems increasingly support water resource monitoring, demand forecasting, infrastructure management, and sustainability planning. These developments present opportunities to enhance resource efficiency while addressing broader environmental challenges. The book therefore examines both sides of the relationship: AI as a driver of water demand and AI as an enabler of smarter water resource management.

The volume also addresses critical questions surrounding governance, policy, and environmental justice. As major technology companies expand data center operations into regions with varying levels of water availability, concerns regarding transparency, accountability, community impacts, and equitable resource allocation have become increasingly prominent. Several chapters investigate legal frameworks, policy responses, and governance models designed to ensure that the benefits of digital transformation do not come at the expense of environmental sustainability or social equity.

The eighteen chapters included in this volume are organized around interconnected themes that collectively provide a comprehensive understanding of the subject.

The opening chapters establish the broader context of AI infrastructure growth and the sustainability challenges associated with increasing energy and water demands. Subsequent chapters examine technological innovations in cooling systems, resource-efficient computing, and predictive analytics for water management. Several contributions focus on freshwater consumption patterns, ecological impacts, and sustainability assessments of AI-driven data ecosystems. Additional chapters explore water scarcity concerns in regions hosting large cloud infrastructures and investigate strategies for balancing technological development with environmental conservation.

The latter part of the volume addresses governance, regulatory frameworks, legal challenges, and policy implications associated with AI data center expansion. These discussions highlight the need for integrated approaches that combine technological innovation, environmental management, stakeholder engagement, and evidence-based policymaking. Together, the chapters provide readers with a multidimensional perspective on one of the most significant sustainability issues emerging from the global AI revolution.

This volume is intended for a broad audience, including researchers, academics, policymakers, environmental professionals, engineers, technology practitioners, sustainability experts, and graduate students interested in the intersection of artificial intelligence, digital infrastructure, and natural resource management. By bridging technical, environmental, and policy perspectives, the book aims to foster interdisciplinary dialogue and support informed decision-making regarding the future of AI development.

As the world enters an era increasingly shaped by artificial intelligence, ensuring the sustainability of the infrastructure that powers these technologies will be among the defining challenges of our time. The future of AI must be built not only on computational efficiency and technological innovation but also on responsible stewardship of the natural resources upon which digital systems depend.

It is our hope that this volume will contribute to ongoing research, informed policy discussions, and practical solutions that advance both technological progress and environmental sustainability. By examining the hidden yet critical relationship between AI data centers and global water resources, this book seeks to encourage a more holistic understanding of what it truly means to build a sustainable digital future.

The Editors

Chapter 1
AI Infrastructure Expansion and Water–Energy Sustainability Challenges

Rampelli Manojkumar
http://orcid.org/0000-0002-8141-0321
BVRIT HYDERABAD College of Engineering for Women, India

ABSTRACT

The rapid expansion of AI-driven data centers has intensified concerns beyond energy consumption, particularly regarding water use for cooling. This chapter examines how high-density AI workloads increase thermal loads, driving demand for water-intensive cooling systems and expanding both direct and indirect water footprints. It analyzes the water–energy nexus, compares cooling technologies, and highlights impacts in water-stressed regions. Regulatory gaps and sustainability metrics such as WUE are discussed alongside emerging solutions, including advanced cooling, water reuse, and AI-based optimization. The chapter provides an integrated perspective to support sustainable and responsible AI infrastructure development.

INTRODUCTION

The rapid proliferation of artificial intelligence (AI) technologies has fundamentally reshaped the architecture, scale, and operational dynamics of modern digital infrastructure. At the core of this transformation lies the exponential expansion of data centers, which serve as the computational backbone for AI model training, inference, and large-scale data processing. Unlike conventional computing

DOI: 10.4018/979-8-3373-8434-4.ch001

workloads, AI-driven applications—particularly deep learning, generative AI, and large language models—require significantly higher computational intensity, leading to increased power densities and continuous operational loads. This shift has elevated data centers from passive digital storage facilities to highly dynamic, resource-intensive infrastructures that operate at the intersection of energy, water, and environmental sustainability.

Recent studies indicate that modern hyperscale data centers typically operate with Power Usage Effectiveness (PUE) values in the range of 1.1–1.6, reflecting improved energy efficiency. However, water consumption remains highly variable, with Water Usage Effectiveness (WUE) ranging from 0.2 to 2.5 liters per kWh depending on cooling technology and climatic conditions. In addition, indirect water consumption associated with electricity generation can range from 1 to 4 liters per kWh, depending on the regional energy mix. These quantitative variations highlight the need to evaluate data center sustainability beyond energy metrics alone.

The sustainability discourse surrounding data centers has historically focused on energy consumption, carbon emissions, and efficiency metrics such as Power Usage Effectiveness (PUE). However, the growing dependence of these facilities on freshwater resources—especially for thermal management—has emerged as a critical but underexamined dimension of sustainability. The increasing thermal loads generated by high-performance computing systems necessitate advanced cooling mechanisms, many of which rely directly or indirectly on water-intensive processes. As a result, the water footprint of data centers is becoming increasingly significant, particularly in the context of AI-driven workloads that demand continuous, high-density computation (de Vries-Gao, 2026; Lei et al., 2025). The total water footprint of a data center can be expressed as the sum of direct water used for on-site cooling and indirect water associated with electricity generation, which depends on the water intensity of the energy mix.

The evolution from traditional enterprise data centers to hyperscale and AI-centric infrastructures has introduced new operational paradigms characterized by higher rack densities, increased server utilization, and continuous processing cycles. Typical AI-oriented facilities now operate at rack densities of 20–80 kW per rack, significantly higher than traditional data centers, thereby intensifying both thermal loads and cooling requirements. These characteristics intensify both direct and indirect water consumption. Direct water use arises from cooling systems such as evaporative cooling towers, chilled water loops, and hybrid cooling architectures, while indirect water consumption is embedded within electricity generation processes. Consequently, the overall water footprint extends beyond on-site usage to include upstream dependencies, forming a complex water–energy nexus that is essential for sustainability assessment (Privette et al., 2026; Hlabisa, 2025).

Figure 1 indicates the overall resource flow in AI-driven data centers, highlighting how increased computational demand leads to higher energy consumption and heat generation, which in turn drives cooling requirements and associated water usage. It also captures the indirect pathway through which energy consumption contributes to additional water use at the power generation stage, emphasizing the interconnected nature of the water–energy nexus.

Figure 1. Resource flow in AI-driven data centers.

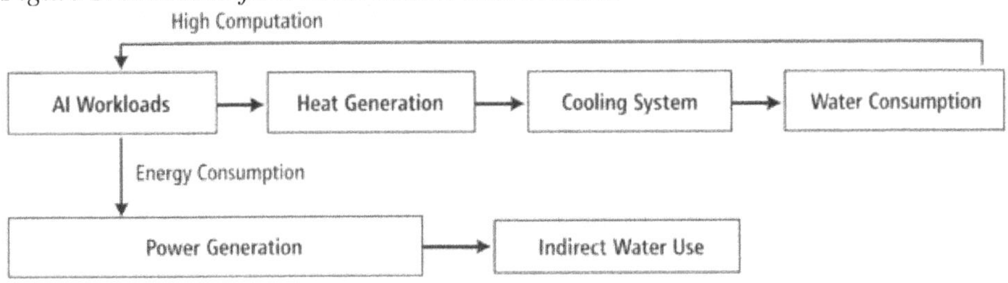

AI workloads further amplify this challenge due to their unique computational characteristics. Training and inference processes often operate under sustained peak conditions, generating substantial heat that must be dissipated continuously to maintain system reliability and performance. This results in persistent cooling demand and increased resource consumption, altering both temporal and spatial patterns of water usage. The continuous nature of AI operations introduces new stresses on infrastructure systems and necessitates more efficient resource management strategies (Ahrabi et al., 2025; Herrera et al., 2025). For example, a 10 MW AI data center operating with evaporative cooling in a hot-arid region may consume 1.5–2.0 liters/kWh of direct water, whereas the same facility in a cool climate using air-based cooling may reduce direct water use to below 0.3 liters/kWh, albeit with higher energy consumption. This illustrates the strong dependence of water footprint on climate and cooling design.

The implications of rising water demand extend beyond technical considerations and enter the domain of environmental and socio-economic sustainability. Data centers are increasingly being located in regions that offer favorable conditions such as low land costs and access to energy resources; however, many such regions are already experiencing water stress. The deployment of water-intensive cooling systems in these areas can exacerbate local scarcity, creating potential conflicts between industrial demand and community water needs. This underscores the importance of integrating environmental constraints into infrastructure planning and decision-making processes (Al Kez et al., 2022; Hiremath, 2024). In water-stressed regions, even moderate-scale data centers can increase local water demand by 5–20%, depending on baseline availability and cooling configuration, thereby intensifying resource competition.

In addition, limited transparency in water usage reporting complicates the evaluation of data center sustainability. While energy-related metrics have become standardized, water-related indicators such as Water Usage Effectiveness (WUE) are not consistently reported. This lack of uniformity restricts benchmarking and delays the development of effective regulatory mechanisms. Addressing this gap requires standardized metrics and improved disclosure practices to support accountability and informed governance (Privette et al., 2026). Despite growing awareness, existing studies largely report water usage descriptively without standardized quantification of direct and indirect components, revealing a critical gap in integrated water–energy assessment frameworks.

The broader ecological footprint of AI-driven data centers encompasses not only water and energy consumption but also material use, land requirements, and lifecycle environmental impacts. The convergence of these factors highlights the need for a systems-level perspective that captures the interdependencies among computational demand, resource utilization, and environmental impact. Such an approach is essential for designing sustainable digital infrastructure capable of supporting long-term technological growth (Hlabisa, 2025).

At the same time, the increasing demand for AI services presents opportunities for innovation in data center design and operation. Advanced cooling technologies, including liquid-based systems, closed-loop water cycles, and waste heat recovery, offer pathways to reduce water consumption while maintaining efficiency. Furthermore, AI-based optimization techniques enable intelligent resource allocation and predictive control, improving overall system performance and sustainability outcomes (Kijak & Woźnicki, 2026; Hiremath, 2024).

From a governance perspective, the integration of Environmental, Social, and Governance (ESG) principles into data center operations is gaining importance. Emerging regulatory frameworks emphasize resource efficiency, environmental responsibility, and transparency in reporting. However, effective implementation requires coordination between technological innovation, policy instruments, and stakeholder engagement to ensure balanced and sustainable infrastructure development (Nahid et al., 2025; Al Kez et al., 2022).

In this context, the water footprint of AI data centers represents a critical research domain that demands interdisciplinary analysis. The interaction between AI workloads, cooling requirements, and water resource availability forms a complex system that must be addressed through integrated engineering, environmental, and policy approaches. Developing sustainable solutions requires not only technological advancements but also strategic planning, regulatory support, and responsible resource management. Unlike prior studies that primarily focus on descriptive assessments of data center resource consumption, this chapter provides a quantified and integrated evaluation of the water–energy nexus in AI-driven data centers, incorporating representative numerical ranges, explicit decomposition of direct and indirect water use, and comparative analysis across cooling technologies and climatic conditions. The main contributions of this chapter are summarized as follows:

1. It provides a structured analysis of water consumption in AI-driven data centers, highlighting the transition from energy-centric to water-inclusive sustainability assessment.
2. It establishes the relationship between AI workload characteristics and increased cooling-driven water demand within modern data center infrastructures.
3. It examines cooling technologies and water–energy trade-offs, along with their environmental implications in water-stressed regions.
4. It presents mitigation strategies and governance perspectives for reducing the water footprint and enabling sustainable AI infrastructure development.

The remainder of this chapter is organized as follows. Section 2 discusses AI-driven workloads and their impact on data center water demand. Section 3 examines cooling technologies and associated water–energy trade-offs. Section 4 analyzes the impacts of data center deployment in water-stressed regions. Section 5 reviews regulatory frameworks and sustainability metrics related to water usage. Section 6 presents water-efficient technologies and mitigation strategies. Finally, Section 7 concludes the chapter with key insights and future directions.

AI-DRIVEN WORKLOADS AND DATA CENTER WATER DEMAND

AI-driven workloads introduce a fundamental shift in computational behavior, characterized by high processing intensity, continuous execution, and extensive reliance on specialized hardware such as GPUs and TPUs. Unlike conventional workloads that exhibit variability and idle cycles, AI applications—particularly deep learning model training and large-scale inference—operate under sustained peak conditions. This results in significantly higher power densities within data centers, often reaching levels that exceed the capabilities of traditional air-cooling systems. In practical deployments, AI-oriented data centers typically operate at rack power densities in the range of 20–80 kW per rack, compared to 5–15 kW in conventional facilities. This increase leads to proportionally higher cooling loads, often exceeding 30–50% of total facility energy consumption in high-density environments. The continuous nature of these workloads leads to persistent heat generation, which must be managed effectively to ensure system reliability and performance. Consequently, thermal management becomes a critical operational requirement, directly influencing both energy and water consumption patterns (Lei et al., 2025; Panchumarthy & Benala, 2025).

The thermal profile of AI workloads is distinct in that it produces consistent and concentrated heat loads over extended periods. High-density server racks used for AI computations can generate several times more heat compared to traditional IT equipment. For instance, AI training clusters can generate heat fluxes exceeding 500–1000 W per processor, necessitating advanced cooling solutions such as liquid or evaporative systems that inherently involve higher water dependency. This shift necessitates the adoption of advanced cooling technologies, many of which rely on water-based mechanisms such as evaporative cooling or liquid cooling systems. As a result, the increase in computational intensity directly translates into higher cooling demand, establishing a strong link between AI workload characteristics and water consumption in data centers (Lu & Wang, 2025; MirhoseiniNejad et al., 2021).

Relationship Between Workloads, Cooling Demand, and Water Consumption

The relationship between AI workloads and water consumption is primarily mediated through cooling requirements. As computational load increases, power consumption rises proportionally, leading to increased heat generation. To maintain optimal operating temperatures, this heat must be dissipated through cooling systems, many of which consume significant amounts of water. In evaporative cooling systems, for instance, water is used to absorb heat through evaporation, making water consumption directly dependent on the thermal load generated by IT equipment. Therefore, AI-driven workloads indirectly drive water demand by increasing the need for efficient heat dissipation (Lei et al., 2025). The relationship between computation and water usage can be described as the sum of direct and indirect water consumption. Direct water use corresponds to the amount of water required for cooling, which depends on the type of cooling technology employed. Indirect water use is associated with electricity consumption and is determined by the water intensity of the regional energy mix. Therefore, the total water use reflects both on-site cooling requirements and the upstream water embedded in energy generation.

Figure 2 illustrates the water–energy nexus in data center operations, showing the interdependent relationship between energy consumption, heat generation, cooling demand, and water use. It reflects how increases in computational load propagate through the system, while also indicating the trade-off where reducing water usage often results in higher energy consumption.

For example, a 5 MW AI workload operating in a hot-arid climate with evaporative cooling may result in direct water usage of 1.5–2.2 L/kWh, whereas the same workload in a cool climate using air cooling may reduce direct water consumption below 0.3 L/kWh, but increase energy consumption by approximately 10–20%. This demonstrates the inherent trade-off between energy and water across climates.

Figure 2. Water–energy nexus in data centers.

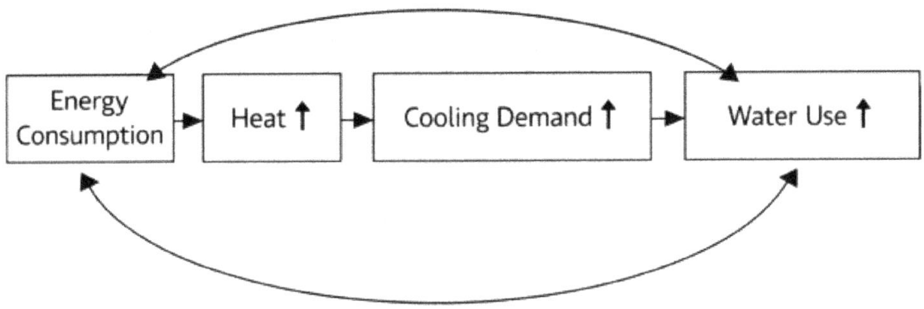

Water Usage Effectiveness (WUE) serves as a key metric for evaluating this relationship, providing a measure of water consumption relative to IT energy usage. However, WUE is influenced by multiple factors, including cooling technology, climatic conditions, and workload characteristics. AI workloads, due to their high and sustained energy consumption, can lead to complex interactions where improvements in energy efficiency do not necessarily correspond to reductions in water usage (Lei & Masanet, 2022).

To better understand the multidimensional nature of performance evaluation, Table 1 presents a comparison of commonly used data center efficiency metrics.

Table 1. Comparison of Data Center Sustainability Metrics

Metric	Definition	Typical range	Limitation
PUE	Total energy / IT energy	1.1-1.6	Does not consider water use
WUE	Water usage / IT energy	0.2-2.5 L/kWh	Highly climate-dependent
CUE	Carbon emissions / IT energy	0.1-0.6 $kgCO_2$/kWh	Depends on energy source

In addition to direct water usage in cooling systems, indirect water consumption associated with electricity generation must also be considered. Data centers rely heavily on grid electricity, and in regions where power generation involves thermoelectric processes, significant volumes of water are used for cooling at power plants. AI workloads, by increasing overall energy demand, indirectly contribute to this upstream water consumption. Ignoring this component leads to an underestimation of the total water footprint, underscoring the importance of adopting a holistic perspective (Lei et al., 2025).

The magnitude of indirect water use varies significantly with the energy mix. For example, thermoelectric power generation may require 2–4 L/kWh, whereas renewable sources such as solar and wind typically require <0.5 L/kWh, thereby reducing the indirect water footprint.

Influence of Climate, Location, and Infrastructure Design

The geographic location of data centers plays a crucial role in determining water consumption patterns. In cooler climates, data centers can utilize free-air cooling or economization techniques, which significantly reduce reliance on water-intensive cooling systems. However, AI workloads often require tightly controlled thermal environments that limit the effectiveness of such approaches. In contrast, data centers located in hot or arid regions depend heavily on evaporative cooling systems, resulting in substantially higher water usage.

In hot-arid regions, WUE values can exceed 2.0 L/kWh, while in temperate climates utilizing economization, WUE can be maintained below 0.5 L/kWh, illustrating strong geographic dependency.

The interaction between workload intensity and climatic conditions creates significant variability in water demand across regions. For instance, the same AI workload may result in different water footprints depending on the local climate and cooling infrastructure. This spatial variability highlights the importance of site selection and infrastructure design in managing water consumption. However, the increasing demand for AI services often prioritizes factors such as connectivity, latency, and energy availability over water sustainability, leading to potential resource imbalances (Lei et al., 2025).

Infrastructure design further influences water consumption through the choice of cooling technology and system configuration. High-density data centers designed for AI workloads often adopt hybrid or liquid-based cooling systems to handle increased thermal loads. While these systems improve cooling efficiency, they may also increase water dependency if not designed with sustainability considerations in mind. Therefore, optimizing infrastructure design requires balancing performance requirements with resource efficiency.

For example, consider a hypothetical 10 MW AI data center: in a cool and humid region using air cooling, the system may operate with a WUE of approximately 0.3 L/kWh and a higher PUE of around 1.5, whereas in a hot and arid region using evaporative cooling, the WUE may increase to approximately 1.8 L/kWh while achieving a lower PUE of around 1.2. This comparison highlights the trade-off between energy efficiency and water consumption driven by location and cooling design.

Workload Management, Optimization, and Resource Efficiency

Efficient workload management plays a critical role in mitigating the resource demands associated with AI-driven data centers. Thermal-aware scheduling techniques aim to distribute computational tasks in a manner that minimizes localized heat generation and reduces peak cooling demand. By balancing workloads across servers and avoiding the formation of thermal hotspots, these strategies can improve cooling efficiency and reduce water consumption.

Studies indicate that thermal-aware workload scheduling and AI-based control can reduce cooling energy by 10–25% and associated water consumption by 5–15%, depending on system configuration.

Machine learning-based approaches further enhance workload management by enabling predictive and adaptive control of data center operations. Predictive models can forecast energy consumption and thermal behavior based on historical data, allowing operators to proactively adjust cooling systems and resource allocation. Such approaches have been successfully applied in building energy systems and industrial processes, demonstrating their potential to improve efficiency and sustainability when extended to data center environments (Pham et al., 2020; Seyedzadeh et al., 2018).

The integration of AI into data center management introduces a dual dynamic: while AI workloads increase resource demand, AI-based optimization techniques offer opportunities to reduce inefficiencies. For example, reinforcement learning algorithms can optimize cooling system parameters in real time, adjusting airflow, temperature setpoints, and water usage to achieve optimal performance. This creates a feedback loop in which AI contributes both to increased demand and improved efficiency, emphasizing the importance of intelligent system design (Panchumarthy & Benala, 2025).

To illustrate the trade-offs between different cooling approaches, Table 2 summarizes the relationship between energy consumption and water usage.

Table 2. Energy–Water Trade-offs in Cooling Technologies

Cooling Type	Energy Consumption	Water Consumption	Typical range
Air Cooling	High	Low	~0–0.2 L/kWh
Evaporative Cooling	Low	High	~1.0–2.5 L/kWh
Liquid Cooling	Moderate	Moderate	~0.2–1.0 L/kWh

The energy–water trade-off complicates data center optimization: reducing energy (e.g., via evaporative cooling) increases water use, while minimizing water (e.g., air cooling) raises energy demand, requiring multi-objective optimization.

Scaling of AI Workloads and Emerging Challenges

The rapid scaling of AI workloads represents a major driver of future water demand in data centers. As organizations develop larger and more complex models, the computational requirements increase exponentially, leading to corresponding increases in energy and cooling demand. This trend is further amplified by the widespread deployment of AI applications across industries, including healthcare, finance, transportation, and communication.

In addition to centralized data centers, the growing adoption of edge computing introduces new challenges in managing resource consumption. Distributed AI workloads are often processed in smaller facilities that may lack advanced cooling infrastructure, resulting in lower efficiency and higher aggregate resource usage. The cumulative impact of these distributed systems can be significant, particularly when considering the global scale of AI deployment.

Furthermore, hybrid and multi-cloud environments complicate the assessment of water consumption, as workloads are distributed across multiple data centers with varying efficiency levels. This fragmentation makes it difficult to accurately measure and manage resource usage, highlighting the need for standardized reporting and coordinated management strategies.

Overall, AI-driven workloads fundamentally reshape the relationship between computation, cooling, and water consumption in data centers. The increasing demand for high-performance computing intensifies thermal loads and drives the adoption of water-intensive cooling technologies. Addressing this challenge requires a comprehensive approach that integrates workload optimization, infrastructure design, and policy frameworks to ensure sustainable operation. Quantitatively, the scaling of AI workloads is expected to increase data center water demand by 20–40% over the next decade under current cooling practices, emphasizing the urgency of integrated optimization strategies.

These workload-induced thermal and resource demands necessitate efficient cooling solutions, making it critical to analyze the available cooling technologies and their associated energy–water trade-offs.

COOLING TECHNOLOGIES AND WATER–ENERGY TRADE-OFFS

Cooling systems form a critical component of data center infrastructure, particularly in the context of AI-driven workloads that generate sustained and high-density thermal loads. As computational intensity increases, the ability to efficiently dissipate heat becomes essential to maintain operational reliability, prevent hardware degradation, and ensure performance stability. Traditional air-cooling systems, which were adequate for earlier generations of data centers, are increasingly insufficient for handling the thermal demands of modern AI-centric infrastructures. This has led to the adoption of advanced cooling technologies that vary in their energy efficiency, water consumption, and operational complexity. In practice, cooling systems can account for 30–50% of total data center energy consumption, with water usage varying from ~0 to 2.5 L/kWh depending on the cooling approach and climatic conditions.

The selection of cooling technology is inherently linked to the trade-off between energy and water usage. For example, evaporative cooling can reduce cooling energy consumption by 20–40% compared to air cooling, but may increase water usage to 1.0–2.5 L/kWh, whereas air cooling typically uses <0.2 L/kWh but at higher energy cost. While some systems minimize energy consumption by leveraging water-intensive processes, others reduce water dependency at the cost of higher energy requirements. This trade-off forms the basis of sustainability challenges in data center cooling, particularly as AI workloads continue to scale (Chu & Huang, 2023; Nadjahi et al., 2018).

Air-Based Cooling Systems

Air cooling remains one of the most widely used methods in data centers due to its simplicity, lower initial cost, and minimal water dependency. In this approach, heat generated by servers is removed through airflow using Computer Room Air Conditioners (CRAC) or Computer Room Air Handlers (CRAH). The air cooling systems exhibit limitations when applied to high-density AI workloads. As rack power densities increase, the volume of air required to effectively dissipate heat grows significantly, leading to higher fan power consumption and reduced overall efficiency. Additionally, air has a relatively low heat capacity compared to liquids, making it less effective for managing concentrated heat loads. As a result, purely air-based cooling systems often require supplementary cooling mechanisms in AI data centers (Ohadi et al., 2012; Jing et al., 2024). Typical air-cooled systems exhibit WUE values below 0.1–0.2 L/kWh, but may result in higher PUE values in the range of 1.4–1.6 under high-density AI workloads.

From a water perspective, air cooling is advantageous because it minimizes direct water usage. However, this benefit is offset by increased energy consumption, which can indirectly contribute to water usage through electricity generation processes. Thus, while air cooling reduces on-site water dependency, it may not necessarily minimize the overall water footprint when indirect effects are considered.

Evaporative Cooling Systems

Evaporative cooling has emerged as a widely adopted solution for improving energy efficiency in data centers. Cooling towers and adiabatic cooling systems are common implementations of this technique.

The primary advantage of evaporative cooling lies in its ability to significantly reduce energy consumption compared to conventional air-cooling systems. By leveraging the latent heat of vaporization, evaporative systems achieve higher cooling efficiency with lower electrical input. However, this efficiency comes at the cost of increased water consumption, as large volumes of water are required to sustain the evaporation process (Chu & Huang, 2023). Evaporative systems typically consume 1.0–2.5 L/kWh, depending on ambient temperature and humidity. In addition, operational factors such as cycles of concentration (typically 3–7), blowdown losses, and chemical treatment (biocides and anti-scaling agents) influence water consumption and wastewater generation.

The dependence on water makes evaporative cooling particularly challenging in regions with limited water availability. In hot and arid climates, where cooling demand is highest, water consumption can become substantial, exacerbating local water scarcity issues. Furthermore, water losses due to drift, blowdown, and evaporation contribute to operational inefficiencies. These factors highlight the inherent trade-off between energy savings and water usage in evaporative cooling systems (Karimi et al., 2022).

Liquid Cooling Technologies

Liquid cooling technologies have gained significant attention as a solution for managing high-density heat loads in AI-driven data centers. Unlike air cooling, liquid cooling involves direct or indirect contact between a cooling fluid and heat-generating components, enabling more efficient heat transfer due to the higher thermal conductivity and heat capacity of liquids.

Two primary types of liquid cooling are commonly employed: direct-to-chip cooling and immersion cooling. In direct-to-chip systems, coolant is circulated through cold plates attached to processors, removing heat at the source. Immersion cooling, on the other hand, involves submerging servers in a dielectric fluid that absorbs heat directly from components.

Liquid cooling systems offer several advantages, including improved thermal efficiency, reduced energy consumption for cooling, and the ability to support higher rack densities. Liquid and immersion cooling systems typically operate with approach temperatures in the range of 5–10°C, which enhances heat transfer efficiency and improves the potential for waste heat recovery compared to air-based cooling systems. However, their impact on water usage depends on system design. Closed-loop liquid cooling systems require minimal make-up water (<0.1–0.3 L/kWh) as heat is recirculated within sealed systems, whereas open-loop configurations that reject heat via cooling towers may still consume 0.5–1.5 L/kWh, depending on design (Hnayno et al., 2022; Liu & Yu, 2021).

Recent advancements in immersion cooling technologies have further enhanced efficiency by enabling uniform heat dissipation and reducing the need for air-based cooling infrastructure. These systems also facilitate waste heat recovery, allowing excess thermal energy to be reused for other applications. Despite these benefits, challenges related to cost, scalability, and maintenance remain barriers to widespread adoption (Wang et al., 2024; Liu et al., 2024).

Comparative Analysis of Cooling Technologies

The choice of cooling technology involves evaluating trade-offs between energy consumption, water usage, cost, and operational complexity. Table 3 provides a comparative overview of major cooling approaches.

Table 3. Comparison of Data Center Cooling Technologies

Cooling Technology	Energy (PUE impact)	Water Use (L/kWh)	Suitable Density	Key Limitation
Air Cooling	High (PUE ~1.4–1.6)	~0–0.2	Low–Moderate	Inefficient at high density
Evaporative Cooling	Low (PUE ~1.1–1.3)	~1.0–2.5	Moderate–High	Water-intensive
Liquid Cooling (Closed-loop)	Low	~0.1–0.3	High	Higher cost
Liquid Cooling (Open-loop)	Moderate	~0.5–1.5	High	Still water dependent
Immersion Cooling	Very Low	~0–0.2	Very High	Deployment complexity

This comparison illustrates that no single cooling technology optimally satisfies both energy and water efficiency requirements. Instead, the selection depends on specific operational priorities and environmental constraints.

Figure 3 presents a comparative view of major cooling technologies used in data centers, emphasizing the trade-offs between energy efficiency and water consumption. It highlights that while air cooling minimizes water use at the cost of higher energy demand, evaporative cooling reduces energy consumption but significantly increases water usage, with liquid and immersion cooling offering intermediate or optimized solutions.

Figure 3. Comparison of major data center cooling technologies.

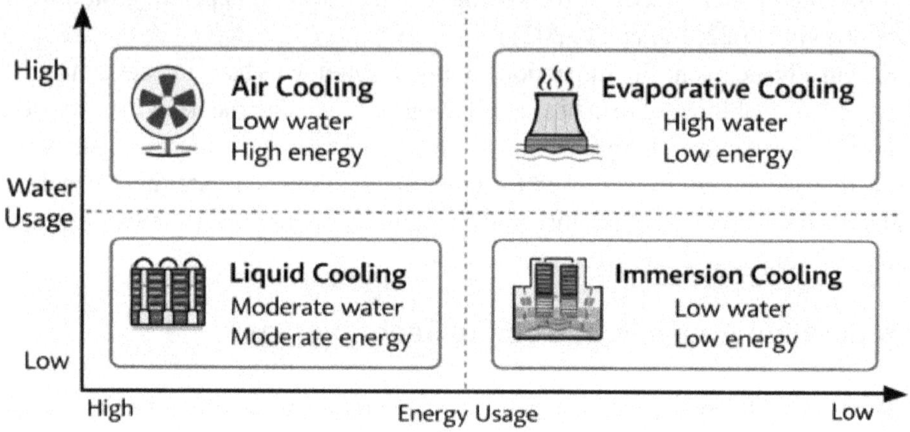

Water–Energy Trade-off Dynamics

The interaction between water and energy consumption in data center cooling systems represents a critical sustainability challenge. Reducing energy consumption often involves the use of water-intensive cooling methods, while minimizing water usage typically requires energy-intensive solutions. This inverse relationship complicates optimization efforts and necessitates a multi-objective approach to system design. Air cooling is characterized by high energy use and very low water consumption, whereas evaporative cooling offers low energy use but high-water consumption. Closed-loop liquid cooling provides a balanced solution with both low energy and low water use, while hybrid systems operate at moderate levels of both energy and water consumption.

The air cooling eliminates water usage but requires higher energy input to achieve the same level of cooling. Liquid cooling systems offer a potential middle ground, balancing energy and water efficiency, but their effectiveness depends on system configuration and operational conditions (Karimi et al., 2025).

The complexity of these trade-offs is further influenced by external factors such as climate, electricity generation mix, and water availability. In regions where electricity is generated using water-intensive processes, reducing energy consumption can indirectly reduce water usage. However, in areas with renewable energy sources, the relationship between energy and water becomes less pronounced, shifting the focus toward direct water usage.

Climate-Specific Performance and Case Insights

Cooling system performance varies significantly across different climatic conditions. In cooler regions, free-air cooling can be used to reduce both energy and water consumption. In such conditions, combined energy–water efficiency improvements of 15–30% can be achieved compared to baseline operation. However, in hot climates, the efficiency of air-cooling decreases, making evaporative or liquid cooling more attractive despite their higher water requirements. For example, consider two scenarios for a 10 MW AI data center.

1. Cool/humid region (air cooling): PUE $\approx$ 1.5, WUE $\approx$ 0.2 L/kWh
2. Hot/arid region (evaporative cooling): PUE $\approx$ 1.2, WUE $\approx$ 1.8 L/kWh

This illustrates that energy-efficient designs in hot climates often come at the expense of significantly higher water consumption.

Case studies in hot-arid regions demonstrate that cooling systems can account for a substantial portion of total water consumption in data centers. In such environments, the reliance on evaporative cooling leads to increased water usage, highlighting the need for climate-specific optimization strategies. Modeling studies further indicate that integrating multiple cooling techniques and dynamically adjusting system operation can improve overall efficiency and reduce resource consumption (Karimi et al., 2022; Karimi et al., 2025).

Emerging Trends and Flexible Cooling Architectures

Recent developments in data center cooling emphasize flexibility and adaptability in managing resource trade-offs. Hybrid cooling systems combine multiple technologies, such as air and liquid cooling, to optimize performance under varying conditions. These systems can dynamically switch between modes based on workload intensity, environmental conditions, and resource availability.

Advanced control strategies, including AI-based optimization, enable real-time adjustment of cooling parameters to achieve optimal energy and water efficiency. Concepts such as cooling flexibility frameworks aim to harmonize trade-offs between water, energy, carbon emissions, and cost, providing a more holistic approach to sustainability (Gnibga et al., 2024). AI-based control strategies have demonstrated potential reductions of 10–20% in cooling energy and 5–15% in water consumption through real-time optimization.

In addition, the integration of energy storage and thermal management systems offers new opportunities for improving cooling efficiency. Technologies such as liquid air energy storage can be combined with cooling systems to enhance performance and reduce resource consumption, further expanding the scope of sustainable data center design (Liu et al., 2024).

Implications for Sustainable AI Infrastructure

The increasing adoption of AI workloads necessitates a re-evaluation of cooling strategies to balance performance with sustainability. The choice of cooling technology must consider not only immediate operational requirements but also long-term environmental impacts, particularly in terms of water usage.

Achieving sustainable cooling requires an integrated approach that combines technological innovation, intelligent control systems, and context-aware design. This includes selecting appropriate cooling technologies based on climate conditions, optimizing workload distribution to reduce thermal stress, and implementing policies that encourage efficient resource use.

Ultimately, the water–energy trade-off in data center cooling represents a central challenge in the development of sustainable AI infrastructure. Addressing this challenge is essential to ensure that the continued growth of AI technologies does not come at the expense of critical water resources.

While cooling technologies determine the operational efficiency of data centers, their water consumption implications become particularly significant when such infrastructures are deployed in regions with limited water availability.

Quantitatively, selecting appropriate cooling strategies can alter total water consumption by a factor of 5–10× for the same computational workload, highlighting the critical importance of informed technology selection.

IMPACTS OF AI DATA CENTERS IN WATER-STRESSED REGIONS

The expansion of AI-driven data centers is increasingly intersecting with regions that already experience water scarcity, creating a critical sustainability concern. Data centers require continuous cooling to maintain operational stability, and when water-based cooling systems are employed, this translates into persistent freshwater demand. In water-stressed regions, even incremental increases in demand can disrupt existing supply–demand balances. In practice, large-scale data centers can increase local water demand by 5–20%, depending on facility size and baseline water availability, significantly intensifying regional water stress. This issue becomes more pronounced with AI workloads, which operate at higher power densities and sustained load conditions, thereby intensifying cooling requirements and water consumption compared to conventional workloads (Ristic et al., 2015; Jiang et al., 2025).

Figure 4 depicts the environmental and socio-economic impacts of data center water consumption in water-stressed regions, illustrating how increased water extraction can lead to groundwater depletion, reduced availability for agriculture, ecological disturbances, and potential social conflicts over resource allocation.

Figure 4. Environmental and socio-economic impacts of AI data center water use.

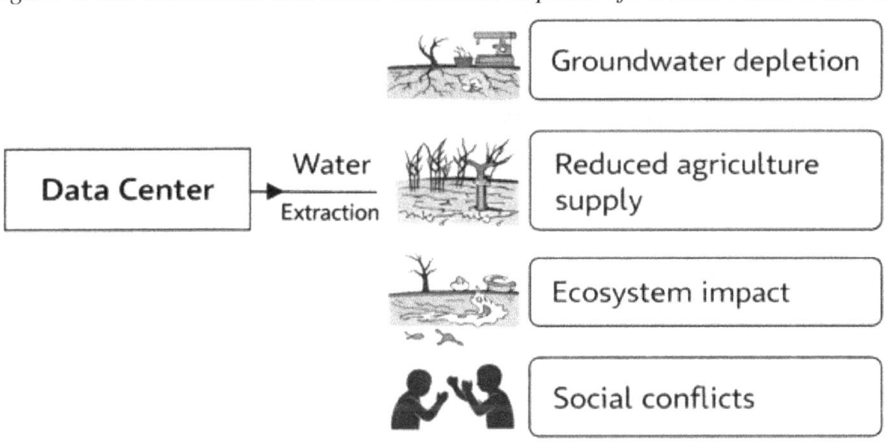

Water demand from data centers is not uniform; it fluctuates with workload intensity, climate conditions, and cooling technology. However, AI-centric facilities tend to operate closer to peak capacity for extended periods, making their water demand more stable but consistently high. In regions already experiencing seasonal water shortages, this steady demand reduces system flexibility and limits the ability to manage drought conditions effectively.

The situation becomes more critical when groundwater sources are used to support data center operations. Continuous extraction without adequate recharge can lead to declining water tables, increased pumping costs, and long-term depletion of aquifers. In heavily exploited regions, groundwater levels can decline by 0.5–1.5 meters per year under sustained industrial withdrawal, including data center operations. Over time, this creates structural water stress that extends beyond the immediate operational footprint of the data center.

Site Selection and Environmental Implications

Data center siting decisions have traditionally prioritized energy access, connectivity, and economic factors, often overlooking water availability as a primary constraint. As a result, several large-scale facilities have been established in regions where water resources are already limited. This creates a structural mismatch between infrastructure expansion and environmental capacity, particularly in arid and semi-arid regions.

The environmental implications extend beyond simple water withdrawal. Continuous extraction for cooling purposes can alter local hydrological cycles, affecting both surface water and groundwater systems. Reduced river flows, lower groundwater recharge rates, and changes in seasonal water availability can impact ecosystems and agricultural activities. In extreme cases, this may contribute to land degradation and reduced ecological resilience.

Thermal discharge from cooling operations introduces additional environmental stress. When heated water is released into natural water bodies, it can disrupt aquatic ecosystems by altering temperature regimes. This affects dissolved oxygen levels and biological processes, potentially leading to biodiversity loss. While modern cooling systems attempt to minimize such impacts, large-scale deployment still presents cumulative environmental risks. Siting decision framework for AI data centers in water-stressed regions based on water availability, competing demand, and regulatory constraints is given in Table 4.

Table 4. Siting decision framework for AI data centers in water-stressed regions.

Factor	Indicator	Decision Criterion
Water Stress	Water Stress Index	Avoid high (>0.6)
Recharge Rate	mm/year	Prefer high recharge zones
Competing Demand	% allocation (agriculture/domestic)	Avoid high competition
Water Source	Fresh vs Reclaimed	Prefer non-potable
Regulatory Constraints	Permits/limits	Ensure compliance

Socio-Economic Conflicts and Resource Competition

In water-stressed regions, water is a shared and often contested resource, essential for domestic use, agriculture, and local industries. The introduction of large data centers adds a new layer of demand that can disrupt existing allocation patterns. This creates competition between industrial consumption and community needs, particularly in regions where water governance systems are already under pressure.

Local communities may experience reduced access to water, increased costs, or declining quality as a result of intensified usage. In some regions, industrial water demand has been observed to increase local water tariffs by 10–30%, affecting both households and agricultural users. Agricultural sectors, which are often highly sensitive to water availability, can be particularly affected. These impacts are not always immediate but may develop gradually as cumulative demand increases.

The perception of imbalance in resource allocation can lead to social tensions. In several instances, concerns over water usage have influenced public opinion and acceptance of data center projects (Agarwal & Beharry, 2026).

At the same time, data centers contribute to economic development through investment, employment, and technological advancement. However, these benefits must be evaluated alongside environmental and social costs to ensure that development remains sustainable and inclusive. A structured community engagement approach includes

1. Use of reclaimed or treated wastewater for cooling
2. Public disclosure of water usage through local dashboards
3. Agreements for shared infrastructure or heat reuse
4. Periodic stakeholder consultations and impact assessments

Indirect Water Footprint and Governance Challenges

The total water footprint of AI data centers extends beyond direct usage in cooling systems. A significant portion of water consumption is embedded in electricity generation, particularly in regions dependent on thermoelectric power plants. These systems require large volumes of water for cooling, making energy consumption a major indirect contributor to water demand. For example, thermoelectric power generation may require 2–4 L/kWh, whereas renewable energy sources such as solar and wind typically require <0.5 L/kWh, significantly reducing indirect water footprint.

AI workloads amplify this effect due to their high energy requirements. As computational demand increases, so does electricity consumption, which in turn raises water usage at the energy production level. This interconnected relationship between water and energy complicates sustainability assessments, as reductions in one domain may lead to increases in another (Jiang et al., 2025). Moreover, indirect water intensity varies temporally with grid conditions, meaning that shifting computational workloads to periods with lower water-intensive generation can reduce overall water footprint.

Regulatory frameworks have not fully adapted to address this complexity. While energy efficiency is widely monitored and regulated, water usage reporting remains inconsistent. Metrics such as Water Usage Effectiveness (WUE) are available but not universally adopted, limiting comparability and transparency across facilities.

The absence of standardized reporting and enforcement mechanisms makes it difficult to assess cumulative impacts at regional and national levels. Addressing these gaps requires coordinated policy efforts, improved disclosure practices, and integration of water considerations into existing sustainability frameworks. Without such measures, the rapid expansion of AI infrastructure risks outpacing the development of effective governance systems.

These environmental and socio-economic challenges highlight the need for effective regulatory frameworks and standardized sustainability metrics to manage and monitor water usage in data center operations.

REGULATORY FRAMEWORKS AND SUSTAINABILITY METRICS

The rapid growth of AI-driven data centers has exposed limitations in existing regulatory frameworks, which have traditionally focused on energy efficiency and carbon emissions while giving relatively less attention to water consumption. As data centers become more resource-intensive, particularly due to AI workloads, water usage is emerging as a critical parameter that requires formal regulation and monitoring. However, regulatory approaches across regions remain fragmented, with significant variation in reporting requirements, enforcement mechanisms, and sustainability standards.

In practice, water usage in data centers can range from 0.2 to 2.5 L/kWh, yet fewer than 30–40% of operators publicly disclose WUE, highlighting a significant regulatory and transparency gap.

In many cases, water consumption is treated as an operational parameter rather than a regulated environmental metric. This creates gaps in accountability, as operators may optimize for energy efficiency while overlooking water impacts. The absence of mandatory disclosure requirements further limits visibility into how much water is being consumed and under what conditions. As a result, policymakers and stakeholders face challenges in assessing cumulative impacts and designing effective interventions (Agarwal & Beharry, 2026). Existing regulatory frameworks largely lack standardized definitions for direct and indirect water usage, and do not mandate unified reporting formats, representing a critical gap in current sustainability governance.

The increasing scale of AI infrastructure necessitates a transition toward integrated regulatory frameworks that explicitly incorporate water usage alongside energy and carbon metrics. This shift is essential to ensure that improvements in one dimension do not lead to unintended consequences in another.

Sustainability Metrics for Data Centers

The assessment of data center sustainability relies heavily on quantitative metrics that capture resource efficiency and environmental impact. Among these, Power Usage Effectiveness (PUE) has been widely adopted as a standard measure of energy efficiency. However, PUE does not account for water consumption, limiting its effectiveness in evaluating overall sustainability. To address this gap, Water Usage Effectiveness (WUE) has been introduced as a complementary metric that measures the amount of water used per unit of IT energy consumption. While WUE provides valuable insights into water efficiency, its interpretation is influenced by several factors, including climate conditions, cooling technologies, and operational practices. As a result, comparing WUE values across different regions or facilities can be challenging (Lei & Masanet, 2022). Table 5 provides key sustainability metrics for data centers.

Table 5. Key Sustainability Metrics for Data Centers

Metric	Definition	Typical range	Limitation
PUE	Total energy / IT energy	1.1-1.6	Ignores water usage
WUE	Water usage / IT energy	0.2-2.5 L/kWh	Climate-sensitive
CUE	Carbon emissions / IT energy	0.1-0.6 $kgCO_2/kWh$	Depends on energy mix

A standardized disclosure framework for data center water performance includes key parameters such as direct WUE, which represents on-site cooling water use measured in L/kWh, and indirect WUE, which accounts for water embedded in electricity consumption, also expressed in L/kWh. It further includes the water source composition, indicating the percentage of fresh, reclaimed, or recycled water used, along with seasonal variation in water usage, typically reported as monthly or seasonal WUE values. Additionally, thermal discharge is reported to capture heat rejection to the environment, measured in terms of temperature (°C) or thermal load (MW).

In addition to these metrics, broader frameworks are being developed to capture multi-dimensional sustainability performance. These include integrated indicators that consider energy, water, carbon emissions, and operational efficiency simultaneously (Murino et al., 2023).

ESG Integration and Governance Perspectives

The incorporation of Environmental, Social, and Governance (ESG) principles into data center operations has gained significant importance in recent years. ESG frameworks provide a structured approach to evaluating sustainability by considering environmental impact, social responsibility, and governance practices. Within this context, water usage is increasingly recognized as a key environmental indicator. For instance, ESG reporting increasingly requires disclosure of water intensity (L/kWh), percentage of non-potable water use, and year-on-year reduction targets, linking operational performance with sustainability commitments.

From an environmental perspective, ESG frameworks emphasize efficient resource utilization, reduction of environmental footprint, and adoption of sustainable technologies. Water management strategies, including recycling, reuse, and reduction of consumption, are integral to achieving these objectives. Social considerations focus on the impact of data center operations on local communities, particularly in water-stressed regions where resource competition may arise.

Governance aspects involve transparency, accountability, and compliance with regulatory standards. Organizations are expected to disclose resource usage data, implement monitoring systems, and align their operations with sustainability goals. The integration of ESG principles into data center management encourages a holistic approach that goes beyond technical efficiency and addresses broader societal and environmental concerns (Miciuła et al., 2026).

At the same time, ESG-driven approaches introduce new challenges, particularly in terms of standardization and measurement. Variations in reporting frameworks and metrics can lead to inconsistencies, making it difficult to compare performance across organizations.

Policy Frameworks and Water Governance

Policy frameworks play a critical role in shaping the sustainability of data center operations. Governments and regulatory bodies are increasingly recognizing the need to incorporate water considerations into infrastructure planning and environmental regulations. This includes the development of policies that promote efficient water use, encourage adoption of sustainable technologies, and require disclosure of resource consumption. Policy levers play a critical role in governing data center water sustainability. Procurement criteria can require the adoption of low-WUE systems, thereby reducing the overall water footprint. Permitting mechanisms can impose limits on water withdrawal to control regional resource impacts. ESG reporting requirements ensure mandatory disclosure of water usage, improving transparency and accountability. Additionally, incentive structures that promote water reuse and recycling can encourage innovation and adoption of sustainable practices.

Water governance frameworks often involve multiple stakeholders, including government agencies, industry operators, and local communities. Effective governance requires coordination among these stakeholders to ensure that water resources are allocated and managed sustainably. In the context of data centers, this may involve setting limits on water usage, establishing guidelines for site selection, and incentivizing the use of alternative cooling technologies (Sharma et al., 2021; Dai et al., 2025). Policies promoting renewable energy adoption can reduce indirect water usage by lowering reliance on thermoelectric power generation.

The integration of smart water management systems further enhances governance capabilities by enabling real-time monitoring and data-driven decision-making. These systems leverage digital technologies to track water usage, detect inefficiencies, and optimize resource allocation. When combined with policy measures, they provide a robust framework for managing water consumption in data centers.

However, the effectiveness of policy frameworks depends on their enforcement and adaptability. Rapid advancements in AI technologies and data center infrastructure require policies that can evolve in response to changing conditions. Static regulations may become outdated, necessitating continuous review and updating to remain relevant.

WATER-EFFICIENT TECHNOLOGIES AND MITIGATION STRATEGIES

The increasing water demand of AI-driven data centers has accelerated the development of advanced cooling technologies aimed at reducing dependence on freshwater resources while maintaining high thermal efficiency. Traditional evaporative cooling systems, although energy-efficient, consume large volumes of water and are less suitable for water-stressed regions. In response, alternative approaches such as liquid cooling, immersion cooling, and hybrid systems are gaining prominence. These technologies can reduce water consumption by 30–70% compared to conventional evaporative cooling, depending on system design and operating conditions.

Figure 5 outlines a framework of mitigation strategies for reducing the water footprint of AI-driven data centers, integrating technological, operational, and planning approaches such as advanced cooling systems, water recycling, AI-based optimization, waste heat recovery, and sustainable site selection.

Figure 5. Framework of water mitigation strategies for AI-driven data centers

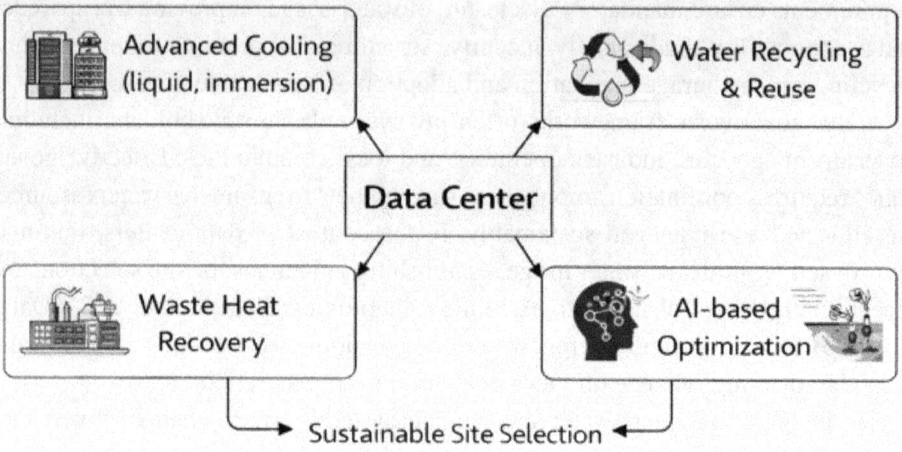

Liquid cooling technologies, particularly direct-to-chip cooling, enable efficient heat removal at the source, significantly reducing the need for large-scale air circulation and evaporative processes. Immersion cooling further enhances thermal performance by submerging electronic components in dielectric fluids, allowing uniform heat dissipation and minimizing thermal resistance. These systems reduce both energy consumption and water usage, especially when implemented in closed-loop configurations. However, their deployment requires higher initial investment and careful system design to ensure reliability and scalability (Liu & Yu, 2021; Wang et al., 2024). Closed-loop liquid cooling systems can reduce direct water usage to below 0.1–0.3 L/kWh, compared to 1.0–2.5 L/kWh in evaporative systems.

Hybrid cooling architectures are also emerging as practical solutions, combining air and liquid cooling methods to optimize performance under varying conditions. By dynamically switching between cooling modes, these systems can balance energy and water usage based on real-time requirements. Such flexibility is particularly valuable in regions with seasonal climate variations or fluctuating workload intensities (Alkrush et al., 2024). Hybrid systems can achieve 10–25% reductions in water consumption while maintaining comparable energy efficiency under varying climatic conditions.

Water Recycling, Reuse, and Closed-Loop Systems

Reducing freshwater dependency requires a shift toward circular water management practices within data center operations. Water recycling and reuse systems enable the treatment and recirculation of water used in cooling processes, significantly lowering net water consumption. Closed-loop cooling systems, in particular, minimize water losses by preventing evaporation and reusing the same water within the system.

In addition to internal recycling, some data centers are adopting the use of non-potable water sources, such as treated wastewater or industrial effluents, for cooling purposes. This approach reduces pressure on freshwater supplies while maintaining operational efficiency. However, it requires robust treatment and monitoring systems to ensure that water quality meets operational requirements and does not compromise equipment performance.

The integration of water reuse strategies also enhances resilience in water-stressed regions by reducing dependence on external water supplies. While these systems may involve higher initial costs and infrastructure complexity, they offer long-term benefits in terms of sustainability and operational stability. Water recycling and reuse strategies can reduce freshwater dependency by 40–80%, depending on system design and availability of non-potable sources.

Waste Heat Recovery and Energy–Water Synergies

An important mitigation strategy involves the recovery and reuse of waste heat generated by data center operations. Instead of dissipating heat through water-intensive cooling systems, waste heat can be captured and utilized for secondary applications such as district heating, industrial processes, or electricity generation.

Technologies such as warm-water cooling systems enable the extraction of usable thermal energy, reducing the need for additional cooling and lowering overall water consumption. The concept of integrating data centers with local energy systems creates opportunities for synergistic resource utilization, where waste heat becomes a valuable resource rather than a byproduct (Zimmermann et al., 2012; Zhu et al., 2020).

The effectiveness of waste heat recovery depends on factors such as temperature levels, proximity to heat demand, and infrastructure availability. While not universally applicable, this approach represents a promising pathway for improving both energy and water efficiency in data center operations. In optimized systems, waste heat recovery can reduce cooling demand by 15–30%, thereby indirectly lowering water consumption.

AI-Driven Optimization and Intelligent Control

The application of AI in data center management provides significant opportunities for improving water and energy efficiency. Advanced algorithms can analyze real-time operational data to optimize cooling system performance, adjust workload distribution, and predict resource demand. This enables dynamic control of cooling parameters, reducing unnecessary water usage while maintaining performance requirements.

A typical control stack for water-efficient operation includes thermal-aware workload placement, rack-level power capping, cooling setpoint optimization, and predictive maintenance with anomaly detection. These control strategies directly influence both energy and water consumption dynamics by enabling coordinated and adaptive system operation.

Predictive models can forecast thermal loads and adjust cooling strategies accordingly, preventing overcooling and minimizing resource wastage. Reinforcement learning techniques further enhance optimization by continuously adapting system behavior based on feedback, enabling more efficient use of both energy and water resources (Omrani & Ghassemi, 2025; Ma et al., 2025).

The integration of AI-based control systems also supports the coordination of multiple subsystems, including cooling, power management, and workload scheduling. This holistic approach improves overall system efficiency and enables more effective management of resource trade-offs. Key performance indicators (KPIs) for continuous monitoring include water usage intensity (L/kWh or L/MWh), cooling tower cycles of concentration, approach temperature (°C), and hourly water consumption per MW load, all of which enable real-time tracking and optimization of water and energy performance in data center operations.

Illustrative Case

A data center transitioning from evaporative cooling (WUE $\approx$ 1.8 L/kWh) to a hybrid liquid cooling system with reclaimed water reduced direct water consumption to 0.6 L/kWh, achieving approximately 65% reduction in freshwater use. Although capital costs increased, operational savings and regulatory compliance resulted in a payback period of 3–5 years, depending on location and water pricing.

Sustainable Design and Infrastructure Planning

Long-term mitigation of water consumption requires incorporating sustainability principles into the design and planning of data center infrastructure. This includes selecting locations based on water availability, climate conditions, and environmental impact, rather than focusing solely on economic or connectivity factors.

Design strategies such as modular architecture, high-efficiency cooling layouts, and optimized airflow management can significantly reduce resource consumption. In addition, integrating renewable energy sources into data center operations can reduce indirect water usage associated with electricity generation, particularly in regions reliant on water-intensive power plants (Murino et al., 2023). This can reduce indirect water consumption by 50–80% when transitioning from thermoelectric to renewable energy sources.

Sustainable design also involves lifecycle considerations, including construction materials, operational practices, and end-of-life management. By adopting a holistic approach, data centers can minimize their overall environmental footprint and improve long-term sustainability.

CONCLUSIONS

The rapid expansion of AI-driven data centers has transformed digital infrastructure into highly resource-intensive systems where water usage emerges as a critical sustainability constraint alongside energy consumption. This chapter demonstrates that AI workloads, characterized by high power densities (20–80 kW/rack) and continuous operation, significantly increase both direct and indirect water consumption through cooling requirements and electricity generation. The resulting water footprint is highly variable, ranging from 0.2 to 2.5 L/kWh (direct) and 1–4 L/kWh (indirect), underscoring the need for integrated water–energy assessment rather than energy-centric evaluation alone.

A key insight is the inherent energy–water trade-off in data center cooling. While air cooling minimizes water use at the cost of higher energy consumption, evaporative cooling reduces energy demand but can increase water usage to 1.0–2.5 L/kWh, with liquid and hybrid systems offering intermediate or optimized solutions. Additionally, the study highlights that data center deployment in water-stressed regions can increase local water demand by 5–20%, intensifying socio-environmental pressures. However, the adoption of advanced mitigation strategies—including liquid cooling, water reuse, AI-based optimization, and waste heat recovery—can reduce overall water footprint by 40–70%, demonstrating significant potential for sustainable operation. Despite these advancements, critical gaps remain in standardized water reporting, regulatory frameworks, and integrated optimization approaches. Future efforts must focus on mandatory disclosure of direct and indirect water usage, climate-aware infrastructure planning, and AI-driven real-time control strategies. A coordinated approach involving technology innovation, policy intervention, and stakeholder engagement is essential to ensure that the continued growth of AI infrastructure is aligned with long-term water sustainability.

ACKNOWLEDGEMENT

The authors acknowledge the use of artificial intelligence (AI) tools for language refinement and structuring of the manuscript. All technical content, analysis, and interpretations were developed and verified by the authors.

REFERENCES

Agarwal, A., & Beharry, E. (2026). A Transparency-Based Approach to Regulating the Resource Footprint of US Data Centers. *GRACE. Global Review of AI Community Ethics*, *4*(1).

Ahrabi, R. R., Mousavi, A., Mohammadi, E., Wu, R., & Chen, A. K. (2025, April). AI-driven data center energy profile, power quality, sustainable sitting, and energy management: A comprehensive survey. In *2025 IEEE Conference on Technologies for Sustainability (SusTech)* (pp. 1-8). IEEE. DOI: 10.1109/SusTech63138.2025.11025802

Al Kez, D., Foley, A. M., Laverty, D., Del Rio, D. F., & Sovacool, B. (2022). Exploring the sustainability challenges facing digitalization and internet data centers. *Journal of Cleaner Production*, *371*, 133633. DOI: 10.1016/j.jclepro.2022.133633

Alkrush, A. A., Salem, M. S., Abdelrehim, O., & Hegazi, A. A. (2024). Data centers cooling: A critical review of techniques, challenges, and energy saving solutions. *International Journal of Refrigeration*, *160*, 246–262. DOI: 10.1016/j.ijrefrig.2024.02.007

Chu, J., & Huang, X. (2023). Research status and development trends of evaporative cooling air-conditioning technology in data centers. *Energy and Built Environment*, *4*(1), 86–110. DOI: 10.1016/j.enbenv.2021.08.004

Dai, Y., Huang, Z., Khan, N., & Labbo, M. S. (2025). Smart water management: Governance innovation, technological integration, and policy pathways toward economic and ecological sustainability. *Water (Basel)*, *17*(13), 1932. DOI: 10.3390/w17131932

de Vries-Gao, A. (2026). The carbon and water footprints of data centers and what this could mean for artificial intelligence. *Patterns (New York, N.Y.)*, *7*(1), 101430. DOI: 10.1016/j.patter.2025.101430 PMID: 41583976

Gnibga, W. E., Chien, A. A., Blavette, A., & Orgerie, A. C. (2024, June). Flex-CoolDC: datacenter cooling flexibility for harmonizing water, energy, carbon, and cost trade-offs. In *Proceedings of the 15th ACM International Conference on Future and Sustainable Energy Systems* (pp. 108-122). DOI: 10.1145/3632775.3661936

Herrera, M., Xie, X., Menapace, A., Zanfei, A., & Brentan, B. M. (2025). Sustainable AI infrastructure: A scenario-based forecast of water footprint under uncertainty. *Journal of Cleaner Production, 526*, 146528. DOI: 10.1016/j.jclepro.2025.146528

Hiremath, R. B. (2024, November). AI-Embedded data centres: promoting sustainability and reducing water footprint. In *2024 first international conference on data, computation and communication (ICDCC)* (pp. 40-44). IEEE. DOI: 10.1109/ICDCC62744.2024.10961316

Hlabisa, S. (2025). The ecology of artificial intelligence: Energy, water, materials, and land limits of digital systems. *Carbon Neutral Systems, 1*(1), 19. DOI: 10.1007/s44438-025-00018-8

Hnayno, M., Chehade, A., Klaba, H., Bauduin, H., Polidori, G., & Maalouf, C. (2022). Performance analysis of new liquid cooling topology and its impact on data centres. *Applied Thermal Engineering, 213*, 118733. DOI: 10.1016/j.applthermaleng.2022.118733

Jiang, F., Duan, C., & Chen, B. (2025). Facility-level energy-driven water footprint and scarcity implications of Chinese data centers: A bottom-up analysis and scenario-based projection. *Applied Energy, 399*, 126522. DOI: 10.1016/j.apenergy.2025.126522

Jing, Y., Xie, L., Li, F., Zhan, Z., Wang, Z., Yang, F., Fan, J., Zhu, Z., Zhang, H., Zhao, C., Shi, Z., Jiang, H., Xie, X., & Jiang, Y. (2024). Field test of cooling systems in two air-cooled data centers: Various regions, air distributions and evaporative cooling technologies. *Applied Thermal Engineering, 248*, 123189. DOI: 10.1016/j.applthermaleng.2024.123189

Karimi, L., Shinh, R., Laisure-Pool, C., Green, M., Yacuel, L., Ashby, J., & Hickenbottom, K. L. (2025). Energy and water dynamics in data center cooling: Insights from a modeling study in hot-arid climates. *Applied Thermal Engineering, 276*, 126802. DOI: 10.1016/j.applthermaleng.2025.126802

Karimi, L., Yacuel, L., Degraft-Johnson, J., Ashby, J., Green, M., Renner, M., & Hickenbottom, K. L. (2022). Water-energy tradeoffs in data centers: A case study in hot-arid climates. *Resources, Conservation and Recycling, 181*, 106194. DOI: 10.1016/j.resconrec.2022.106194

Kijak, R., & Woźnicki, J. (2026). Solutions for the Data Centers' Sustainability. In *The Palgrave Encyclopedia of Sustainable Resources and Ecosystem Resilience* (pp. 1–10). Springer Nature Switzerland. DOI: 10.1007/978-3-030-67776-3_81-1

Lei, N., Lu, J., Shehabi, A., & Masanet, E. (2025). The water use of data center workloads: A review and assessment of key determinants. *Resources, Conservation and Recycling*, *219*, 108310. DOI: 10.1016/j.resconrec.2025.108310

Lei, N., & Masanet, E. (2022). Climate-and technology-specific PUE and WUE estimations for US data centers using a hybrid statistical and thermodynamics-based approach. *Resources, Conservation and Recycling*, *182*, 106323. DOI: 10.1016/j.resconrec.2022.106323

Liu, C., Hao, N., Zhang, T., Wang, D., Li, Z., & Bian, W. (2024). Optimization of data-center immersion cooling using liquid air energy storage. *Journal of Energy Storage*, *90*, 111806. DOI: 10.1016/j.est.2024.111806

Liu, C., & Yu, H. (2021). Evaluation and optimization of a two-phase liquid-immersion cooling system for data centers. *Energies*, *14*(5), 1395. DOI: 10.3390/en14051395

Lu, R., & Wang, D. (2025). A Thermal-aware Workload Scheduler for High-performance LLM Inference in Cooling-regulated Datacenters. *ACM SIGENERGY Energy Informatics Review*, *5*(2), 98–104. DOI: 10.1145/3757892.3757906

Ma, S., Liu, Y., Liu, Y., Wang, J., Fang, Q., & Huang, Y. (2025). Artificial intelligence-enabled predictive energy saving planning of liquid cooling system for data centers. *Advanced Engineering Informatics*, *65*, 103283. DOI: 10.1016/j.aei.2025.103283

Miciuła, I., Wojtaszek, H., Mastalerz, M., Wysocki, W., Plecka, P., Czaplewski, M., Doskocz, J., & Raba-Schulze, A. (2026). Optimization of Electricity Consumption by Information Technology Devices in Accordance with Environmental, Social, Governance and Sustainable Development Principles. *Energies*, *19*(5), 1139. DOI: 10.3390/en19051139

MirhoseiniNejad, S. M., Badawy, G., & Down, D. G.MirhoseiniNejad. (2021). Holistic thermal-aware workload management and infrastructure control for heterogeneous data centers using machine learning. *Future Generation Computer Systems*, *118*, 208–218. DOI: 10.1016/j.future.2021.01.007

Murino, T., Monaco, R., Nielsen, P. S., Liu, X., Esposito, G., & Scognamiglio, C. (2023). Sustainable energy data centres: A holistic conceptual framework for design and operations. *Energies*, *16*(15), 5764. DOI: 10.3390/en16155764

Nadjahi, C., Louahlia, H., & Lemasson, S. (2018). A review of thermal management and innovative cooling strategies for data center. *Sustainable Computing : Informatics and Systems, 19*, 14–28. DOI: 10.1016/j.suscom.2018.05.002

Nahid, F. A., Islam, M. M., Kamruzzaman, M., & Al Mansur, A. (2025, December). The Rising Energy Footprint of Data Centers: A Review of Global Trends, Challenges, and Opportunities. In *2025 IEEE 7th International Conference on Sustainable Technologies For Industry 5.0 (STI)* (pp. 1-6). IEEE.

Ohadi, M. M., Dessiatoun, S. V., Choo, K., Pecht, M., & Lawler, J. V. (2012, March). A comparison analysis of air, liquid, and two-phase cooling of data centers. In *2012 28th Annual IEEE Semiconductor Thermal Measurement and Management Symposium (SEMI-THERM)* (pp. 58-63). IEEE. DOI: 10.1109/STHERM.2012.6188826

Pham, A. D., Ngo, N. T., Truong, T. T. H., Huynh, N. T., & Truong, N. S. (2020). Predicting energy consumption in multiple buildings using machine learning for improving energy efficiency and sustainability. *Journal of Cleaner Production, 260*, 121082. DOI: 10.1016/j.jclepro.2020.121082

Privette, A. P., Barros, A., & Cai, X. (2026). Data centers water footprint: The need for more transparency. *AGU Advances, 7*(2), e2025AV002140.

Ristic, B., Madani, K., & Makuch, Z. (2015). The water footprint of data centers. *Sustainability (Basel), 7*(8), 11260–11284. DOI: 10.3390/su70811260

Seyedzadeh, S., Rahimian, F. P., Glesk, I., & Roper, M. (2018). Machine learning for estimation of building energy consumption and performance: A review. *Visualization in Engineering, 6*(1), 5. DOI: 10.1186/s40327-018-0064-7

Sharma, S. K., Seetharaman, A., & Maddulety, K. (2021). Framework for sustainable urban water management in context of governance, infrastructure, technology and economics. *Water Resources Management, 35*(12), 3903–3913. DOI: 10.1007/s11269-021-02916-1

Wang, H., Yuan, X., Zhang, K., Lang, X., Chen, H., Yu, H., & Li, S. (2024). Performance evaluation and optimization of data center servers using single-phase immersion cooling. *International Journal of Heat and Mass Transfer, 221*, 125057. DOI: 10.1016/j.ijheatmasstransfer.2023.125057

Zhu, X., Jiang, W., Liu, F., Zhang, Q., Pan, L., Chen, Q., & Jia, Z. (2020, May). Heat to power: Thermal energy harvesting and recycling for warm water-cooled datacenters. In *2020 ACM/IEEE 47th Annual International Symposium on Computer Architecture (ISCA)* (pp. 405-418). IEEE.

Zimmermann, S., Meijer, I., Tiwari, M. K., Paredes, S., Michel, B., & Poulikakos, D. (2012). Aquasar: A hot water cooled data center with direct energy reuse. *Energy*, *43*(1), 237–245. DOI: 10.1016/j.energy.2012.04.037

KEY TERMS AND DEFINITIONS

Artificial Intelligence (AI) Workloads: Computational tasks such as deep learning, model training, and inference that require high processing power and continuous operation.

Carbon Usage Effectiveness (CUE): A sustainability metric representing carbon emissions produced per unit of IT energy consumption in a data center.

Closed-Loop Cooling System: A cooling configuration in which coolant or water is recirculated within the system, minimizing water losses and freshwater demand.

Data Center Water Footprint: The total amount of water consumed directly for cooling and indirectly through electricity generation in data center operations.

Environmental, Social, and Governance (ESG): A framework used to evaluate sustainability performance based on environmental responsibility, social impact, and governance practices.

Evaporative Cooling: A cooling technique that removes heat through water evaporation, reducing energy consumption but increasing water usage.

Immersion Cooling: An advanced cooling method in which servers or components are submerged in dielectric fluid to improve heat dissipation efficiency.

Liquid Cooling: A thermal management approach where liquid coolants directly absorb heat from electronic components for efficient cooling.

Power Usage Effectiveness (PUE): A metric used to evaluate energy efficiency in data centers, defined as the ratio of total facility energy consumption to IT equipment energy consumption.

Thermal-Aware Workload Scheduling: An optimization strategy that distributes computational workloads to minimize thermal hotspots and reduce cooling demand.

Water Usage Effectiveness (WUE): A metric that measures the amount of water consumed per unit of IT energy usage, usually expressed in liters per kilowatt-hour (L/kWh).

Water–Energy Nexus: The interconnected relationship between water consumption and energy use in data center operations and cooling systems.

Chapter 2

AI Data Centers in Cooling Mechanisms and Freshwater Consumption: Challenges, Technologies, and Sustainable Pathways

Oiendri Mukhopadhyay
http://orcid.org/0009-0008-0694-261X
Kalinga Institute of Industrial Technology, India

Banasree Maji
http://orcid.org/0009-0000-1828-4767
Kalinga Institute of Industrial Technology, India

Hitesh Mohapatra
http://orcid.org/0000-0001-8100-4860
School of Computer Engineering, Kalinga Institute of Industrial Technology (Deemed to be University), Bhubaneswar, India

ABSTRACT

The rapid growth of AI infrastructure has increased freshwater demand in global data centers due to the cooling requirements of high-density GPU clusters. While carbon emissions from AI systems receive major regulatory attention, water consumption remains largely overlooked despite its impact on water-stressed communities and ecosystems. This chapter examines the hidden water footprint of AI data centers across three scopes: direct cooling water use (Scope 1), indirect water use from electricity generation (Scope 2), and embodied water in hardware manufacturing (Scope 3). It reviews cooling technologies such as air cooling, liquid cooling, and

DOI: 10.4018/979-8-3373-8434-4.ch002

immersion cooling. It also analyzes intelligent optimization methods, including reinforcement learning controllers, digital twins, and IoT-based predictive maintenance. Regional disparities are explored with emphasis on water-stressed regions and Africa. Policy measures and future research directions are discussed for sustainable AI infrastructure.

INTRODUCTION

Artificial intelligence has emerged as one of the most transformative technological forces of the twenty-first century, reshaping industries from healthcare and manufacturing to finance, education, and national security. The computational infrastructure underpinning this transformation a global network of large-scale data centers housing tens of thousands of graphics processing units (GPUs) and tensor processing units (TPUs) has grown at an unprecedented rate in both size and energy density. Unlike the general-purpose server infrastructure of the 2000s and 2010s, which benefited from decades of efficiency improvements through virtualization, cloud consolidation, and hardware miniaturization, the modern AI data center presents a qualitatively different thermal challenge. Systems such as the NVIDIA H100 GPU operate at thermal design power (TDP) values exceeding 700 watts per device, and integrated AI training clusters such as the GB200 NVL72 rack are projected to draw up to 600 kilowatts per rack by 2027 (Latif et al., 2025). These power densities far exceed the cooling capacity of conventional air-based systems, which have practical thermal limits of approximately 15 to 30 kilowatts per rack. The consequence is a structural and irreversible transition toward liquid-based cooling strategies a transition that has direct and substantial implications for the consumption of freshwater, a resource whose global scarcity is simultaneously being intensified by climate change, population growth, and competing industrial demands.

The timeliness of this problem cannot be overstated. Freshwater constitutes only approximately 2.5 percent of all water on Earth, and of that, a mere 0.5 percent is accessible and safe for human and industrial use (EESI, 2025). Yet data centers, which already account for approximately one to two percent of global electricity consumption, are increasingly competing with households, agriculture, and natural ecosystems for this finite resource. A typical large hyperscale data center consumes up to 5 million gallons of water per day equivalent to the daily needs of a town of up to 50,000 people (Brookings Institution, 2025). The training of a single large language model such as GPT-3 directly evaporated an estimated 700,000 liters of freshwater at Microsoft's United States data centers, a figure that rises to approximately 5.4 million liters when the water consumed at off- site power generation facilities is included (Li et al., 2025). At the inference level, a single ChatGPT response is estimated to require between three and ten times more water than a conventional web search, and as global AI adoption scales to billions of daily interactions, the aggregate water demand becomes staggering. Researcher Alex de Vries estimated in 2026 that total AI water consumption could reach between 312 and 764 billion liters in that year alone a volume comparable to the entire annual global consumption of bottled water. Against this backdrop, the convergence of rapidly expanding AI infrastructure and intensifying global water scarcity represents one of the defining sustainability challenges of the current decade, one that demands urgent, interdisciplinary attention from engineers, policymakers, and the research community alike.

The role of freshwater in data center cooling is rooted in thermodynamic necessity. Computing hardware converts electrical energy into computational work, but a significant portion of that energy is dissipated as heat. Without continuous and effective heat removal, processors rapidly exceed safe operating temperatures, leading to performance degradation, hardware failure, and ultimately system shutdown. Historically, data centers managed this thermal load through air- based cooling: chilled air was circulated through server rows via computer room air conditioning (CRAC) and computer room air handler (CRAH) units, and waste heat was rejected to the outside environment. This approach is relatively water-efficient, achieving Water Usage Effectiveness (WUE) values near zero, but it is energetically expensive, driving Power Usage Effectiveness (PUE) values to 1.4 and above. As rack densities increased with the adoption of AI accelerators, the limitations of air cooling became increasingly apparent as air simply does not have sufficient thermal capacity to remove heat at the densities required by GPU clusters. The industry response has been the adoption of liquid cooling in its various forms: direct-to-chip cooling, which circulates water through cold plates mounted on heat-generating components; rear-door heat exchangers; and full immersion cooling, in which servers are submerged in dielectric fluids. Of these, direct-to- chip and open-loop evaporative cooling systems remain the most widely deployed at scale. Evaporative cooling in particular is thermodynamically efficient water absorbs heat approximately 3,000 times more effectively than air per unit volume, but this efficiency comes at the direct cost of freshwater consumption, as the evaporative process permanently removes water from the local watershed. The result is a fundamental tension at the heart of data center sustainability: the most energy-efficient cooling methods are also the most water-intensive, forcing operators, policymakers, and researchers to grapple with a trade-off that cannot be resolved by optimizing for a single environmental metric.

Beyond this water-energy trade-off, the geographic dimension of data center water consumption raises profound questions of equity and resource governance. Approximately two-thirds of data centers constructed since 2022 have been located in water-stressed regions, driven by the availability of cheap land, favorable tax regimes, and access to electricity infrastructure, rather than by proximity to abundant water resources (Lincoln Institute, 2025). S&P Global analysis of over 9,000 data center assets worldwide found that 43 percent currently operate in areas of high- water stress, with approximately 30 percent of those under construction projected to face intensifying scarcity by 2050. In the United States, projections by the Houston Advanced Research Center estimate that Texas data centers alone will consume 49 billion gallons of water in 2025 and up to 399 billion gallons annually by 2030. Community backlash has already materialized in multiple jurisdictions: in Newton County, Georgia, a Meta data center was permitted to consume over 500,000 gallons per day, leading residents to report declining water pressure and supply issues; in The Dalles, Oregon, Google's data centers consumed roughly one-third of all city water; and in Chile, Santiago's environmental tribunal partially revoked Google's data center construction permit over water impact concerns. These cases illustrate that the freshwater footprint of AI is not an abstract environmental metric it translates directly into competition for a finite shared resource between industrial infrastructure and the communities surrounding it. This chapter therefore addresses the problem at multiple levels: the physics of heat generation and cooling in AI data centers, the engineering landscape of available cooling technologies and their water implications, the intelligent optimization systems capable of reducing both water and energy consumption, the geographic and climatic variability that shapes water intensity across deployment regions, and the policy frameworks that are beginning to govern water disclosure and efficiency in the sector.

This study is motivated by three principal objectives. First, it seeks to establish a rigorous, scope- inclusive accounting of the freshwater footprint of AI data centers, moving beyond simplistic on- site WUE metrics to encompass the full Scope 1, 2, and 3 water consumption chain. Second, it aims to provide a comprehensive comparative analysis of available cooling technologies from conventional air systems to advanced immersion approaches evaluating their respective thermal performance, water implications, energy efficiency, and operational suitability for AI workloads. Third, it investigates the role of intelligent systems including reinforcement learning-based cooling controllers, digital twins, predictive IoT maintenance frameworks, and workload scheduling strategies in simultaneously reducing the water and energy overhead of AI infrastructure. In doing so, this chapter contributes to the growing body of evidence that water sustainability must be elevated to co-equal status with carbon reduction in the design, operation, and governance of AI data centers. The structure of the chapter proceeds as follows: Section 2 presents a critical literature review of past and contemporary research on data center energy and water consumption. Section 3 details the proposed analytical framework and methodology. Section 4 presents the main findings and discussions. Section 5 integrates the conclusion and future research directions. References, additional reading, and key term definitions follow in the concluding sections.

LITERATURE REVIEW

Historical Evolution of Data Center Energy Research

The academic study of data center energy consumption has a well-established lineage. Koomey (2011) produced one of the landmark analyses of global and United States data center electricity use between 2005 and 2010, employing a scenario-based approach across four projections. His findings contradicted earlier alarmist projections: electricity consumption growth had slowed considerably compared to the 2000-2005 period, with 2010 global data center electricity use estimated at 1.1 to 1.5 percent of total global electricity and 1.7 to 2.2 percent for the United States. Koomey attributed this deceleration primarily to a slower-than-expected growth in the installed server base, particularly volume servers rather than to dramatic efficiency improvements alone. This finding was significant because it challenged the narrative that digital expansion necessarily implies proportional energy growth, establishing instead that server utilization rates, infrastructure efficiency (measured through PUE), and virtualization adoption were the primary moderating variables.

Masanet et al. (2020), writing in Science, substantially extended this line of work with a recalibration of global data center energy use estimates through 2018. Despite an approximately 550 percent increase in global computing output and a six-fold increase in internet traffic between 2010 and 2018, total data center energy consumption grew only modestly from approximately 194 TWh in 2010 to approximately 205 TWh in 2018. The authors attributed this remarkable efficiency to improvements in hardware energy performance, the rapid transition to hyperscale architectures with PUE values below 1.2, and software optimizations including virtualization. Critically, Masanet et al. (2020) identified AI workloads as an emerging factor likely to break the historical efficiency-demand balance, given their substantially higher computational density and energy requirements per unit of useful output. This prescient observation now constitutes the central challenge addressed by the present study. While energy efficiency research dominated the data center sustainability literature through the 2010s, water consumption received comparatively little systematic attention. Shehabi et al. (2016) in the United States Data Center Energy Usage Report produced some of the earliest estimates of direct and indirect water use by data centers, situating water consumption within the broader PUE and energy framework. Their work highlighted the significant water demands of thermoelectric power generation the predominant electricity source for data centers establishing the conceptual foundation for what would later be formalized as the Scope 1/2/3 water accounting framework. Figure.1 presents sustainable ai data center with integrated cooling and water management system.

Figure 1. Sustainable AI Data Centre with Integrated Cooling and Water Management System

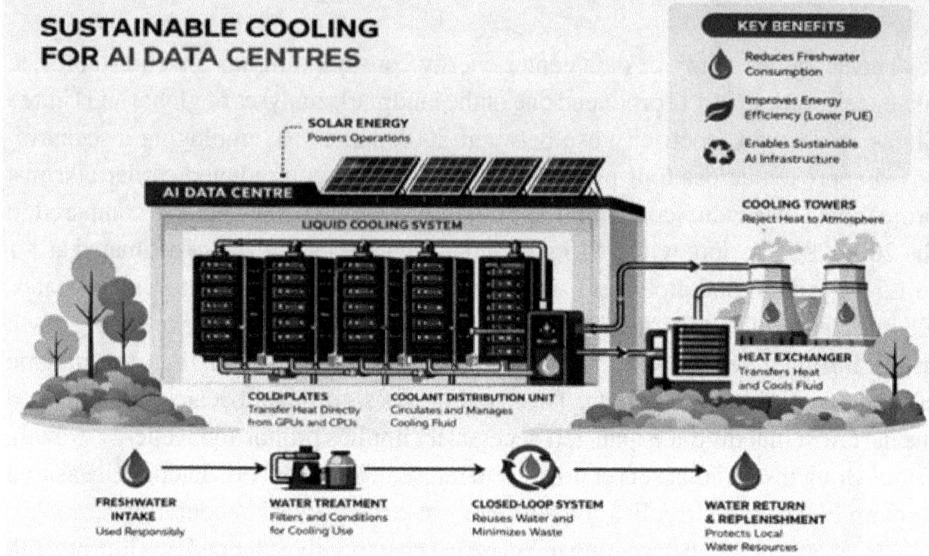

The Emergence of the AI Water Footprint

The publication of Li et al. (2025), originally circulated as a preprint in 2023, represented a watershed moment in the data center water literature. Titled 'Making AI Less Thirsty,' the study provided the first systematic quantification of the water footprint of large language models using GPT-3 as a case study. The authors demonstrated that training GPT-3 in Microsoft's U.S. data centers directly evaporated approximately 700,000 liters of freshwater (Scope 1), rising to approximately 5.4 million liters when Scope 2 electricity-related water consumption was included and projected to be roughly three times higher had training occurred in less water-efficient facility locations. At the inference scale, the study estimated that each set of 10 to 50 ChatGPT responses consumes approximately 500 milliliters of freshwater equivalent, a finding widely cited in policy and media discussions. Critically, Li et al. (2025) formalized the three-scope water accounting framework and argued that water and carbon footprints are complementary but non-substitutable metrics, requiring simultaneous optimization rather than trade-off management. This framing has proven influential in shaping subsequent research and regulatory responses.

Lei et al. (2025) extended the water efficiency literature significantly through a workload-level analysis as opposed to the facility-level WUE metrics that had dominated prior work. Their study demonstrated that workload-level water consumption can vary by more than four orders of magnitude across different data center configurations, primarily due to differences in server efficiency, electricity water intensity, server utilization rates, cooling configuration, and the proportion of inactive servers. This finding has important practical implications: it means that facility-level WUE metrics are insufficient for comparing the water impact of different AI deployments, and that a workload-based assessment framework integrating both computational performance and water consumption is necessary for accurate benchmarking. The authors also identified hardware refresh cycles as a significant but underappreciated determinant of water efficiency, noting that older, less efficient hardware can dramatically increase per-workload water consumption even in otherwise well-operated facilities.

Shumba et al. (2025) provided a critical geographic extension of the water footprint literature by developing the first comprehensive hourly WUE dataset for 41 African countries across five climate zones. Their work highlighted the inadequacy of global or developed-world-centric WUE frameworks for contexts where infrastructure inefficiencies particularly water leakage from aging distribution systems represent a major and typically unaccounted source of water loss. The study demonstrated that some African countries exhibit lower overall water intensity than global averages due to their energy mix, while others face compounding challenges of high water stress, poor infrastructure, and limited alternative energy options. The application of the framework to estimate LLM inference water footprints for Llama-3-70B and GPT-4 across multiple African countries provided a concrete illustration of how identical AI workloads can generate radically different water footprints depending on deployment geography. Figure.2 presents the cooling technology performance comparison (PUE vs WUE).

Figure 2. Cooling Technology Performance Comparison (PUE vs WUE)

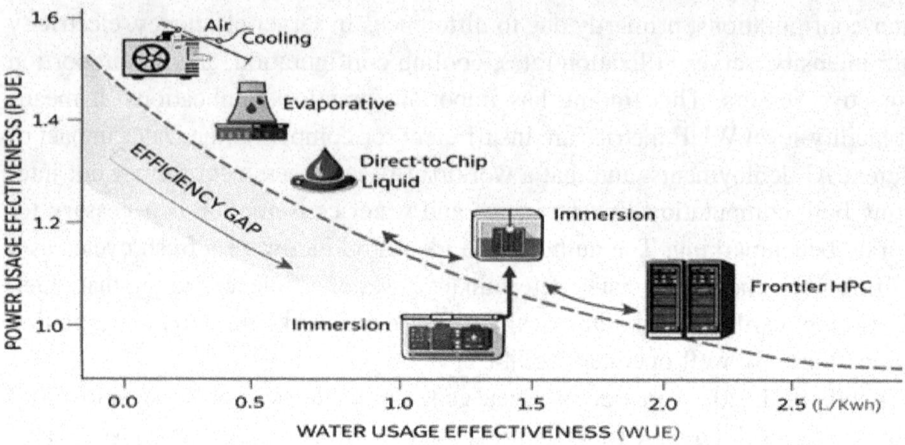

Cooling Technologies: Performance and Water Implications

The engineering literature on data center cooling technologies has expanded substantially in response to the AI-driven density challenge. The review by Chang et al. (2024) in Sustainability provided a comprehensive bibliometric and technical analysis of cooling system optimization strategies, documenting the shift from PID and model predictive control (MPC) approaches toward reinforcement learning-based systems. The authors noted a growing emphasis in the literature on intelligent, adaptive control as the preferred paradigm for managing the dynamic and uncertain thermal profiles characteristic of AI workloads. Importantly, their analysis identified WUE as a metric receiving increasing but still insufficient attention relative to PUE in optimization studies, reflecting the historical lag in water sustainability research relative to energy research.

Latif et al. (2025) contributed a rigorous empirical benchmarking study comparing liquid-cooled and air-cooled systems using dual 8x NVIDIA H100 GPU nodes under both synthetic stress and real-world LLM and vision-language model training workloads. Their results established that liquid-cooled systems maintain GPU temperatures 15 to 22°C lower than equivalent air-cooled systems, achieving up to 17 percent higher computational throughput per GPU and reducing node- level power consumption by 1 to 1.5 kW during intensive training. The study represents one of the first empirically grounded demonstrations that the transition to liquid cooling is not merely a sustainability choice but a performance necessity for modern

AI infrastructure. This finding critically undermines any argument that air cooling can be retained as an adequate solution for high-density AI clusters.

The literature on immersion cooling has similarly advanced. Comparisons documented by ScienceDirect (2024) and in arXiv (2022) demonstrate PUE values of 1.02 to 1.04 for immersion- cooled systems, with rack power densities up to 250 kW values unachievable through any other commercially available cooling approach. Single-phase immersion systems using dielectric oils present near-zero direct water consumption, representing the most water-efficient high- performance cooling solution currently available. However, the literature consistently identifies high capital costs, compatibility issues with standard server form factors, and long-term fluid reliability concerns as barriers to widespread adoption. Two-phase immersion systems, while thermally superior, face additional environmental challenges related to the use of fluorinated fluids with significant global warming potential a problem the engineering community has yet to fully resolve. Figure.3 presents reinforcement learning-based cooling control loop (LC-Opt Framework).

Figure 3. Reinforcement Learning-Based Cooling Control Loop (LC-Opt Framework)

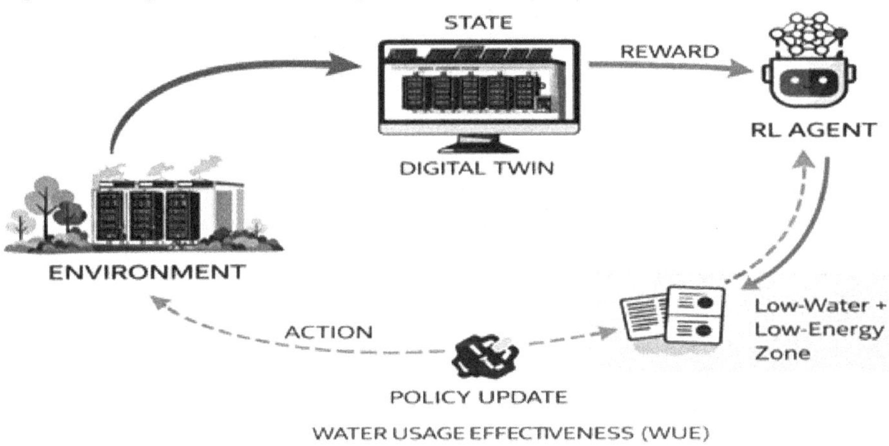

Intelligent Optimization and Leak Management

The application of reinforcement learning to data center cooling optimization has emerged as an active and productive research stream. Naug et al. (2025) introduced LC-Opt, a benchmarking framework built on a digital twin of the Frontier exascale supercomputer's cooling infrastructure, enabling systematic evaluation of RL-based end-to-end liquid cooling control. Their experimental results demonstrated that centralized multi-agent RL systems with multi-head architectures substantially outperform ASHRAE guideline-based control, with heat recovery unit integration yielding an additional 21 percent energy savings. The incorporation of policy distillation to produce LLM-based, human-readable control explanations addresses a key practical barrier to real-world deployment: operator trust. This line of work represents a significant advance over earlier rule-based and MPC approaches, which struggle to adapt to the non-linear and rapidly changing thermal dynamics of AI workload execution.

Jadhav and Liu (2026) demonstrated the power of physics-based digital twins for data center cooling optimization through a Modelica-based model of the Frontier supercomputer validated against one full year of operational data. Their layered optimization analysis is particularly instructive: flow-only optimization achieves 20.4 percent energy savings, joint flow and temperature optimization reaches 30.1 percent, and a practically constrained ramp-rate-limited strategy retains 27.8 percent — preserving over 92 percent of the theoretical maximum under real- world operational constraints. The identification of baseline inefficiencies with the Frontier system operating at nearly three times the thermally minimum flow rate highlights the large, low-cost efficiency gains available in existing facilities through improved control alone, without capital investment in new hardware.

Sunkara and Konakanchi (2025) addressed an underexplored operational risk of liquid cooling adoption through their IoT-LSTM-Random Forest predictive leak detection framework. The system achieves 96.5 percent detection accuracy and 87 percent forecasting accuracy within a 30- minute window, with the capacity to predict imminent leaks up to four hours in advance. For a 47- rack liquid-cooled facility, the prevention of emergency shutdowns enabled by this system is estimated to save approximately 1,500 kWh of annual energy loss. From a technology adoption perspective, this work is significant because coolant leakage risk is one of the most frequently cited barriers to liquid cooling deployment, particularly in organizations without prior liquid infrastructure experience. Demonstrating that this risk can be proactively managed through intelligent sensing systems removes a critical obstacle to wider adoption.

Critical Synthesis and Identified Research Gaps

A critical reading of the existing literature reveals several persistent structural gaps that constrain both the quality of current assessments and the progress of practical interventions. The most significant of these is the near-universal reliance on facility-level, annual aggregate water metrics particularly WUE as the primary basis for sustainability evaluation. As demonstrated by Lei et al. (2025), facility-level WUE obscures variation of over four orders of magnitude at the workload level, rendering it inadequate for comparing the water impact of specific AI deployments across different facilities or geographies. Despite this documented inadequacy, WUE remains the metric mandated by the most advanced regulatory frameworks currently in force, including the EU's Delegated Regulation 2024/1364. A workload-level metric that is both technically sound and practically measurable without access to proprietary operational data has yet to be standardized, and developing one constitutes an urgent research priority.

A second critical gap concerns the temporal dimension of water consumption. The existing literature predominantly reports annual aggregate water use, which conceals the highly variable, time-of-day and season-dependent nature of data center cooling demands. Cooling water requirements peak during summer months when ambient temperatures are highest and evaporative efficiency is lowest, precisely the period when water stress is most acute in many regions. Real- time, high-resolution WUE data would enable both more accurate risk assessment and more sophisticated adaptive management strategies, yet such data are almost entirely absent from the public domain. The Africa WUE dataset of Shumba et al. (2025), which provides hourly estimates, is a notable exception and a methodological model for expanding temporal granularity in future datasets. Third, the literature has largely treated data center water consumption as a facility-level engineering problem, underemphasizing its social and governance dimensions. Questions of water rights allocation, prior appropriation doctrines, community consent, and equitable access are central to whether AI infrastructure expansion can be considered socially sustainable yet these dimensions appear rarely in the engineering-focused literature that dominates this field. Bridging the engineering and social science perspectives on AI data center water consumption represents both a research gap and a necessary condition for developing governance frameworks that are both technically effective and socially legitimate.

PROPOSED STUDY

Research Framework and Scope

This study proposes an integrated analytical framework for understanding, quantifying, and reducing the freshwater footprint of AI data centers through the combined lens of cooling technology assessment, intelligent optimization system design, and policy analysis. The framework is structured around three interconnected analytical pillars: (1) a comprehensive scope- inclusive water accounting methodology spanning Scope 1, 2, and 3 water consumption; (2) a comparative technical evaluation of cooling technologies evaluated against both thermal performance and water efficiency criteria relevant to AI-specific workloads; and (3) a synthesis of intelligent control and optimization strategies capable of simultaneously reducing water and energy consumption across the operational lifecycle of a liquid-cooled AI data center. The proposed framework is grounded in the critical insight from Li et al. (2025) and Lei et al. (2025) that neither facility-level metrics alone nor single-scope accounting provides a sufficient basis for sustainable AI infrastructure decision-making, and that a holistic, workload-aware, and geography-sensitive assessment is essential.

Water Accounting Methodology: A Three-Scope Model

The proposed study adopts and extends the three-scope water footprint accounting framework formalized by Li et al. (2025). Scope 1 quantifies direct on-site water consumption associated with cooling tower evaporation, chiller blowdown, and liquid cooling loop makeup water. This is the component captured by the standard WUE metric (WUE = total annual site water use in liters / IT equipment energy consumption in kWh) as standardized in ISO/IEC 30134-9:2022. However, the present study argues that Scope 1 alone is deeply misleading: as demonstrated by the GPT-3 case study, Scope 1 water use accounts for only approximately 13 percent of the total three-scope water footprint, with Scope 2 electricity-related water constituting the dominant share at approximately 87 percent.

Scope 2 water consumption is calculated by multiplying the total electricity consumed by the data center by the water intensity factor (WIF) of the electricity grid serving that facility, expressed in liters of water per kilowatt-hour of electricity generated. This factor varies significantly by energy source and region: coal-fired power plants require approximately 19,185 gallons of water per MWh, natural gas plants approximately 2,800 gallons per MWh, while wind and solar photovoltaic generation require negligible cooling water. The implication is profound: a data center powered by renewable energy from wind or solar sources carries a Scope 2 water footprint close to zero, while an identical facility powered by coal generates a Scope 2 footprint orders of magnitude larger. This creates a direct and quantifiable sustainability dividend from renewable energy transition that extends beyond carbon reduction to water conservation a dual benefit that is underappreciated in current policy discussions. Table.1 presents water footprint scopes for AI data centers.

Table 1. Water footprint scopes for AI data centers: definitions and illustrative values

Scope	Source	Description	Estimated Contribution (GPT- 3 Training)
Scope 1 (Direct)	On-site cooling systems	Water evaporated by cooling towers and chillers physically located at the data center	~700,000 liters directly evaporated on-site
Scope 2 (Indirect)	Electricity generation	Water consumed at off-site thermoelectric power plants supplying electricity to the DC	~4.7 million additional liters
Scope 3 (Embodied)	Hardware supply chain	Water embedded in semiconductor fabrication, server manufacturing, and component logistics	Significant; currently unreported industry-wide

Scope 3 water accounting remains the least developed component of the framework. Semiconductor fabrication is among the most water-intensive manufacturing processes in the electronics industry: a typical 300mm wafer fabrication facility consumes between 3 and 9 million gallons of ultra-pure water per day, and the production of a single advanced GPU involves numerous photolithographic, etching, and chemical mechanical planarization steps that collectively require thousands of liters of purified water. When this embodied water is amortized across a GPU's operational lifetime and attributed to the AI workloads it executes, it adds a meaningful but currently unaccounted contribution to the total water footprint of AI operations.

The present framework recommends the development of standardized life-cycle assessment (LCA) methodologies for AI hardware water intensity as a critical priority for future work and explicitly flags Scope 3 as a significant gap in current corporate water reporting.

Comparative Analysis of Cooling Technologies for AI Workloads

The cooling technology landscape for AI data centers spans a wide range of approaches with fundamentally different water and energy implications. The present study evaluates seven primary technologies against four criteria: direct water consumption (WUE), energy efficiency (PUE), maximum sustainable rack power density, and primary operational barriers to adoption. This comparative framework is designed to assist data center designers and operators in selecting cooling strategies appropriate to their specific workload profiles, climate contexts, and sustainability constraints. Table.2 presents comparative evaluation of cooling technologies for AI data centers.

Table 2. Comparative evaluation of cooling technologies for AI data centers

Technology	Water Use (WUE)	PUE Range	Max Rack Density	Primary Limitation
Air Cooling (CRAC/CRAH)	~0 L/kWh	1.4 - 1.6	≤15 kW	Thermal ceiling for AI GPUs
Evaporative (Open-loop Tower)	1.0 - 2.5 L/kWh	1.2 - 1.4	≤30 kW	High freshwater consumption
Direct-to-Chip Liquid Cooling	0.1 - 0.5 L/kWh	1.1 - 1.3	≤80 kW	Plumbing complexity; leak risk
Single-Phase Immersion	~0 L/kWh	1.02 - 1.05	≤200 kW	High CAPEX; fluid cost
Two-Phase Immersion	~0 L/kWh	~1.02	≤250 kW	Fluorinated fluid GWP concerns
Closed-Loop Hybrid	0.2 - 0.8 L/kWh	1.1 - 1.3	≤60 kW	Design complexity
Free Cooling (Economizer)	Low-Moderate	1.1 - 1.2	Moderate	Climate-dependent effectiveness

Air cooling, while near-zero in direct water use, is fundamentally unsuitable for the rack densities demanded by modern AI GPU clusters. The thermal design power of current AI accelerators exceeds 700 watts per device, and a fully populated 8-GPU server node draws in excess of 5 to 6 kW; scaling this to a 10-server rack exceeds 50 kW, and full AI clusters with networking and memory infrastructure routinely reach 80 kW and above beyond the practical ceiling of forced-air systems. The retention of air cooling for AI infrastructure therefore necessitates severe underutilization of available rack density, translating directly into greater physical footprint, higher capital costs, and lower computational efficiency per unit of floor area.

Evaporative cooling towers have served as the workhorse of hyperscale data center thermal management for over a decade, offering a favorable PUE of 1.2 to 1.4 and the ability to support moderate rack densities. However, the water penalty is substantial: open-loop evaporative systems using freshwater sources consume 1.0 to 2.5 liters per kilowatt-hour of IT energy, and approximately 80 percent of withdrawn water is permanently lost to evaporation rather than returned to the watershed. For a 100 MW AI data center operating at full load, this translates to approximately 100 to 250 million liters of freshwater consumed per month a volume that constitutes a meaningful fraction of municipal water supply in many regions. The proposal of closed-loop evaporative systems, which can reduce freshwater use by up to 70 percent through water recirculation and heat rejection to the ambient environment, represents the most immediately deployable water conservation measure within the evaporative paradigm, though it introduces additional capital cost and design complexity.

Direct-to-chip liquid cooling represents the current sweet spot for AI data centers seeking to balance performance, water efficiency, and operational familiarity. Cold plates are mounted directly on GPU die surfaces, circulating water or a water-glycol mixture to absorb chip-level heat fluxes that air cooling cannot accommodate. The coolant is then pumped to a Coolant Distribution Unit (CDU), where heat is transferred to a building water loop and ultimately rejected via a dry cooler or cooling tower. When implemented as a closed-loop system, direct-to-chip cooling achieves WUE values of 0.1 to 0.5 L/kWh a reduction of 70 to 90 percent compared to open evaporative systems while supporting rack densities up to 80 kW. The primary operational barrier is the introduction of water infrastructure into the server environment, with attendant leak risk. This risk, as demonstrated by Sunkara and Konakanchi (2025), is manageable through predictive IoT-based monitoring, but requires organizational investment in sensing infrastructure and maintenance protocols that many operators have not previously needed.

Immersion cooling, in both single-phase and two-phase variants, represents the technological frontier of data center thermal management. By eliminating the air-coolant thermal interface entirely and ensuring direct contact between the dielectric fluid and all heat-generating surfaces, immersion systems achieve PUE values approaching 1.02 within two percent of theoretical perfection. The direct water footprint is near zero, as dielectric fluids are recirculated in closed systems with minimal makeup requirements. Maximum rack power densities of 200 to 250 kW place immersion cooling as the only viable option for the most extreme AI compute clusters. Despite these advantages, adoption has been limited by high capital cost, the absence of standardized server form factors compatible with immersion deployment, and in the case of two- phase systems the environmental profile of fluorinated dielectric fluids, which are persistent in the environment and carry global warming potentials orders of magnitude greater than carbon dioxide.

The development of low-GWP or bio-based alternative dielectric fluids is therefore identified as a critical enabling research area for sustainable immersion cooling at scale. Figure.4 presents water–carbon trade-off in data center cooling technologies.

Figure 4. Water-Carbon Trade-off in Data Centre Cooling Technologies

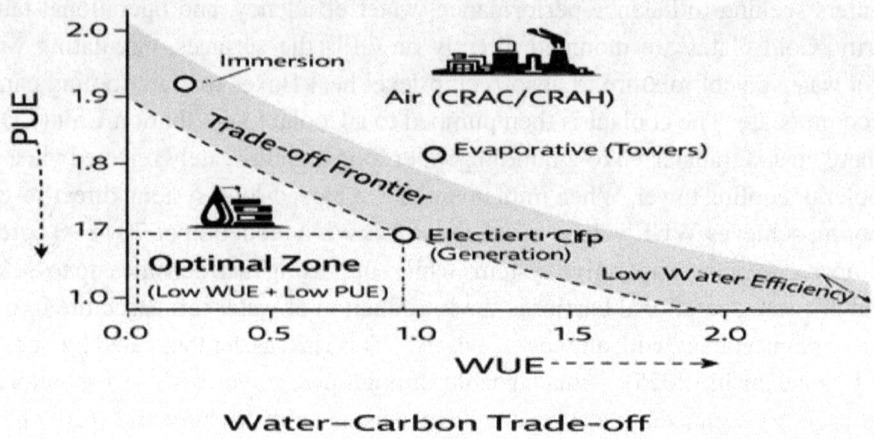

Intelligent Optimization Framework for Water-Efficient AI Data Centers

The proposed study synthesizes four categories of intelligent optimization systems applicable to water-efficient AI data center operation. These are not mutually exclusive but rather form a layered, complementary architecture spanning the physical, control, workload, and operational dimensions of data center management. The first layer is physics-based digital twin modelling, following the approach of Jadhav and Liu (2026). A digital twin a high-fidelity computational model calibrated against real operational data enables safe and comprehensive exploration of optimization strategies, including scenarios that would be impractical or risky to test in live facilities. The demonstrated accuracy of Modelica- based digital twins (CV-RMSE below 2.7%) provides sufficient fidelity for meaningful control optimization, and the finding that over 92% of theoretical energy savings can be retained under practical operational constraints makes digital twin-based optimization highly actionable for real-world facilities.

The second layer is reinforcement learning-based cooling control, exemplified by the LC-Opt framework of Naug et al. (2025). RL-based controllers offer a fundamentally different paradigm from rule-based or PID control: they learn optimal control policies through repeated interaction with the environment (or its digital twin), adapting dynamically to changing workload profiles, ambient conditions, and system degradation without requiring explicit system modeling. The demonstrated superiority of centralized multi-agent RL over ASHRAE guideline-based control, combined with the interpretability improvements afforded by LLM-based policy distillation, positions RL control as the leading candidate for intelligent liquid cooling management in next- generation AI data centers.

The third layer is temperature-aware and geography-aware workload scheduling, drawing on the insights of Moore et al. (2005) and the Green-DCC framework of Sarkar et al. (2025). The strategic temporal and spatial distribution of AI training workloads scheduling computationally intensive tasks to times and locations with lower WUE and carbon intensity can meaningfully reduce both water and carbon footprint without requiring any hardware modifications. At the facility level, workload placement policies such as Zone-Based Discretization (ZBD) and Minimize Heat Recirculation (MinHR) can reduce annual cooling costs by up to 50 percent in simulation by maintaining balanced thermal profiles and minimizing heat recirculation between server zones.

The fourth layer is predictive IoT-based maintenance, targeting the operational risks introduced by liquid cooling infrastructure. The LSTM-Random Forest framework of Sunkara and Konakanchi (2025) demonstrates that coolant leaks can be predicted with 87 percent accuracy up to four hours in advance, enabling preventive intervention before costly emergency shutdowns occur. The integration of this predictive maintenance capability with the broader digital twin and RL control architecture creates a unified intelligent management system that addresses thermal optimization, energy efficiency, water conservation, and operational resilience simultaneously. Figure.5 presents IoT-based leak detection pipeline for liquid-cooled data centers.

Figure 5. IoT-Based Leak Detection Pipeline for Liquid-Cooled Data Centers

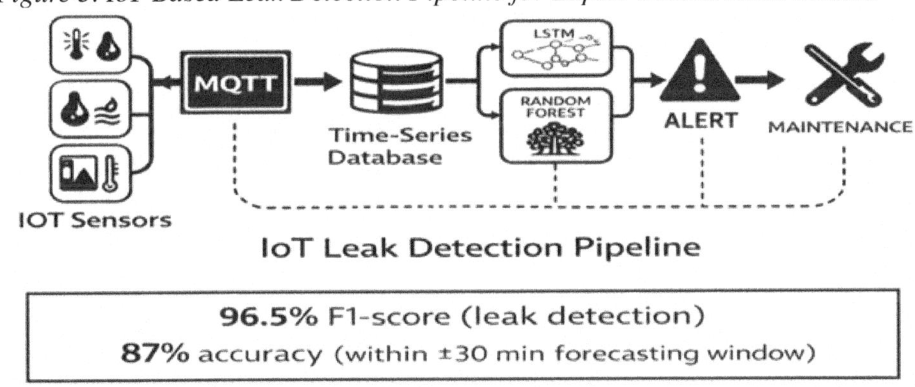

Integrated Water-Energy Co-Optimization: Proposed Methodology

The preceding four optimization layers are most powerful when integrated into a unified co- optimization architecture that treats water consumption and energy efficiency as joint objective functions rather than separate considerations. The proposed methodology for this co-optimization proceeds as follows. First, a facility-level digital twin is constructed and validated against at least six months of operational data to achieve CV-RMSE below three percent across key thermal and hydraulic variables. Second, a multi-objective RL controller is trained within the digital twin environment using a reward function that simultaneously penalizes WUE excess (direct and Scope 2), PUE excess, and thermal constraint violations. Third, the trained controller is distilled into an interpretable decision tree or LLM-based policy following the approach of Naug et al. (2025) and validated in shadow mode against live facility operations before full deployment. Fourth, the workload scheduling layer is integrated with the facility-level controller through a hierarchical control architecture: the higher-level scheduler allocates workloads across facilities based on real- time WUE and carbon intensity signals, while the facility-level controller manages local thermal dynamics within those allocations.

This four-layer architecture is designed to be modular and incrementally deployable. Operators who cannot immediately implement all four layers can begin with digital twin construction and flow optimization for the lowest-cost intervention, capturing approximately 20 percent of energy savings before any RL or IoT infrastructure is in place. The IoT leak detection layer can be deployed independently as an operational risk mitigation measure with a short payback period. The RL control layer requires the digital twin as a prerequisite but can be deployed alongside existing control systems in shadow mode with zero operational risk during the validation phase. This incremental deployment pathway lowers the barrier to adoption and allows operators to capture tangible benefits at each stage, building organizational confidence and capability in advance of full system integration.

The proposed framework is also designed to be climate-adaptive: by incorporating real-time ambient temperature and humidity forecasts into the RL controller's state representation, the system can proactively shift between cooling modes for example, increasing free cooling utilization during cool periods and switching to more closed-loop recirculation during peak summer heat in ways that minimize both water consumption and energy use as external conditions vary. This adaptive capability is particularly valuable in regions with high seasonal variability in temperature and humidity, where static cooling configurations may be well-optimized for one season while being significantly suboptimal in another.

The proposed study also explicitly addresses the under-examined relationship between water consumption and the temporal scheduling of AI training workloads. Large-scale model training unlike inference, which is driven by unpredictable user demand is inherently schedulable: training runs can, within limits, be timed to coincide with periods of lower ambient temperature, higher renewable energy availability, and lower grid water intensity. The Green-DCC framework of Sarkar et al. (2025) operationalizes precisely this insight at the multi-facility level, demonstrating that hierarchical RL-based workload allocation across geographically distributed data center clusters can meaningfully reduce both carbon emissions and, by extension through the Scope 2 linkage, water consumption. The present framework extends this logic by incorporating real-time WUE forecasting derived from ambient temperature, humidity, and grid water intensity projections as an explicit input to the workload scheduling optimization. This enables operators not merely to minimize energy cost or carbon intensity in workload placement decisions, but to minimize the combined environmental footprint across all three scopes simultaneously. Implementing this capability requires close integration between the facility-level cooling management system and the higher-level workload orchestration layer an architectural integration that represents a departure from the currently prevalent practice of treating cooling management and compute scheduling as separate, independently optimized systems. Bridging this architectural gap is therefore identified as a key implementation challenge for the proposed framework, and one whose resolution would yield significant and measurable environmental benefits at the scale of large AI training clusters. The proposed framework is designed to align with and support the emerging global regulatory landscape for data center water reporting and efficiency. As summarized below, a growing number of jurisdictions are beginning to mandate water disclosure and set WUE performance targets, though the overall regulatory framework remains fragmented and largely reactive relative to the pace of AI infrastructure expansion. Table.3 presents selected global policy measures governing data center water efficiency (2022-2025).

Table 3. Selected global policy measures governing data center water efficiency (2022-2025)

Region / Jurisdiction	Policy / Regulation	Year	Key Water Provision
European Union	Energy Efficiency Directive (2023/1791) + Delegated Regulation 2024/1364	2024	Mandatory annual WUE and water input reporting for DCs with installed IT power ≥500 kW; first reports due September 2024
Netherlands	Hyperscale DC Moratorium	2024	Permanent ban on new hyperscale data centers above 70 MW capacity
Singapore	Green Data Center Roadmap	2023-ongoing	WUE target of 2.0 m³/MWh; PUE ≤1.25 mandatory for new builds
China	National DC Building Code	2022	Sole country with mandatory WUE performance standards embedded in national building code (IEA, 2025)
USA - California	Senate Bill 58	2025	Tax credit for data centers adopting sustainable practices including use of recycled water in cooling systems
USA - Minnesota	DC Water Supply Regulation	2025	Developers required to partner with Dept. of Natural Resources to verify adequate water supply prior to permitting
USA - Arizona	Municipal Water Caps	2025	Water consumption limits forcing developers to commit to zero-water cooling designs in new projects

The EU's Delegated Regulation 2024/1364 represents the most comprehensive binding framework currently in force, requiring annual reporting against 24 sustainability indicators including total water input and WUE for all data centers with installed IT power of 500 kW or more. First reports were due in September 2024, and the European Commission's planned Data Center Energy Efficiency Package, expected in April 2026, will assess reported data and explore minimum performance standards. Singapore's requirement of WUE ≤ 2.0 m³/MWh for new builds provides a concrete, enforceable efficiency target against which the technologies and frameworks described in this study can be benchmarked. The proposed analytical framework is designed to be directly compatible with these reporting requirements, enabling operators to use it both as an internal optimization tool and as a basis for regulatory compliance documentation.

DISCUSSION

The synthesis of findings across the water accounting, cooling technology, intelligent optimization, and policy dimensions reveals several cross-cutting insights that merit extended discussion. The first and most fundamental is the inadequacy of single-metric sustainability evaluation for AI data center operations. The persistent dominance of PUE as the primary operational efficiency metric in the industry has contributed to a systematic undervaluation of water consumption: data center operators have rationally optimized toward lower PUE through evaporative cooling, while simultaneously increasing their freshwater footprint. The WUE metric, introduced by The Green Grid in 2011 and standardized as ISO/IEC 30134-9:2022, was designed to address this gap, but its adoption remains inconsistent and its scope-exclusive (Scope 1 only) definition means it captures only a fraction of the true water impact. The workload-based framework proposed by Lei et al. (2025) represents a more meaningful advance: by integrating computational performance with both direct and indirect water consumption, it enables cross- facility comparisons that account for the radically different energy mixes and server efficiency profiles that drive four-orders-of-magnitude variation in per-workload water intensity. The adoption of this framework or a formally standardized equivalent as the basis for regulatory reporting would constitute a significant improvement over current WUE-based disclosure.

The water-carbon trade-off identified in the literature is more nuanced than a simple binary between water-intensive and carbon-intensive cooling choices. The analysis in this study reveals that the trade-off is resolvable not through technological compromise, but through systemic redesign. Closed-loop liquid cooling powered by renewable electricity can achieve simultaneously low WUE and low carbon intensity: renewable energy sources such as wind and solar carry near-zero Scope 2 water footprints, meaning that a liquid-cooled facility powered by renewables achieves best-in-class performance on both dimensions simultaneously. This convergence is not merely theoretical: it represents the design target toward which the industry is moving, as evidenced by the major cloud providers' commitments to achieve 100 percent renewable energy matching and water positivity by 2030. The implication for policymakers is that carbon transition incentives and water conservation mandates are, in many cases, mutually reinforcing rather than competing objectives, and that joint policy frameworks addressing both simultaneously would be more efficient than separate regulatory tracks.

The geographic dimension of the freshwater challenge presents a more intractable problem. The concentration of AI data center construction in water-stressed regions reflects market forces cheap land, favorable tax treatment, proximity to electricity infrastructure that are not automatically corrected by technology improvements or corporate sustainability commitments. Even the most water-efficient immersion-cooled facility, if sited in a severely water-stressed region, will contribute to the depletion of a shared resource that local communities depend on for domestic and agricultural use. The finding of MSCI (2025) that approximately 30 percent of data centers currently under construction are located in regions projected to face intensifying water scarcity by 2050 suggests that current siting decisions are not adequately accounting for long-term water risk. Addressing this requires policy interventions beyond the facility level: mandatory water availability impact assessments in planning and permitting processes, zoning restrictions on water- intensive industrial development in stress-prone watersheds, and economic instruments that internalize the true social cost of freshwater consumption in industrial tariff structures.

The equity dimension of AI data center water consumption deserves explicit recognition in the research and policy discourse. Water-intensive AI infrastructure is disproportionately concentrated in regions including the southwestern United States, parts of southern Europe, Chile, and sub- Saharan Africa where freshwater is already a constrained resource shared among agricultural communities, municipal users, and natural ecosystems. The economic benefits of data center investment tax revenues, employment, digital infrastructure rarely accrue to the same communities that bear the costs of reduced water availability. In Newton County, Georgia, residents reported declining water pressure attributable to Meta's permitted consumption of over 500,000 gallons per day from a shared municipal supply. In Chile, environmental regulators partially revoked Google's data center construction permits following community opposition rooted in water rights concerns. These cases represent a pattern of structural inequity in which the economic gains of AI infrastructure accrue globally while the environmental costs are highly localized. Addressing this inequity requires not only technological solutions but governance reforms that give affected communities meaningful standing in data center permitting decisions and enforceable rights to information about water consumption impacts. The development of community water impact assessment frameworks analogous to environmental impact statements but specifically focused on freshwater resource competition represents a critical governance innovation that the current literature has not yet systematically addressed.

The role of intelligent systems in water efficiency improvement deserves critical evaluation against realistic deployment timelines and barriers. The RL-based and digital twin-based optimization frameworks reviewed in this study demonstrate impressive theoretical and simulation-validated performance improvements, but their real-world deployment at scale remains limited. The validation of Jadhav and Liu (2026) against one year of Frontier supercomputer operational data is among the most rigorous available, yet it represents a single facility. The generalizability of RL control policies trained on one data center's thermal characteristics to other facilities with different equipment configurations, workload profiles, and cooling architectures is an open research question. The IoT-based predictive leak detection framework of Sunkara and Konakanchi (2025), while achieving strong performance metrics, was validated exclusively on synthetic data a significant limitation that must be addressed through real-world pilot deployments before the system can be recommended for operational use. These gaps do not diminish the promise of intelligent optimization, but they underscore the importance of staged, empirically grounded validation before large-scale deployment, and highlight the need for open benchmark datasets and shared testbeds in the data center research community.

CONCLUSION AND FUTURE SCOPE

This chapter has demonstrated that the freshwater footprint of AI data centers is a problem of immediate and growing severity, one that demands a fundamental reorientation of how sustainability is conceptualized and operationalized in the AI infrastructure sector. The analysis establishes three core conclusions. First, water and carbon are complementary and non- substitutable sustainability metrics: the historical tendency to optimize exclusively for energy efficiency through evaporative cooling has systematically worsened freshwater consumption, and genuine sustainability requires the simultaneous minimization of both dimensions through integrated design choices. Second, technologically viable pathways to water-efficient AI data center operation exist today: closed-loop liquid cooling and immersion systems can reduce direct water consumption to near zero, RL-based intelligent control can recover over 90 percent of theoretical energy savings under real-world constraints, and predictive maintenance systems can manage the operational risks of liquid infrastructure.

Third, technology alone is insufficient: the concentration of AI data centers in water-stressed regions, the absence of comprehensive multi- scope water disclosure requirements, and the persistence of market incentives that prioritize cheap land and energy over water availability all require policy interventions that internalize the true social cost of freshwater consumption in industrial operations. The urgency of these interventions is heightened by the pace of AI expansion: the decisions made in the next five years about data center siting, cooling technology selection, and water governance frameworks will shape the freshwater implications of AI infrastructure for decades to come. Addressing AI's water footprint is therefore not merely an environmental imperative it is a prerequisite for the long-term social license of artificial intelligence as a global technology.

Table 4. Future research directions for AI data center water sustainability

Research Direction	Objective	Methodology Suggested	Expected Impact
Scope 3 Embodied Water Accounting	Quantify water embedded in AI hardware manufacturing and supply chains	Life-cycle assessment (LCA) combined with semiconductor fab water- use datasets	Complete total water footprint model for AI infrastructure
Waterless Cooling for AI Chips	Develop dry or near-dry cooling technologies viable at rack densities >120 kW	Materials science R&D; microfluidic chip-level heat exchangers	Eliminate direct freshwater dependency for new AI data centers
AI-Adaptive Cooling Control	Deploy real-time RL controllers trained on live facility data across multiple DC types	Multi-site benchmarking using LC-Opt and Green-DCC frameworks	20-30% additional reduction in cooling energy and water use
Reclaimed Water Integration	Standardize use of treated municipal wastewater as cooling water source	Pilot studies at hyperscale facilities; water quality and biofouling analysis	Decouple data centers from potable water competition
Region-Specific WUE Frameworks	Expand WUE datasets to Southeast Asia, South Asia, and Latin America	Hourly climate-energy-infrastructure modelling following Shumba et al. (2025) methodology	Globally equitable water sustainability benchmarks for AI deployment
Low-GWP Dielectric Fluids	Identify or synthesize immersion cooling fluids without global warming potential	Chemical screening and thermal performance testing	Enable safe, large-scale immersion cooling without climate co-harms
Community Water Rights Framework	Develop governance models balancing DC water use with community and ecosystem needs	Interdisciplinary policy, legal, and hydrology research	Equitable and legally defensible water allocation for AI infrastructure

The path toward water-sustainable AI infrastructure is navigable, but it requires coordinated action across four dimensions simultaneously. Technologically, the industry must accelerate the transition from open evaporative cooling to closed-loop liquid and immersion systems, supported by investment in low-GWP dielectric fluid development and standardized server form factors compatible with immersion deployment. Operationally, intelligent management systems integrating digital twins, reinforcement learning control, and predictive IoT maintenance must be developed, validated on real-world facilities, and made accessible beyond the small number of hyperscale operators currently capable of building such infrastructure independently.

Metrologically, the WUE metric must be upgraded from a facility-level, Scope-1-only measure to a workload-level, multi-scope framework that captures the true freshwater impact of AI operations and enables meaningful regulatory and corporate accountability. And institutionally, water governance frameworks must evolve to treat freshwater as a shared public resource whose allocation to industrial uses including AI data centers is subject to democratic oversight, community participation, and enforceable equity safeguards. The convergence of these four dimensions is not merely aspirational: it reflects the minimum standard of stewardship that the global freshwater challenge demands from an industry whose growth trajectory is one of the most rapid in the history of industrial infrastructure. Future research should prioritize the seven directions identified in the table below, spanning technical, methodological, and governance dimensions. Each direction addresses a specific gap identified in the current literature and is positioned to generate actionable outcomes relevant to both the research community and industry practitioners.

REFERENCES

Chang, Q., Huang, Y., Liu, K., Xu, X., Zhao, Y., & Pan, S. (2024). Optimization control strategies and evaluation metrics of cooling systems in data centers: A review. *Sustainability (Basel), 16*(16), 7222. DOI: 10.3390/su16167222

de Vries-Gao, A. (2026). The carbon and water footprints of data centers and what this could mean for artificial intelligence. *Patterns (New York, N.Y.), 7*(1), 101430. DOI: 10.1016/j.patter.2025.101430 PMID: 41583976

Jadhav, S. A., & Liu, Z. (2026). Digital Twin-Based Cooling System Optimization for Data Center. *arXiv preprint arXiv:2603.01198.*

Koomey, J. (2011). Growth in data center electricity use 2005 to 2010. *A report by Analytical Press, completed at the request of The New York Times, 9*(2011), 161.

Latif, I., Shafique, M. A., Ullah, H., Newkirk, A. C., Yu, X., & Munir, A. (2025). Cooling Matters: Benchmarking Large Language Models and Vision-Language Models on Liquid-Cooled Versus Air-Cooled H100 GPU Systems. *arXiv preprint arXiv:2507.16781.*

Lei, N., Lu, J., Shehabi, A., & Masanet, E. (2025). The water use of data center workloads: A review and assessment of key determinants. *Resources, Conservation and Recycling, 219*, 108310. DOI: 10.1016/j.resconrec.2025.108310

Li, P., Yang, J., Islam, M. A., & Ren, S. (2025). Making ai less' thirsty'. *Communications of the ACM, 68*(7), 54–61. DOI: 10.1145/3724499

Masanet, E., Shehabi, A., Lei, N., Smith, S., & Koomey, J. (2020). Recalibrating global data center energy-use estimates. *Science, 367*(6481), 984–986. DOI: 10.1126/science.aba3758 PMID: 32108103

Masanet, E., Shehabi, A., Lei, N., Smith, S., & Koomey, J. (2020). Recalibrating global data center energy-use estimates. *Science, 367*(6481), 984–986. DOI: 10.1126/science.aba3758 PMID: 32108103

Moore, J. D., Chase, J. S., Ranganathan, P., & Sharma, R. K. (2005, April). Making Scheduling" Cool": Temperature-Aware Workload Placement in Data Centers. In *USENIX ATC, General Track* (pp. 61-75).

Naug, A., Guillen, A., Kumar, V., Greenwood, S., Brewer, W., Ghorbanpour, S., . . . Sarkar, S. (2025). Lc-opt: Benchmarking reinforcement learning and agentic ai for end-to-end liquid cooling optimization in data centers. *arXiv preprint arXiv:2511.00116.*

Sarkar, S., Naug, A., Guillen, A., Gundecha, V., Gutiérrez, R. L., Ghorbanpour, S., & Bash, C. (2025, April). Hierarchical multi-agent framework for carbon-efficient liquid-cooled data center clusters. *Proceedings of the AAAI Conference on Artificial Intelligence, 39*(28), 29694–29696. DOI: 10.1609/aaai.v39i28.35370

Shehabi, A., Smith, S., Sartor, D., Brown, R., Herrlin, M., Koomey, J., ... & Lintner, W. (2016). United states data center energy usage report.

Shumba, N., Tshekiso, O., Li, P., Fanti, G., & Ren, S. (2025, July). A water efficiency dataset for African data centers. In *Proceedings of the ACM SIGCAS/SIGCHI Conference on Computing and Sustainable Societies* (pp. 453-460). DOI: 10.1145/3715335.3735483

Sunkara, K. C., & Konakanchi, R. (2025). Smart IoT-Based Leak Forecasting and Detection for Energy-Efficient Liquid Cooling in AI Data Centers. arXiv preprint arXiv:2512.21801.

KEY TERMS AND DEFINITIONS

Coolant Distribution Unit (CDU): A mechanical unit that manages the distribution of liquid coolant throughout a direct-to-chip or immersion cooling system, typically incorporating pumps, heat exchangers, and flow regulation valves to maintain optimal coolant temperature and pressure.

Digital Twin: A high-fidelity computational model that replicates the physical and thermal behavior of a real data center or cooling infrastructure, enabling simulation-based optimization and controller training without risk to live production systems.

Immersion Cooling: A data center cooling technique in which servers and electronic components are fully submerged in a dielectric fluid that directly absorbs heat, achieving PUE values near 1.02 and near-zero direct water consumption.

Power Usage Effectiveness (PUE): A metric equal to total data center facility energy divided by IT equipment energy; a value of 1.0 represents perfect efficiency. The global average for hyperscale facilities is approximately 1.2.

Reinforcement Learning (RL): A machine learning paradigm in which an agent learns to optimize a policy through interaction with an environment, applied in this context to data center cooling control for dynamic, workload-adaptive energy and water optimization.

Scope 1 / 2 / 3 Water Footprint: A three-tier framework: Scope 1 covers direct on-site cooling water; Scope 2 covers indirect water from electricity generation; Scope 3 covers embodied water in hardware manufacturing and supply chains.

Water Usage Effectiveness (WUE): A metric standardized as ISO/IEC 30134-9:2022, calculated as total annual site water usage in liters divided by IT equipment energy in kWh; lower WUE indicates greater water efficiency. The industry average is approximately 1.8 L/kWh.

Water-Carbon Trade-off: The inverse relationship between water efficiency and energy/carbon efficiency in data center cooling: evaporative systems improve PUE but worsen WUE, while dry systems conserve water but consume more electricity and produce greater carbon emissions.

Chapter 3
Cooling Technologies and Comparative Sustainability Studies:
AI Workloads, High-Performance Computing, and Energy-Water Relationships

Riadh Hanachi
http://orcid.org/0000-0003-1657-1652
Laboratory Desalination and Natural Water Valorization, Center of Water Researches and Technologies (CERTE), Tunisia

Emna Melliti
http://orcid.org/0000-0001-8539-7315
Laboratory Desalination and Natural Water Valorization, Center of Water Researches and Technologies (CERTE), Tunisia

Hamza Elfil
Laboratory Desalination and Natural Water Valorization, Center of Water Researches and Technologies (CERTE), Tunisia

ABSTRACT

This chapter examines the sustainability challenges of AI-intensive and high-performance computing (HPC) data centers, focusing on cooling technologies and their impacts on energy and water use. It situates these infrastructures within the broader energy–water nexus and compares major cooling approaches—air, water-based, and liquid—using key metrics such as PUE, WUE, and CUE. The chapter highlights how the rapid growth of AI workloads, marked by high power densities

DOI: 10.4018/979-8-3373-8434-4.ch003

and dynamic thermal constraints, increases resource demand and complicates thermal management. It also explores regional variations, showing how climate, water availability, and electricity mix affect sustainability outcomes. By integrating technical, environmental, and policy perspectives, it identifies trade-offs between energy efficiency and water consumption and emphasizes context-aware design. Ultimately, it argues that combining thermal modeling, energy optimization, hydrological analysis, and carbon assessment supports more informed infrastructure and location decisions.

1. GENERAL INTRODUCTION: AI, HPC AND GROWING ENVIRONMENTAL PRESSURE

1.1 Expansion of AI and HPC workloads

Over the past decade, the rapid growth of artificial intelligence (AI) and high-performance computing (HPC) has transformed the scale and architecture of digital infrastructures. Large deep learning models, foundation models and generative AI systems require substantial computing power, accelerating the development of hyperscale and specialized data centers with high power densities and near-continuous operation (Masanet et al., 2020; Jones, 2018; Yadav, K.D. 2024).

Unlike traditional IT workloads, AI and HPC applications rely heavily on accelerators such as GPUs, TPUs and ASICs, which process simultaneously large volumes of data. Training large models can involve thousands of accelerators operating simultaneously, producing rack power densities of some tens of kilowatts and generating temporary thermal peaks. These conditions confront conventional cooling systems, which were initially designed for more stable and less concentrated loads (Patterson et al., 2021; Li et al., 2025; ASHRAE, 2021).

Large-scale inference is generally less energy-intensive per task than training, but its geographic distribution near end users limits the ability to locate workloads only in cold climates regions or low-carbon electricity places. This makes the environmental management of AI infrastructures more complex, especially when cooling demand, energy availability and carbon intensity vary across regions (Javier Jareño et al., 2026).

1.2 Intensification of the environmental footprint of digital infrastructures

The development of AI and HPC data centers is also accompanied by a marked increase in their ecological footprint. Energy efficiency gains have slowed data center electricity growth, but AI could reverse the trend. (Masanet et al., 2020; IEA, 2023 ; Weike Zhang et al; 2023).

In addition to electricity, data centers also consume water, both directly and indirectly, and this is an ecological footprint, which is becoming increasingly worrying, especially in recent years. "Data centers directly consume water, mostly in evaporative cooling systems, and indirectly, in generating their electricity, especially if they receive power from a thermal power plant with a water cooling system" (Van Vliet et al., 2012). Water withdrawals for data centers have, become, in certain regions, a matter of public debate, especially in regions with water stress situation (Mytton, 2021).

Recent projections suggest that water demand linked to AI infrastructures could reach some billion cubic meters annually in the second half of the decade, depending on the deployment scenarios and technological strategies adopted (Li et al., 2025; IEA, 2023). This dynamic raises questions about the long-term sustainability of digital growth in regions where competition for water resources is already intense.

1.3 Objectives and positioning of the chapter

While previous studies have provided important estimates of global data-center electricity demand and efficiency trends, the present chapter focuses on the comparative sustainability performance of cooling technologies under emerging AI and HPC workloads. Masanet et al. (2020) and other studies, recalibrated global data-center energy-use estimates and showed that data centers accounted for approximately 1-2% of global electricity use (approximately 300-400 TWh annually). The International Energy Agency (IEA) projects that data center electricity consumption could double by 2030, potentially reaching 600-800 TWh annually. However, energy use alone does not capture the full environmental trade-off of cooling-system selection, particularly where water availability, electricity carbon intensity and high-density AI accelerators impose additional constraints.

The main objective is twofold:

1. Provide a structured comparison of major cooling architectures using PUE, Water Usage Effectiveness WUE, and Carbon Usage Effectiveness CUE.
2. Examine the environmental trade-offs associated with these technologies in different climatic, energy and institutional contexts.
3. Examine AI infrastructure sustainability by combining technical analysis, environmental assessment, and governance considerations.

2. THERMAL CHARACTERISTICS OF AI AND HPC WORKLOADS

The design of servers for AI and HPC is changing rapidly, creating new challenges for modern data centers, particularly in terms of thermal management. Unlike traditional computing workloads, AI and HPC applications consume large quantity of power, with energy demand fluctuating significantly over short periods. These workloads are also highly sensitive to temperature variations, making the selection of an appropriate cooling system essential. As a result, data centers supporting AI and HPC need advanced cooling solutions capable of adapting to rapid changes in power consumption and heat generation. Moreover, these infrastructures consume important amounts of energy and water (SeyedMorteza et al., 2021).

2.1 Thermal constraints of HPC systems

Modern HPC and AI accelerators impose severe thermal constraints because most of their electrical input power is converted into heat inside a compact package. While conventional server CPUs commonly operate in the range of 200–400 W per socket, recent AI/HPC accelerators often operate between 300 and 750 W per device, and next-generation modules may reach approximately 1000 W or more. For example, AMD reports a default TDP of 360 W for the 128-core EPYC 9754 CPU, with a configurable TDP range of 320–400 W. For accelerators, AMD reports a typical board power of 750 W peak for the Instinct MI300X, while NVIDIA H100 SXM-class accelerators are commonly specified around 700 W.

At the rack level, these values rapidly translate into high thermal density. An 8-GPU server using 700 W accelerators dissipates about 5.6 kW from GPUs alone, before accounting for CPUs, memory, networking, storage and power-conversion losses. A rack containing several such nodes can therefore exceed 30–40 kW/rack, which is the range where airflow management becomes increasingly difficult and liquid-assisted cooling is often considered.

Thermal throttling occurs when the processor reduces frequency, voltage, or boost behavior to remain within its power and temperature envelope. For CPUs, this is generally controlled through the thermal control circuit and package temperature limits. For GPUs, thermal management is controlled through firmware and telemetry tools such as nvidia-smi, which expose power limits, temperature limits, and clock-throttling reasons. NVIDIA documentation and developer discussions indicate that GPU thermal actions include software throttling near the maximum operating temperature and hardware slowdown at higher thermal limits; exact values depend on the GPU, server platform, firmware, and OEM thermal design. Therefore, the manuscript should avoid claiming a universal throttling temperature, and instead state that throttling is triggered when device junction or package temperature approaches the vendor-defined thermal limit.

2.2 Distributed inference and geographic constraints

Inference workloads, although generally less energy intensive per computing unit than training, present another complexity: their geographic distribution. To minimize latency and ensure quality of service, inference workloads are often deployed in data centers located close to end users.

This constraint reduces the possibility of locating all AI infrastructures in regions with cold climates or low carbon intensity. Training can be relocated for efficiency, but it is not the case for inference which complicate environmental optimization. Table 1 summarizes representative TDP or board-power values for modern CPUs and accelerators used in HPC and AI data centers.

Table 1. Representative TDP/TBP values for modern HPC CPUs and accelerators

Device	Type	Representative power rating	Thermal relevance
AMD EPYC 9754	Server CPU	360 W default TDP; 320–400 W configurable TDP	High-core-count CPU; contributes substantially to node-level heat load
Intel Xeon Max / high-end Xeon class	Server CPU	~300–350 W class	Typical upper range for modern HPC CPU sockets
NVIDIA A100 SXM	GPU accelerator	~400 W	Previous-generation AI/HPC accelerator; air or liquid cooling depending on system density
NVIDIA H100 SXM	GPU accelerator	~700 W	High-density AI/HPC accelerator; often drives rack densities above conventional air-cooling limits
NVIDIA H200 SXM	GPU accelerator	~700 W	Similar power envelope to H100, with larger HBM3e memory capacity
AMD Instinct MI300X	GPU accelerator	750 W peak TBP	High-density AI/HPC accelerator; strong liquid-cooling relevance
NVIDIA B200 / Blackwell-class modules	GPU accelerator	~1000 W class	Next-generation AI accelerator power level; generally strengthens need for direct liquid cooling

2.3 Thermal specificities of artificial intelligence workloads

AI workloads, especially those linked to training large models (large language models, generative models, multimodal systems), exhibit distinct thermal characteristics.

2.3.1 Training large-scale models

AI workloads differ not only in their total energy demand but also in their temporal power profile. Training workloads generally operate accelerators at high and sustained utilization for long periods because forward propagation, backpropagation, gradient synchronization and optimizer steps are repeated over many iterations. As a result, training tends to have a relatively low peak-to-average power ratio, typically around 1.1–1.3 at the accelerator or node level under stable distributed training. In contrast, inference workloads are more bursty because request arrival rates, sequence lengths, batching efficiency and latency constraints vary over time. Consequently, inference can show a higher peak-to-average power ratio, commonly around 1.5–3.0, and potentially higher under low-utilization or highly bursty serving conditions. Patterson et al. (2021).

A useful metric to quantify this variability is: $\mathbf{PAR = P_{avg} / P_{peak}}$

where PAR is the peak-to-average power ratio, Ppeak is the short-duration maximum power demand, and Pavg is the average power over the measurement interval.

Table 2 summarizes indicative peak-to-average power characteristics for common AI workloads. Training typically produces a sustained thermal load close to the accelerator power envelope, whereas inference can generate more variable power demand because request rates and batching conditions fluctuate. Therefore, thermal design for AI data centers should consider both average energy consumption and short-duration peak power.

Table 2. Indicative power-variability characteristics of AI workloads

AI workload type	Typical operating pattern	Indicative peak-to-average power ratio	Thermal implication
Large-scale training	Sustained accelerator utilization over long jobs	1.1–1.3	High continuous heat load; cooling must support sustained near-peak operation
Fine-tuning	Shorter training phases; moderate utilization variability	1.2–1.6	Moderate variability; thermal cycling may occur depending on job scheduling
Batch inference	High batching efficiency; throughput-oriented	1.2–1.8	More stable than online inference, but load depends on batch size and queue depth
Online / real-time inference	Bursty request arrivals and latency constraints	1.5–3.0	Larger transient peaks; cooling and power delivery must handle short-term load spikes
Low-utilization inference	Idle periods interrupted by bursts	>3.0 possible	High variability despite lower average energy use

2.4 Transition to high energy density architectures

The combined rise of HPC and AI is leading to a structural transformation of data center architectures. The massive integration of specialized accelerators increases the energy density per unit area, modifying internal thermal balances.

2.4.1 Air cooling technology

Air-cooling relies on the circulation of cold airflows to dissipate heat generated by IT equipment. However, the volumetric heat capacity of air being relatively low, its efficiency decreases when the thermal density increases. At levels above 30–40 kW per rack, maintaining homogeneous air distribution becomes complex and energy-intensive, requiring high flow rates and fine management of hot and cold aisles.

These constraints lead to an increase in the energy consumption of auxiliary systems (fans, compressors), which can degrade the Power Usage Effectiveness (PUE). In addition, the proliferation of ventilation equipment increases the risks of localized thermal imbalance (hot spots).

2.4.2 Growing Adoption of Liquid Cooling

Faced with these limitations, many operators are adopting liquid cooling solutions, including direct-to-chip cooling and immersion cooling. Liquids have a higher heat capacity and thermal conductivity than air, which allows high heat fluxes to be efficiently dissipated over small surface areas.

This transition not only modifies the configuration of racks, but also the overall architecture of data centers, including hydraulic circuits, heat exchangers and heat recovery systems. It also opens up the possibility of operating at higher fluid temperatures, facilitating free cooling and the recovery of waste heat. However, liquid cooling improves thermal management but introduces new environmental impacts, requiring a systemic assessment of energy, water, and materials.

2.5 Sustainability assessment implications

The specific thermal characteristics of AI and HPC workloads directly influence data center environmental performance indicators. Higher energy density can improve relative efficiency (e.g., by reducing distribution losses), but it can also increase absolute energy and water consumption due to the increase in total computing volume.

Load variability and reliance on specialized accelerators reinforce the importance of adaptive, real-time optimized cooling solutions. They also highlight the need to evaluate cooling technologies not only based on their thermal efficiency, but also their interactions with local energy and water systems.

3. OVERVIEW OF AI AND HPC DATA CENTER COOLING TECHNOLOGIES

Due to the increasing demand from workloads using AI and HPC, the methods used to cool data centers have drastically changed. Cooling methods used today can be divided into 3 major categories: (i) air cooling; (ii) water-based systems which may also include evaporative systems; and (iii) direct/immersion liquid cooling solutions (Figure 1). The majority of the time, cooling technologies will be deployed in a combined hybrid fashion, where the cooling methods are chosen based on local climate, energy availability, and their respective costs.

Figure 1. Cooling methods selection by Rack Heat Density

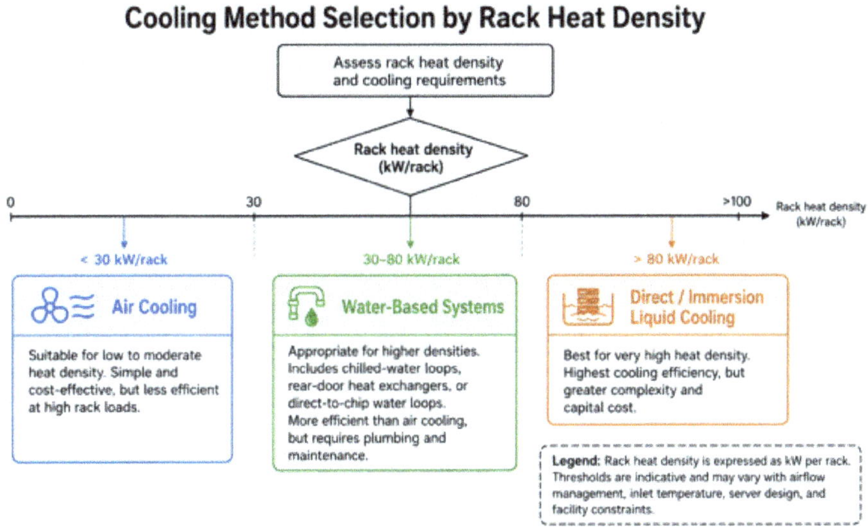

The comparative evaluation of these solutions requires an analysis not only of their thermal efficiency, but also of their energy, water and environmental implications.

3.1 Air cooling: technological maturity and physical limits

3.1.1 Operating principles

Air-cooling has historically been the dominant technology in data centers. It relies on removing heat from IT equipment using conditioned air via Computer Room Air Conditioner CRAC or Computer Room Air Handler CRAH systems. Typical configurations include a raised floor to provide cold air distribution, Hot/cold aisle separation and Ventilation assisted by fans integrated into the servers as illustrated in Figure 2.

Figure 2. Components of efficient cooling systems

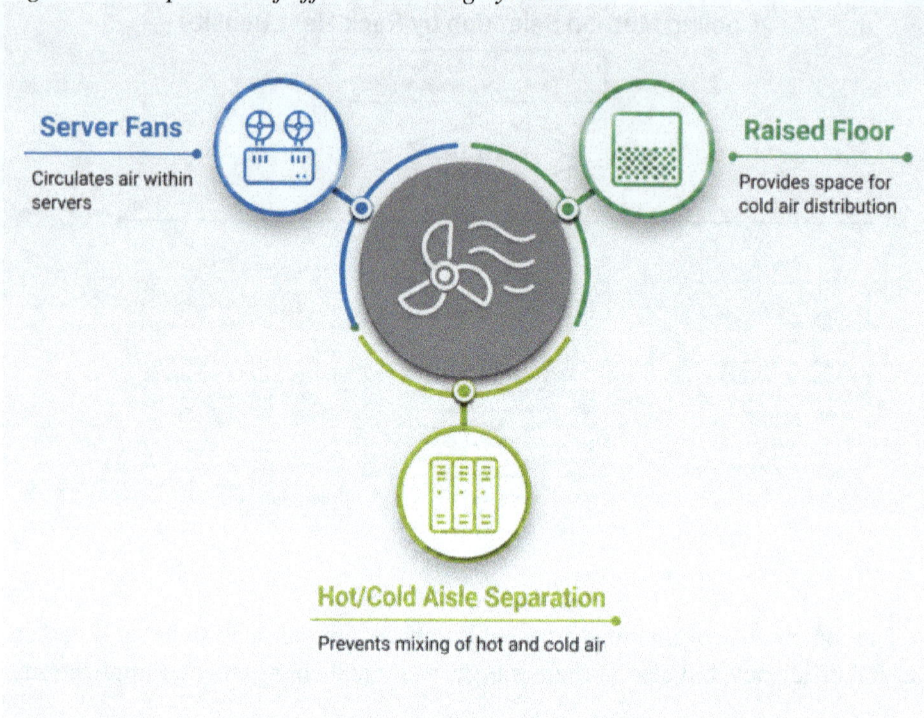

Air cooling works based on the principle of forced convection; cold air travels through the servers to absorb heat from them, and then leaves the room via a heat exchanger that is typically fed by a chilled water source or through a refrigeration system.

3.1.2 Advantages

There are multiple benefits to using air-cooling systems as shown in Figure 3:

Figure 3. Advantages of air cooling systems.

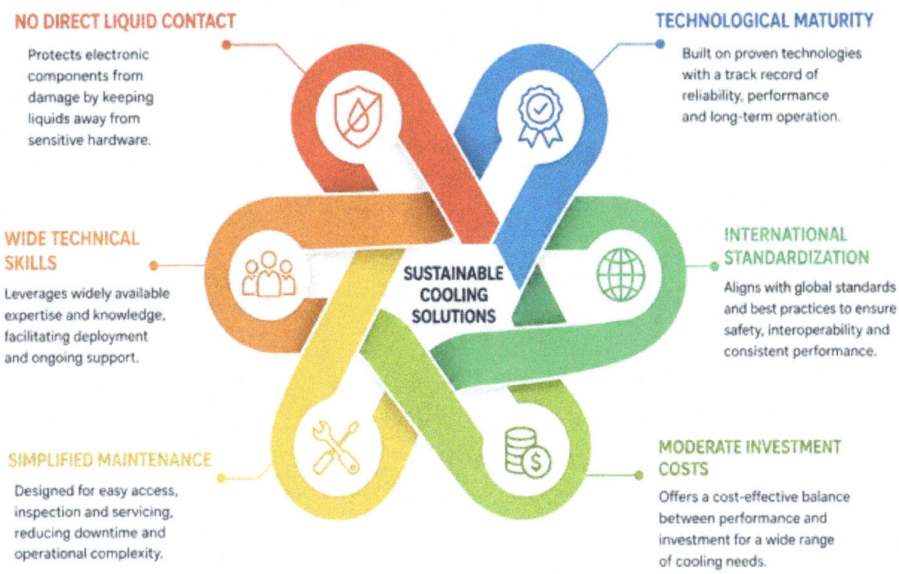

In temperate or cold climates, the use of outside air (free cooling) can significantly reduce energy consumption linked to refrigeration compression.

3.1.3 Limitations in the AI/HPC context

However, air has a relatively low volumetric heat capacity. As power densities per rack exceed 30–40 kW, several limitations have been shown in Figure 4.

Figure 4. Air cooling limitations I AI/HPC

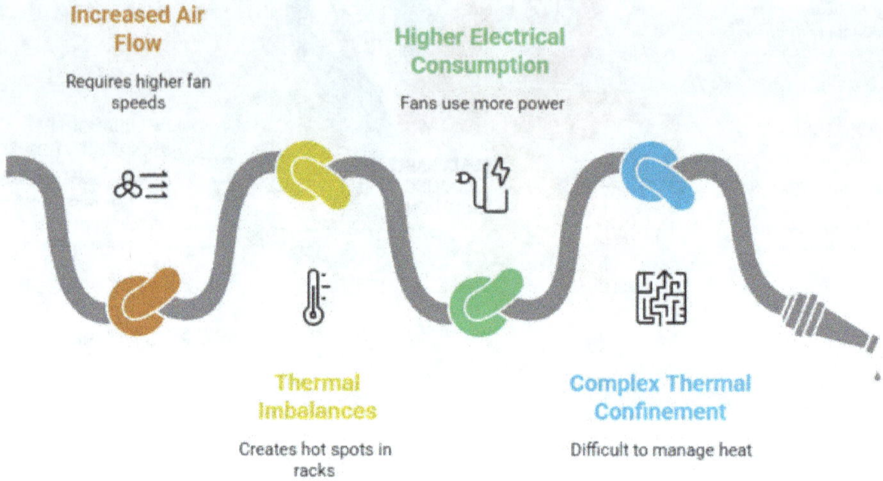

The cooling capacity of an air-based system can be approximated using the convective heat-transfer relation: $\mathbf{q=h\,.A\,.\,\Delta T}$

Where q is the heat removed from the rack, h is the convective heat-transfer coefficient, A is the effective heat-transfer area, and ΔT is the temperature difference between the server surface or exhaust stream and the cooling air. At high rack densities, increasing q requires either higher airflow rates, larger temperature gradients, or improved heat-transfer efficiency.

In racks exceeding 40 kW/rack, CFD analysis can help evaluate whether the supplied airflow can maintain inlet temperatures within acceptable limits or whether liquid-assisted cooling is required. The Figure 5 highlights the progressive increase in cooling-energy demand with increasing rack density.ASHRAE. (2021).

Figure 5. Conceptual effect of rack heat density on PUE and hotspot risk in air-cooled data centers

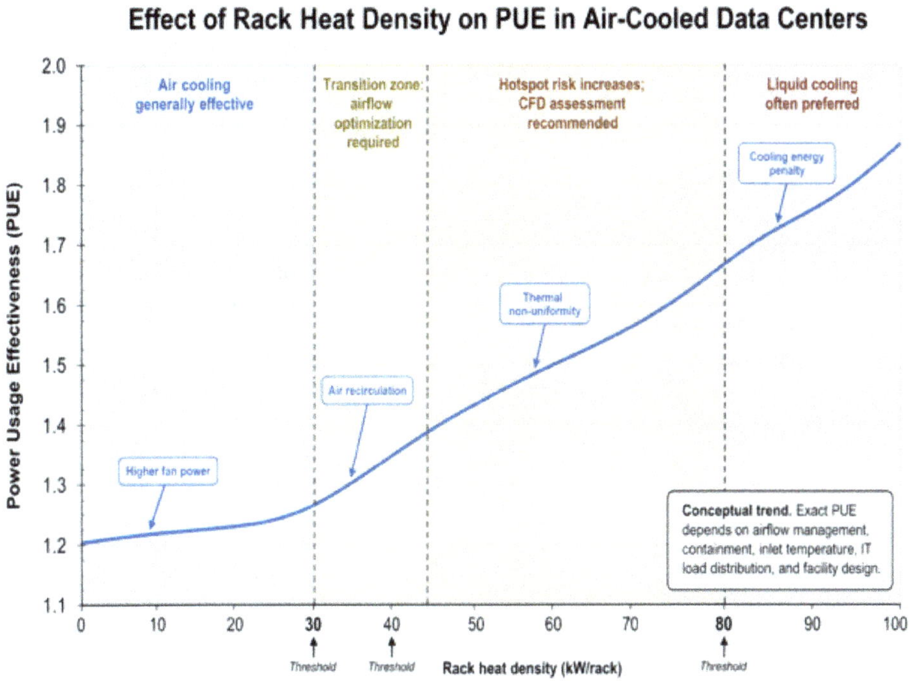

3.2 Water-based systems: energy efficiency and water dependence

Hydraulic systems constitute a major development enabling thermal performance to be improved while reducing the electrical consumption of cooling systems.

3.2.1 Cooling by chilled water systems

In these systems, chilled water circulates through heat exchangers to absorb heat from the internal air. The coolers ensure heat dissipation to the outside.

Figure 6. Benefits of chilled water-cooling

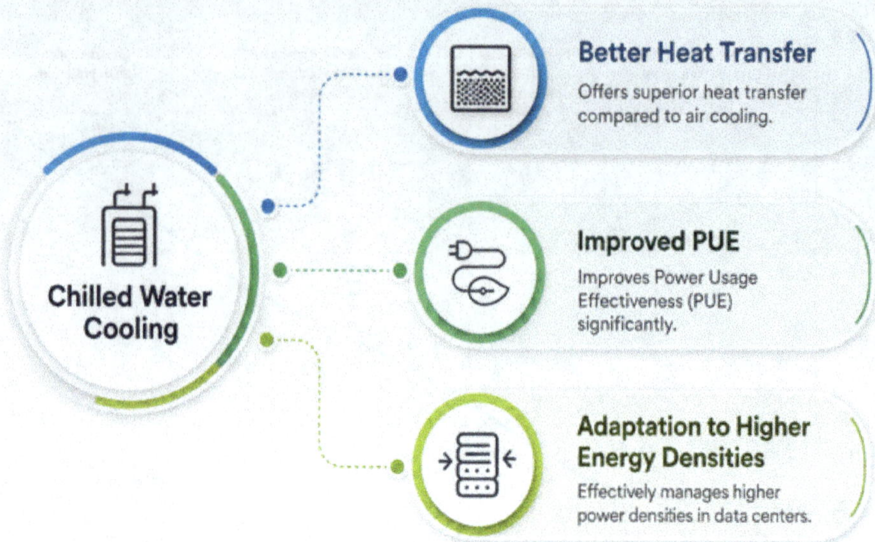

Figure 6 cites different Benefits in relation with chilled water-cooling. However, refrigeration compression systems associated with chillers can be energy intensive, particularly in hot climates.

3.2.2 Evaporative cooling

Evaporative cooling exploits the physical principle that evaporating water absorbs heat (Cuce, P. M. 2025). Cooling towers are frequently used in this context. Figure 7 presents the benefits and boundaries of Evaporative cooling technology.

Figure 7. Dimensions of evaporative cooling

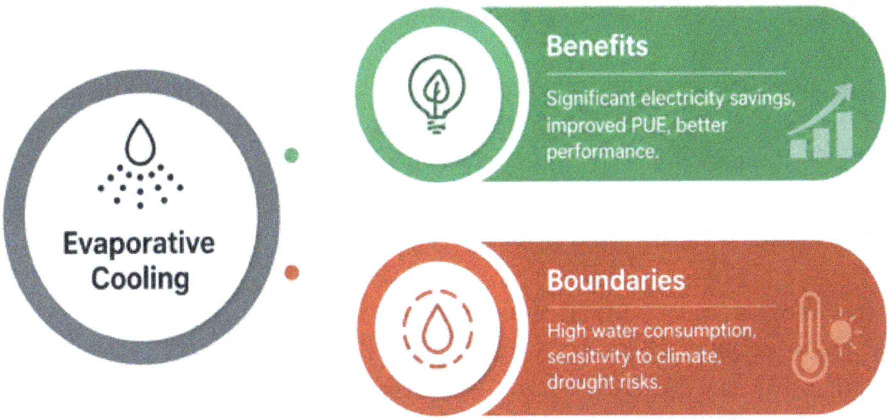

In arid regions or those experiencing water stresses, this technology can generate tensions between industrial, agricultural and municipal uses of water.

Following Lei and Masanet's climate- and technology-specific modelling framework, WUE was considered as a climate-dependent indicator controlled by cooling technology, outdoor dry-bulb temperature, relative humidity, economizer operation, and evaporative cooling availability.

Table 3 presents indicative WUE ranges for evaporative cooling under different climatic conditions. The classification is based on wet-bulb depression, $T_{db}-T_{wb}$, which controls the evaporative cooling potential.

Table 3. Climate-dependent wet-bulb depression and indicative WUE ranges for evaporative cooling in data centers

Climate type	Typical wet-bulb depression, (T_db-T_wb)	Evaporative cooling potential	Indicative WUE range, L/kWh
Hot-dry / arid	> 12 °C	High	1.5–2.0
Semi-arid / warm-dry	8–12 °C	Moderate to high	1.0–1.5
Temperate / mixed	4–8 °C	Moderate	0.5–1.0
Hot-humid	< 4 °C	Low; hybrid/mechanical backup often required	0.5–0.8

3.3 High-performance computing (HPC) and Artificial Intelligence (AI): an example of high-energy density thermal management

Advanced liquid cooling technology is experiencing increasingly faster adoption rates due to their capability in accommodating the increasing need to accommodate HPC and AI workloads.

3.3.1 Direct-to-chip liquid cooling

In a direct-to-chip configuration, a heat transfer fluid can be circulated through cold plates in physical contact with the critical components (CPU/GPU) of the computer. The heat is then transferred from the cold plates, to a second heat transfer circuit and dissipated via a heat exchanger. A variety of benefits can be obtained from this application, as indicated in Figure 8.

Figure 8. Benefits of direct to chip cooling

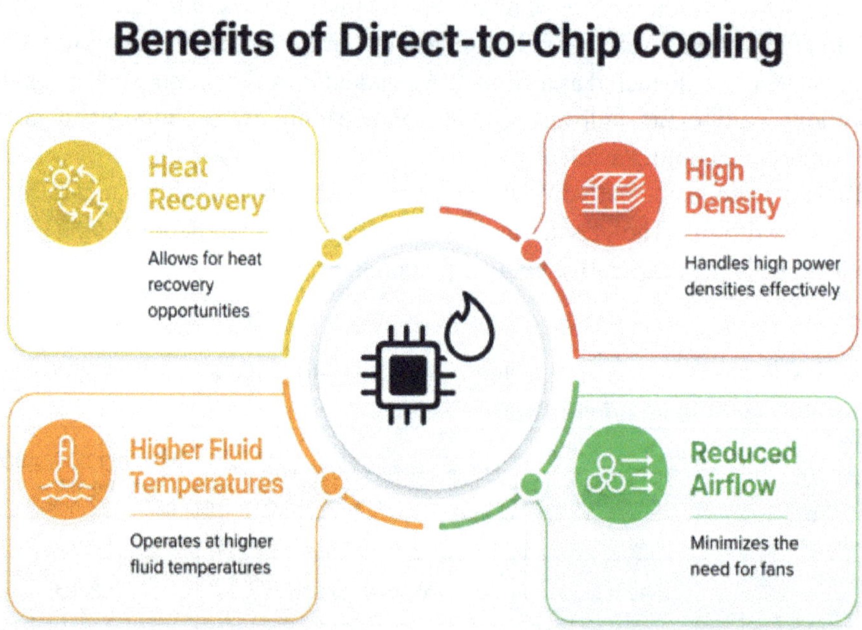

This technology makes it possible to support racks exceeding 80 kW, or even more in certain HPC configurations.

3.3.2 Immersion cooling

Immersion consists of immersing the servers in a non-conductive dielectric fluid. Two variants exist: Single-phase immersion: the fluid remains in the liquid state and two-phase immersion: the fluid evaporates on contact with hot components then condenses. Benefits and Constraints of Immersion cooling have been presented in Figure 9.

Figure 9. Immersion cooling

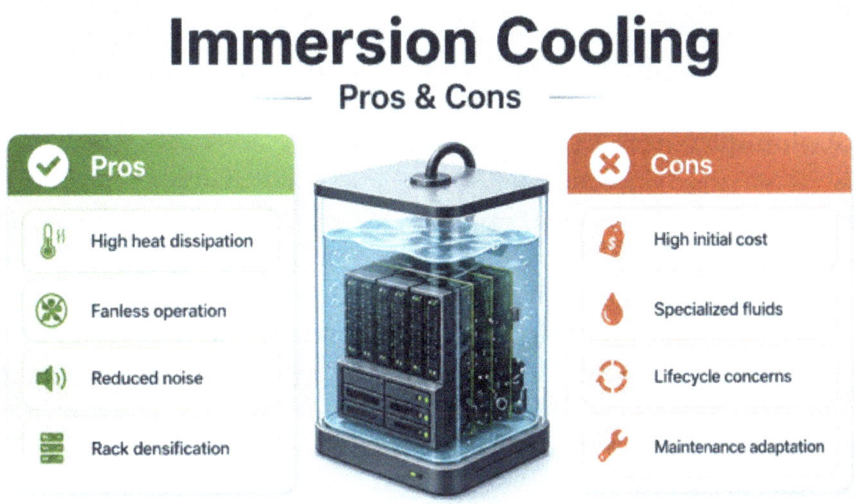

3.3.3 Environmental implications

Liquid cooling can reduce reliance on refrigeration compressors and limit auxiliary energy consumption. However, its overall environmental impact depends on:

- From the source of electricity,
- The type of fluid used,
- End-of-life equipment management,
- Possibilities for thermal recovery and valorization.

3.4 Hybrid configurations and contextual adaptation

A hybrid approach to cooling is used by many data centers, which typically utilize a combination of different technologies to maximize the effectiveness and flexibility of cooling. Air cooling is commonly used for low to moderate load requirements while direct to chip cooling is used for computing with the greatest density or accelerator-type workloads. In some cases, evaporative cooling towers may provide seasonal cooling or additional support as well. Similarly, dry coolers may be also deployed in temperate regions. The use of hybrid cooling technologies adds flexibility to help support the variability of AI workloads and to optimise the ability to adapt to the varying climatic conditions of the location/pathway. The final determination regarding which cooling technology to use is often context-dependent and is impacted by a number of factors, including the climate specific to that region/city, availability and cost of fresh water, energy sources, regulations, and funding constraints (both Capital Expenditure CAPEX and Operating Expenditure OPEX).

3.5 Preliminary comparative synthesis

Currently, there are numerous major themes developing in data center cooling strategies. Air-based cooling is still appropriate for mid/high power densities and geographies that are well suited to use it. Evaporative cooling methods can improve energy performance considerably; however, these methods exert additional pressure on water resources. Advanced liquid cooling technologies are now being developed as a streamlined approach for dealing with high-density AI/HPC workloads and need to undergo a complete environmental evaluation at the system level.

4. METHODOLOGICAL FRAMEWORK FOR COMPARING COOLING TECHNOLOGIES

Evaluating AI and HPC cooling technologies requires a robust methodology combining standard metrics with holistic energy–water–carbon considerations. While some indicators such as Power Usage Effectiveness (PUE) have become the established industrial reference to reflect upon, their individual use alone does not provide enough of a basis to contextualize the true sustainability of digital infrastructure.

This section introduces an organized approach for assessing cooling architectures through the application of a multi-criteria methodology. The methodology combines measurement of energy efficiency, water usage, carbon emission levels as well as contextual measurements into a comprehensive framework for performance assessment.

4.1 Standardized operational indicators

4.1.1 Power Usage Effectiveness

The PUE, developed by The Green Grid, measures the ratio between the total energy consumption of a data center and the energy consumed by IT equipment (IT load):

$$PUE = \frac{\text{Total facility energy}}{\text{IT equipment energy}(kwh)}$$

PUE measures data center energy efficiency, with an ideal value of 1; in practice, top facilities reach ~1.1, while older ones exceed 1.6–2.0. PUE has limitations as a metric; for example, measuring PUE does not determine the overall efficiency of the computing processes and it does not factor in the carbon intensity of the power used to compute. PUE also does not capture the water load that will be consumed in producing the energy consumed by the computing processes and PUE can be artificially improved with high IT utilization. As the total demand for AI workloads increases, improvement in PUE does not necessarily reduce the total impact on the environment.

4.1.2 Water Usage Effectiveness

WUE measures the amount of water consumed directly by the data center per unit of IT energy (The Green Grid. 2021):

$$WUE = \frac{\text{Annual water usage}}{\text{Annual IT energy}(kwh)}$$

It is generally expressed in liters/kWh. The WUE allows cooling technologies to be compared according to their water dependence:

- Air cooling in cold climates: WUE close to zero.
- Evaporative cooling: WUE up to 1–2 L/kWh or more depending on climate.

The WUE Limitations appear especially in distinguish lack between drinking, industrial or recycled water, ignores local water stress and indirect water linked to electricity production. From an energy–water nexus perspective, the WUE must be supplemented by an estimate of the indirect water footprint (The Green Grid. 2021).

4.1.3 Carbon Usage Effectiveness (CUE)

CUE relates CO_2 emissions associated with total data center energy consumption to IT energy:

$$CUE = \frac{\text{CO emissions } (kg)}{\text{IT energy } (kWh)}$$

It mainly depends on PUE, the regional energy mix and the integration of renewable sources. A facility driven by a high-carbon network can maintain a high CUE even if it has an efficient PUE.

4.2 Expanded indicators and systemic approach

Operational metrics need to be incorporated into a more comprehensive framework to prevent a disjointed perspective on sustainability.

4.2.1 Indirect water footprint

Thermal electricity production (coal, gas, nuclear) consumes significant quantities of water for cooling power plants. Thus, a data center with low direct WUE may have a high indirect water footprint if its electricity comes from water-cooled thermal power plants.

The indirect water footprint associated with electricity consumption by IT infrastructure was estimated as: $WF_{\text{indirect}} = WUE_{\text{grid}} \times E_{\text{IT}}$

where WF_{indirect} is the indirect water footprint of electricity use (L), E_{IT} is the electricity consumed by IT equipment (kWh), and WUE_{grid} is the grid-level water-use intensity (*L/kWh*). For a grid supplied by several generation technologies, WUE_{grid} can be calculated as:

$$WUE_{\text{grid}} = \Sigma_{\text{i}}\, s_{\text{i}} \times WUE_{\text{i}}$$

where s_{i} is the electricity share of generation technology iii, and WUE_{i} is its operational water-consumption factor. This formulation allows the indirect water footprint to be estimated for coal-, gas-, and nuclear-dominated electricity grids. Macknick et al. (2011). Table 4 summarizes the water-use intensity factors associated with coal-, natural gas-, and nuclear-based electricity generation. These factors were used to convert IT electricity demand into an indirect water footprint.

Table 4. Indirect water footprint factors for IT electricity consumption under coal-, gas-, and nuclear-based electricity grids

Grid / plant assumption	Median water-consumption factor from NREL	Converted (WUE_i)	Indirect WF equation
Coal, generic tower cooling	687 gal/MWh	2.60 L/kWh	$WF_{indirect} = 2.6 \times E_{IT}$
Coal, subcritical tower cooling	471 gal/MWh	1.78 L/kWh	$WF_{indirect} = 1.78 \times E_{IT}$
Coal, supercritical tower cooling	493 gal/MWh	1.87 L/kWh	$WF_{indirect} = 1.87 \times E_{IT}$
Natural gas, combined cycle tower cooling	198 gal/MWh	0.75 L/kWh	$WF_{indirect} = 0.75 \times E_{IT}$
Natural gas, steam tower cooling	826 gal/MWh	3.13 L/kWh	$WF_{indirect} = 3.13 \times E_{IT}$
Nuclear, generic tower cooling	672 gal/MWh	2.54 L/kWh	$WF_{indirect} = 2.54 \times E_{IT}$
Nuclear, once-through cooling	269 gal/MWh	1.02 L/kWh	$WF_{indirect} = 1.02 \times E_{IT}$

The total water footprint can be formulated as:

Total Water = Direct Water + Indirect Water

This distinction is particularly relevant in regions dependent on thermal power plants.

4.2.2 Integrating local water scarcity

WUE measures water efficiency, but its environmental impact depends on local water availability. A facility that utilizes 1 L/kWh in a water-rich environment will have very different outcomes than one that utilizes 1 L/kWh in a water-stressed environment. Therefore, the inclusion of water stress indicators such as per capita renewable water availability and/or scarcity indicators will help provide context for the evaluation (The Green Grid. 2021):.

4.2.3 Life Cycle Assessment (LCA)

Liquid cooling technologies and immersion technologies both require unique components (i.e. plate coolers, heat exchangers), heat transfer fluids or dielectric fluids, and additional hydraulic structures. The incorporation of LCA allows for the inclusion of: (1) Emissions created through the manufacture of the facility and all components; (2) Environmental impacts resulting from raw material extraction;

and (3) Environmental impacts associated with the disposal of fluids and equipment after useful life.

4.2.4 Potential for heat recovery and valorization

The ability of a cooling system to operate at high outlet temperatures determines its heat recovery potential. Liquid cooling architectures facilitate supplying urban heating networks and Industrial recovery of low temperature heat. This dimension can transform a thermal externality into a secondary energy resource, improving overall systemic efficiency.

4.3 Comparative multi-criteria approach

To compare cooling technologies in a structured way, it is necessary to adopt a multi-criteria as mentioned in Figure 10.

Figure 10. Cooling technologies multi-criteria.

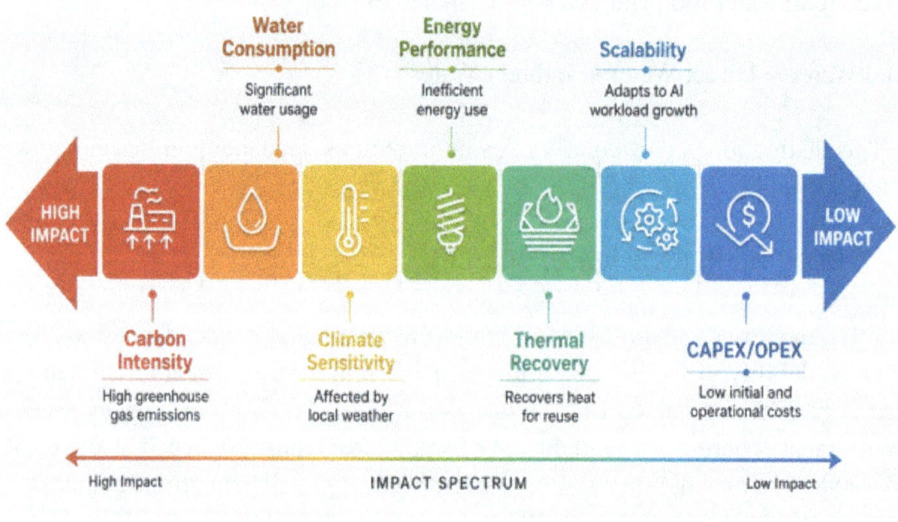

A comparative matrix is structured as follows:

continued on following page

Table 5. Continued

Table 5. Performance Comparison of Cooling Technologies in Data Centers Dlzar et al. (2025)

Cooling technology	Typical PUE range	Water use / WUE profile	CUE variability by electricity mix	Supported rack density	Heat capture / removal performance	Climate sensitivity	Heat reuse potential	Main limitations
Conventional air cooling (CRAC/ CRAH with containment)	1.5–2.0 for CRAC; 1.2–1.5 for CRAH	Low direct water use for DX/air systems; higher water demand when chilled-water CRAH or cooling towers are used	High variability: CUE remains strongly dependent on grid carbon intensity because fan and chiller loads remain significant	Typically <15 kW/rack; up to ~25–40 kW/rack with strong airflow containment	Limited for high-density AI/HPC loads; prone to hotspots and bypass airflow	High; performance declines in hot climates and under variable thermal loads	Low to moderate; exhaust air is usually low-grade, around 30 °C	Limited scalability for AI/HPC workloads; high fan/chiller energy; airflow management complexity
Free air / indirect free cooling	1.05–1.4 depending on climate and configuration	Very low water use for direct free cooling; indirect systems may use moderate water depending on heat exchanger and evaporative support	Low to moderate where paired with low-carbon electricity; still climate-dependent because mechanical backup may be needed	Usually 10–40 kW/rack	Good in suitable climates; less effective for dense AI clusters without liquid assistance	Very high; best in cool, dry or temperate climates	Moderate; exhaust heat can reach ~30–50 °C in favourable designs	Requires filtration, humidity control and backup cooling; unsuitable in polluted or humid regions
Evaporative / adiabatic cooling	~1.1–1.35 in favourable dry climates; higher in humid climates	Medium to high; reported water use ranges from about 0.5 to 8.4 million L/MW/ year depending on climate and design	Variable: lower energy-related CUE than mechanical cooling, but environmental benefit depends on water availability and electricity mix	Up to ~40 kW/rack in optimised deployments	Effective pre-cooling; can reduce cooling energy by ~30–40% in warm dry regions	Very high; performance constrained by wet-bulb temperature and humidity	Low to moderate	Water consumption, scaling, microbial growth, humidity control and limited performance in humid climates
Direct-to-chip liquid cooling / cold plates	~1.07–1.15; broader deployments may fall around 1.1–1.5 depending on residual air cooling and facility design	Low direct water loss in closed-loop systems; indirect water use depends on heat rejection system	Lower and more stable than air cooling when powered by low-carbon electricity; CUE increases proportionally with grid carbon intensity	>50 kW/ rack; up to 300+ kW/rack in advanced cold-plate or microchannel systems	Typically 70–75% heat capture in commercial DLC; cold plates may remove 60–70% of server heat, requiring residual air cooling	Low to moderate; less climate-sensitive than air cooling	High; warm-water output around 40–45 °C can support district heating or industrial reuse	Higher CapEx, CDU requirement, leakage risk, coolant monitoring, retrofit complexity
Immersion cooling, single- or two-phase	~1.03–1.10	Very low direct cooling water use; can eliminate cooling towers in some designs	Lowest CUE potential when combined with low-carbon electricity due to very low cooling overhead; still affected by embodied/fluid impacts and electricity mix	100–368 kW/ rack depending on fluid, tank design and heat rejection	Very high; near-complete heat capture, around 95–100%	Low; suitable for extreme density and less dependent on ambient air	Very high; heat output can reach ~55 °C, suitable for district heating, greenhouses or industrial reuse	High infrastructure cost, custom tanks/racks, dielectric-fluid management, material compatibility and maintenance complexity

Cooling technology	Typical PUE range	Water use / WUE profile	CUE variability by electricity mix	Supported rack density	Heat capture / removal performance	Climate sensitivity	Heat reuse potential	Main limitations
Rear-door heat exchanger (RDHx) / hybrid air–liquid cooling	~1.25–1.33	Low to moderate; WUE reported around 0.01–0.635 L/kWh depending on temperature difference and heat rejection design	Moderate to low; reduces cooling energy compared with CRAC, but CUE still depends on pumps, fans and grid mix	Typically 40–100 kW/rack; some configurations up to ~200 kW/rack	High rack-level heat capture; up to 100% exhaust-air heat capture in suitable configurations	Moderate; more robust than room-level air cooling but still linked to chilled-water/free-cooling availability	Moderate to high; heat output around 40 °C supports low-grade reuse	CDU and pipe routing needed; transient response tuning; trained maintenance staff required

Case-study benchmark: PUE and WUE variation with cooling technology

To strengthen the quantitative comparison, Lei and Masanet developed a physics-based benchmark model for U.S. data centers, covering 10 data-center archetypes and 16 ASHRAE/IECC climate zones. Their study reports thermodynamically compatible PUE and WUE values, meaning that energy and water performance are calculated under the same cooling technology, operating conditions, and climate assumptions. The model includes airside economizers, waterside economizers, water-cooled chillers, air-cooled chillers, direct expansion systems, adiabatic cooling, cooling towers, and humidification systems.

A key result is that PUE decreases in cooler climates and differs strongly with data-center scale and cooling technology. Median PUE values range from 1.12–1.25 for large-scale data centers, 1.39–1.98 for midsize data centers, and 1.71–2.22 for small data centers. This shows that hyperscale or large-scale facilities generally achieve better energy efficiency than smaller facilities, mainly because of optimized operation, higher UPS efficiency, wider allowable temperature ranges, and greater use of free cooling.

The same study also shows that WUE varies more strongly with cooling technology than PUE. Table 6 summarizes benchmark results from Lei and Masanet's physics-based data-center model. Lei, Eric Masanet. (2021).

Table 6. Benchmark comparison of PUE and WUE by data-center cooling technology

Data-center / cooling configuration	Typical PUE behavior	Typical WUE behavior	Main interpretation
Large-scale airside economizer + adiabatic cooling + backup water-cooled chiller	Low; large-scale median PUE about 1.12–1.25	Low to moderate; climate-dependent	Best combined energy–water performance in favorable climates because free cooling reduces mechanical cooling demand
Large-scale waterside economizer + water-cooled chiller	Low; large-scale median PUE about 1.12–1.25	Moderate to high	Efficient in energy terms, but still requires cooling-tower water for heat rejection
Midsize water-cooled chiller without economizer	Moderate; midsize median PUE about 1.39–1.98	High	Lower PUE than air-cooled systems, but high onsite water use due to cooling towers
Midsize air-cooled chiller	Moderate to high; within midsize PUE range 1.39–1.98	Very low direct WUE	Reduces onsite water use but increases electricity demand relative to water-cooled systems
Small air-cooled chiller or direct expansion system	High; small data-center median PUE about 1.71–2.22	Very low direct WUE	Minimal onsite water use, but poorer energy efficiency; indirect water footprint may increase through electricity consumption

4.4 Integration into the energy–water–carbon framework

The proposed framework is based on an integrated vision where cooling technologies are analyzed as interfaces between three interdependent systems as shown in Figure 11:

Figure 11. The synergy of integrated cooling systems

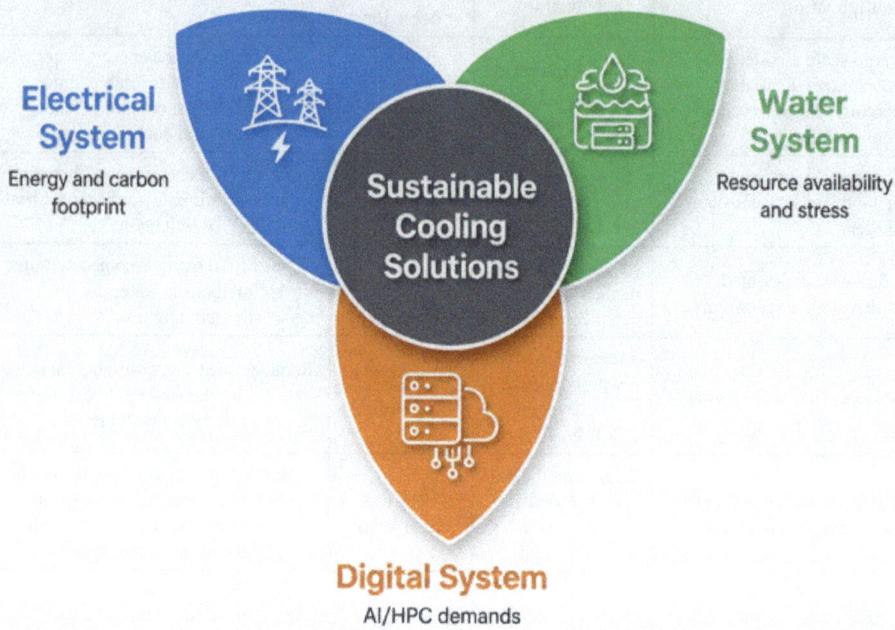

A comprehensive method will help ensure that trade-offs will not occur through performance indicators (PUE and WUE) that result in suboptimal solutions through competing resource types. For example, PUE could be improved without considering the impact on WUE, thereby exacerbating local water scarcity. Similarly, WUE could be reduced without considering its impact on CUE, for instance by shifting from water-based cooling to more energy-intensive air cooling in a carbon-intensive grid, thereby increasing overall carbon emissions.

4.5 Methodological implications for AI workloads

In the specific context of AI workloads, an integrated approach that combines thermal modeling, energy optimization, regional hydrological analysis, and an assessment of the carbon intensity of the electricity mix is essential. Such a holistic framework supports more informed infrastructure design and location decisions, aligning technological performance with sustainability objectives as presented in Figure 12.

Figure 12. Methodological implications for AI workloads

5. COMPARATIVE ANALYSIS OF ENERGY–WATER–CARBON TRADE-OFFS

The rise of AI and HPC infrastructures highlights complex trade-offs between energy performance, water consumption and carbon emissions. Cooling technologies should not be evaluated in isolation, but as interface devices between three interconnected systems: the electrical system, the water system and the digital system. This section analyzes the fundamental tradeoffs between these dimensions and highlights the environmental pressure transfer risks associated with different cooling architectures.

5.1 Fundamental trade-offs between energy and water

5.1.1 Evaporative cooling: energy reduction, water intensification

The very principle of evaporation involves a continuous withdrawal of fresh water to compensate for losses. In some hyperscale centers, WUE can reach or exceed 1–2 liters per kWh of IT energy, depending on climate conditions and operating strategies. Under favorable climatic conditions (hot and dry climate), these systems can improve PUE by limiting the use of mechanical cooling (Cuce, P. M. 2025).

5.1.2 Air cooling in temperate climates: water conservation, energy dependence

Air cooling, particularly when it uses free cooling in cold or temperate climates, can significantly reduce direct water consumption. The WUE can then be close to zero. However, when climatic conditions do not allow permanent free cooling, the use of refrigeration compression systems increases auxiliary electricity consumption. If the electricity comes from a high carbon mix, the improvement in WUE can be accompanied by increased carbon intensity (high CUE). This second arbitration highlights that minimizing the direct use of water does not guarantee a reduction in overall impacts if the carbon footprint of electricity is high.

5.1.3 Liquid cooling: thermal efficiency and systemic complexity

Liquid cooling solutions allow a reduction in ventilation requirements and in dependence on compressors, more efficient use of free cooling and an increase in outlet temperatures facilitating heat recovery as shown in Figure. 13a.

Operationally, these technologies can simultaneously improve PUE and reduce direct water consumption if combined with closed circuits or dry coolers. However, their environmental performance depends on contextual factors as presented in Figure 13b.

Figure 13. **a)** *Liquid cooling advantages* **b)** *Factors influencing environmental performance of cooling Technologies.*

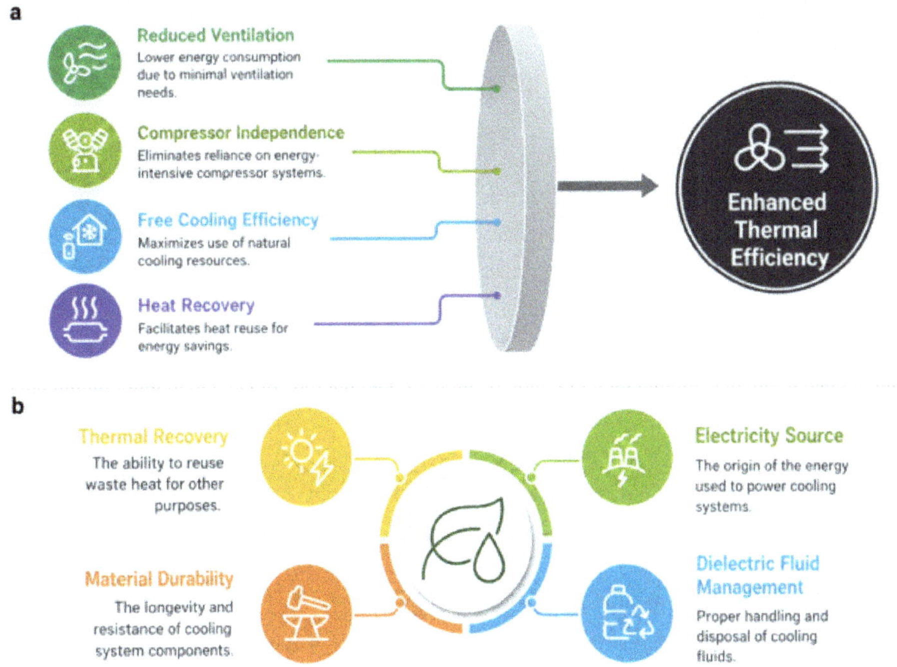

Liquid cooling therefore does not eliminate energy–water–carbon trade-offs, but reconfigures them in a more complex framework.

5.2 Risk of transfer of environmental pressure

5.2.1 Energy → water transfer

Energy optimization via evaporative cooling can shift pressure towards water resources. This transfer is particularly problematic in arid regions or regions subject to recurrent droughts, where industrial withdrawals compete with agriculture and municipal supplies. In these contexts, an improvement in PUE can paradoxically increase local socio-environmental vulnerability.

5.2.2 Water → carbon transfer

The adoption of low-water consumption technologies in regions dependent on water-cooled thermal power plants can shift the water footprint upstream of the energy system. Indirect water used to produce electricity can exceed water consumed directly on site. Thus, a center with a low WUE can nevertheless contribute to a high total water footprint in case its electricity supply is highly temperature-dependent.

5.2.3 Geographic transfer of impacts

The partial mobility of AI workloads, particularly for training, theoretically makes it possible to optimize geographic location based on Climate, energy mix, the water availability. However, this relocation may result in a spatial transfer of environmental impacts to regions with weaker regulations or limited water governance. The risk is then to shift environmental pressures without reducing them overall.

5.3 Absolute growth and volume effect of AI workloads

Even when relative metrics (PUE, WUE) improve, the exponential growth of AI workloads can lead to an increase in absolute energy and water consumption.

This phenomenon is analogous to the rebound effect:

- Improvement in unit efficiency,
- Rapid expansion of global demand,
- Increase in total impacts.

Thus, comparative analysis cannot be limited to relative technical performances; it must integrate growth scenarios in the volume of AI calculation.

6. COMPARATIVE REGIONAL CASE STUDIES: COOLING TECHNOLOGIES AND THE ENERGY–WATER NEXUS

The analysis of AI and HPC data center cooling technologies cannot be separated from their territorial context. The environmental performances observed result from a complex interaction between climatic conditions, energy mix, water availability, infrastructural maturity and regulatory frameworks. Table 7 summarizes indicative PUE, WUE and CUE ranges for three representative regional contexts: Nordic, Gulf and hot-arid regions. The comparison reflects the fact that cooling performance is controlled by local climate, cooling technology and electricity carbon intensity. This is consistent with Lei and Masanet's physics-based framework, which shows that PUE and WUE should be evaluated together because both vary with climate, cooling-system configuration and operational setpoints. Their model covers 10 data-center archetypes and 16 climate zones, and explicitly links PUE/WUE to cooling technology and local weather conditions.

Table 7. Regional benchmark comparison of PUE, WUE and CUE for data-center cooling strategies

Region / climate context	Typical cooling strategy	Indicative PUE	Indicative WUE, L/ kWh IT	Indicative CUE, kg CO_2/kWh IT	Main interpretation
Nordic / cold climate	Free cooling, airside economizer, limited mechanical cooling	1.05–1.20	0.0–0.3	0.02–0.10	Best energy performance due to low ambient temperature and high free-cooling potential; low CUE if electricity is hydro, wind or nuclear dominated
Gulf / hot-humid climate	Mechanical cooling, chilled water, limited free cooling	1.35–1.70	0.2–1.0	0.55–1.10	High cooling energy demand; humidity limits evaporative/free-cooling benefits; CUE can be high if grid electricity is fossil-fuel dominated
Hot-arid region	Direct/indirect evaporative cooling, adiabatic systems, hybrid cooling	1.20–1.50	1.0–2.0	0.45–0.90	Evaporative cooling can reduce PUE but increases WUE; water scarcity makes WUE a critical decision criterion

This section examines several contrasting regional configurations to illustrate the variability of energy–water–carbon trade-offs and the implications for the sustainability of digital infrastructures.

6.1 Developed economies with temperate or cold climates: favorable convergence of indicators

6.1.1 Northern Europe and Nordic countries

Sweden, Finland, Norway and Iceland as the Nordic countries are often cited as favorable environments for the establishment of hyperscale data centers. Numerous factors converge:

- Cold climate that provides free cooling for a large part of the year,
- A mostly carbon-free electricity mix that nuclear, wind, and hydroelectricity power),
- Reliable energy infrastructure,
- Increasing integration into urban heat networks.

In these contexts, it is possible to simultaneously achieve a low PUE (≈1.1–1.2), low WUE (limited use of direct water), and reduce CUE thanks to a low carbon mix.

Waste heat recovery constitutes an additional strategic lever. Liquid or tempered water cooling architectures facilitate the injection of heat into district heating networks, improving systemic energy efficiency.

6.1.2 Structural limits

There are some restrictions that prevent making the model broadly applicable as follows: local electric capacity is limited; there are also restrictions on land availability; the reliance on interconnectivity through networks; and lastly, there are geopolitical issues involved with localising data. On top of this, though training for AI may occur in those areas, inference will typically be done as close to the end-user as possible.

6.2 Geographies of hot climate and carbon energy source; trade-offs are mandated

6.2.1 Southern United States, Middle East region and parts of Asia

The regions having both high ambient temperature and fluctuating/seasonal humidity will have many complex trade-offs with their electricity generation. Air conditioning systems require a large electrical draw to offset the ambient heat in the atmosphere. Increased demand for electric energy will increase PUE and CUE. PUE can be improved by evaporative cooling; however, these systems tend to introduce the potential for increased water use. In areas like the southwest of the United States, where droughts are common, this reliance on water has raised public concerns about how to allocate water resources among data centers, agriculture, and domestic needs.

6.2.2 Case of Gulf countries

Energy-intensive systems are primarily what drive cooling. Most freshwater supplies come from desalination, another energy-intensive process. The energy-water cycle is very circular: energy is generated from water, and water is used to cool the data center; both the cooling and generation processes use the same energy to operate as presented in Figure 14.

Figure 14. Gulf countries situation

The ability to efficiently combine renewable and thermal-based technologies in cooling systems is key to improving environmental performance.

6.3 Regions under structural water stress

6.3.1 Arid and semi-arid zones

In regions subject to high hydrological variability or chronic water scarcity, the establishment of water-intensive data centers can exacerbate social and environmental tensions.

Although the volumes of water consumed by a data center may seem modest on a national scale, their local concentration can be significant. The impact is determined by the watershed concerned, level of existing water stress, and local water governance. In these contexts, WUE must be interpreted in light of water scarcity indicators and not as a decontextualized absolute value.

6.3.2 Strategic arbitrage

The comparative analysis suggests that, in regions with high water stress, priority should be given to water-lean technologies:

- Optimized air cooling,
- Direct-to-chip with closed circuits,
- Dry coolers rather than evaporative towers.

Although these solutions may involve slightly higher energy consumption, they reduce the risk of water use conflicts.

6.4 Africa and emerging economies: infrastructural constraints and governance

6.4.1 Digital growth and infrastructure maturity

The expansion of data centers in Africa and other emerging economies is driven by the rapid growth of digital services and mobile connectivity. However, these regions often present: unstable electricity networks, limited production capacity, unequal water infrastructure, and variable environmental governance.

Advanced cooling technologies (immersion, direct-to-chip) can be expensive to deploy in these contexts, due to high initial investments and lack of specialist skills.

6.4.2 Dependence on diesel generators

In some areas, the power supply relies partially on diesel generators, greatly increasing the CUE. In these cases, the environmental priority may be decarbonization of energy supply rather than marginal optimization of PUE.

6.4.3 Risk of environmental inequalities

Locating data centers in regions with weaker environmental governance can result in: an under-evaluation of water impacts, low transparency of indicators, and a lack of consultation with local communities.

This raises questions of environmental justice and equity in the distribution of the benefits and burdens of the digital transition.

6.5 Synthetic comparison of regional contexts

The regional analysis highlights three typical configurations is presented in Figure 15.

Figure 15. Comparative Analysis of Regional Constraints in Data Center Deployment

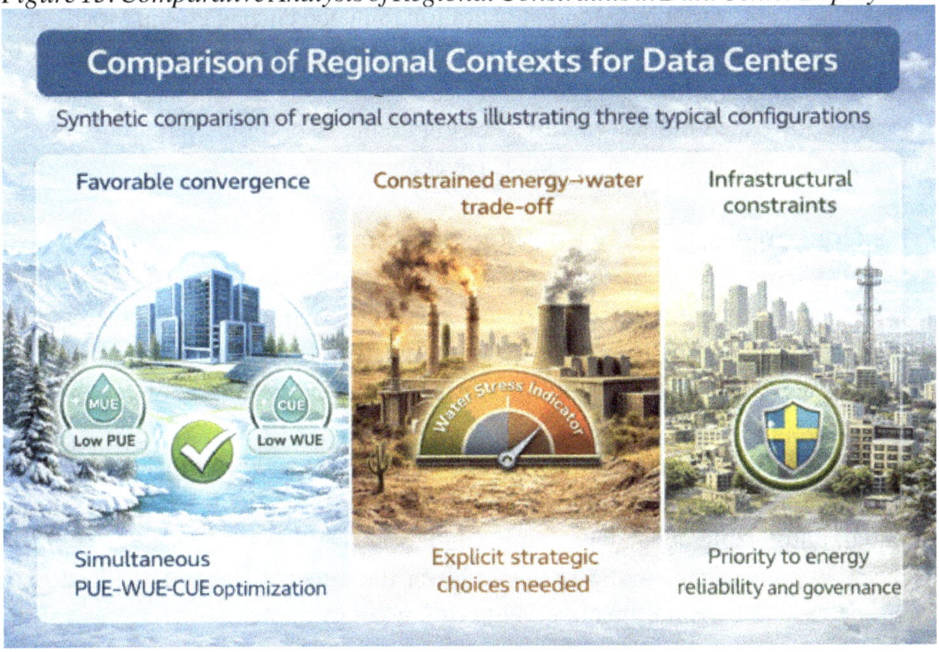

This diversity shows that no cooling technology can be considered universally optimal. Environmental performance depends on a fine adjustment between technology and territorial context.

7. TECHNOLOGICAL INNOVATIONS AND OPTIMIZATION OF AI INFRASTRUCTURES

Because of the advanced computational density of both AI and HPC workloads, data centre operators are faced with competing resources for energy, water and carbon; therefore, they are using a dual approach to lower the environmental footprint while maintaining performance and framework reliability through new uses of technologies and operational optimisations.

7.1 Water-lean cooling technologies

7.1.1 Advanced air cooling

Cooling systems based on air will continue to evolve with improved performance through the application of many technology and operational advances. Examples of these improvements include better engineered airflow management for hot and cold aisle containment, variable speed fan operation to match current cooling loads and indirect free cooling systems all assist in reducing mechanical cooling which will optimize overall performance. The controlled rise in operating temperatures makes it possible to reduce the use of refrigeration compressors, thus reducing energy consumption without resorting to intensive evaporation.

7.1.2 Dry coolers and closed circuit systems

Dry coolers dissipate heat via air-liquid exchangers without direct consumption of water by evaporation. They constitute an attractive alternative in regions experiencing water stress, although their performance depends strongly on the outside temperature.

Closed circuit hydraulic systems reduce water losses and limit make-up needs. In certain configurations, water can be replaced by fluids with low evaporation potential, minimizing dependence on local resources.

These approaches make it possible to lower the direct WUE, but can lead to a slight increase in PUE if climatic conditions are unfavorable.

7.1.3 Alternative refrigerants and thermodynamic innovation

Research also targets low-impact dielectric fluids, low-GWP (Global Warming Potential) refrigerants, and optimized thermodynamic cycles. These innovations aim to reduce the impacts linked to refrigerants, some of which have a high global warming potential.

7.2 Advanced liquid cooling and thermal recovery

7.2.1 Direct-to-chip optimization

Direct-to-chip cooling allows heat to be extracted as close as possible to the source, reducing intermediate heat losses. Recent advances include high conductivity cold plates, modular circuits facilitating maintenance, and the integration of high-resolution thermal sensors. These developments improve thermal stability and make it possible to operate at higher fluid temperatures (warm water cooling), facilitating heat recovery.

7.2.2 Immersion cooling and extreme densification

Immersion cooling, especially biphasic, enables very high densities without server ventilation, reducing fan energy use and improving thermal uniformity. However, challenges remain linked to the cost of fluids, their recyclability and industrial standardization.

7.2.3 Recovery and reuse of waste heat

Thermal recovery transforms an energy residue into a secondary resource, illustrated un Figure 16, improving the overall efficiency of the local energy system. However, its feasibility depends on the proximity of compatible infrastructure and incentive regulatory frameworks.

Figure 16. Benefits of Liquid Cooling in Data Centers

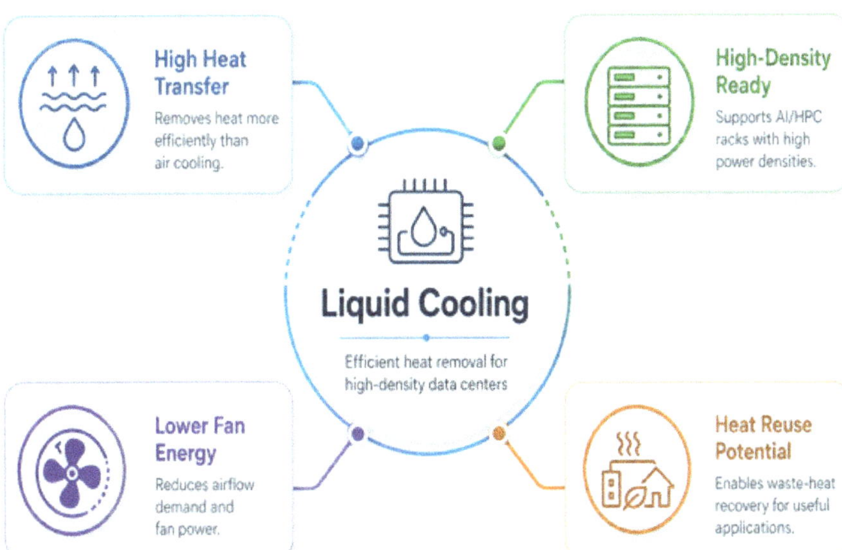

7.3 Optimization by artificial intelligence and predictive control

Figure 17 illustrates a closed-loop AI-based fan-speed control framework for data-center thermal management. Real-time sensor data are used for thermal-state estimation, short-term IT load forecasting, and MPC-based optimization of fan-speed setpoints. The objective is to reduce fan and cooling energy while maintaining rack inlet temperatures within allowable limits. The framework also highlights latency constraints, since delayed sensing, prediction, or actuation can reduce hotspot mitigation and control stability.

Figure 17. Flowchart of AI-optimized fan-speed control in data centers.

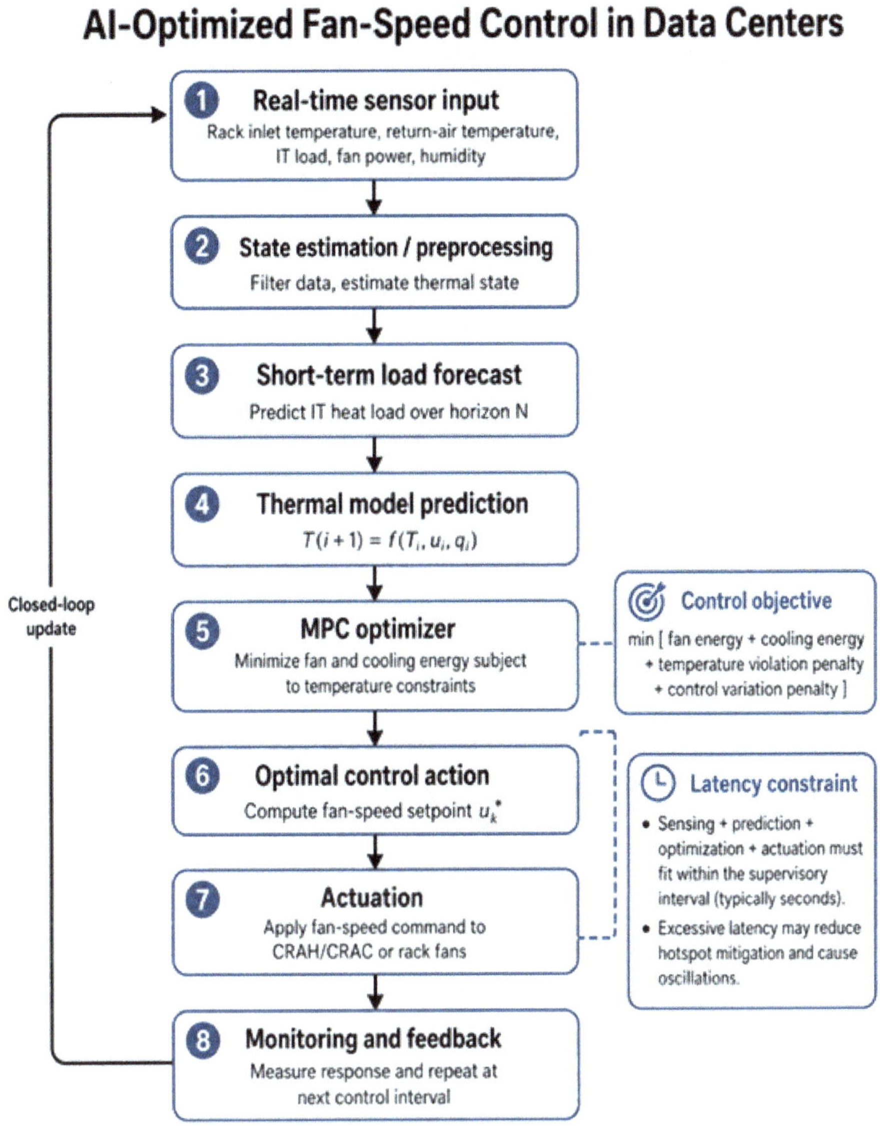

7.3.1 Thermal control based on machine learning

AI optimizes cooling by predicting thermal loads and dynamically adjusting systems, reducing auxiliary energy use and improving PUE.

7.3.2 Workload orchestration and carbon-aware scheduling

Beyond the physical infrastructure, software optimization constitutes a major lever. Strategies include:

- Shifting training loads to low-carbon regions,
- Temporal planning based on the availability of renewable energy,
- Dynamic adjustment of processor frequency.

This multi-objective orchestration makes it possible to reduce the CUE and, indirectly, the water footprint linked to electricity production.

7.3.3 Multi-criteria energy–water–carbon optimization

Advanced approaches combine several factors as shown in figure 18.

Figure 18. Key Dimensions of Multi-Criteria Optimization in Data Centers

The goal is to simultaneously minimize PUE, WUE and CUE, rather than optimizing a single metric. This approach requires high-resolution data and advanced analytical capabilities.

7.4 Integration of renewable energies and storage

7.4.1 Direct coupling with renewables

The integration of renewable sources (solar, wind, hydroelectricity) reduces the carbon intensity of the electricity consumed. In some cases, data centers are developing power purchase agreements (PPAs) or investing directly in renewable capacity.

7.4.2 Energy storage and operational flexibility

Storage systems (batteries, thermal storage) allow: smooth electrical demand, absorb renewable surpluses, and reduce dependence on periods of high carbon intensity. Thermal storage, in particular, can improve management of cooling peaks.

7.5 Limits and barriers to deployment

Despite their potential, these innovations face barriers such as high CAPEX, limited standardization, skill requirements, regulatory constraints, and uncertain long-term profitability.

7.6 Summary: towards systemic optimization

Technological and operational innovations offer real opportunities to reduce the combined footprint of AI data centers. However, their effectiveness depends on their integration into a systemic strategy taking into account: Climatic constraints, the energy mix, Water availability, and projected workload growth as presented in Figure 19.

Figure 19. Systemic strategy optimizes AI data centers

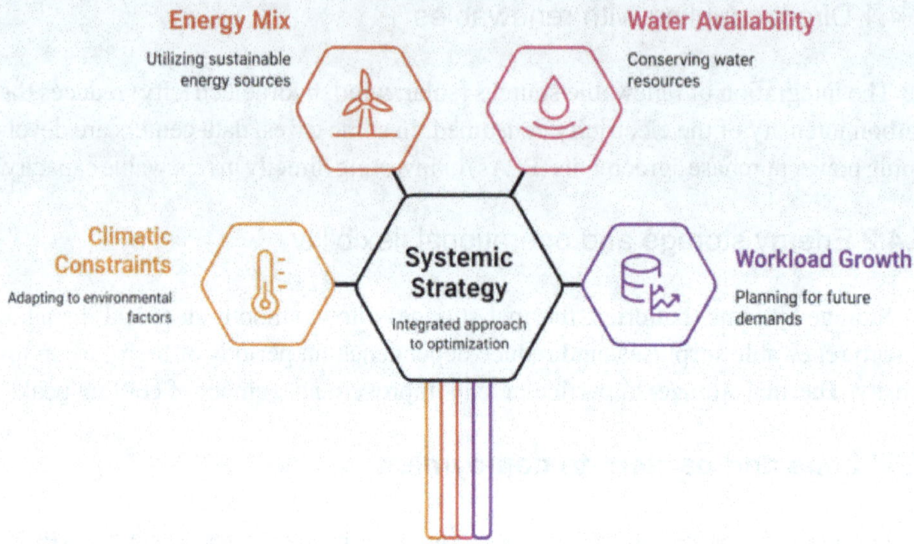

Future optimization of AI infrastructures will likely rely on a combination of several foctors as shown in Figure 20.

Figure 20. Pathways to sustainable AI

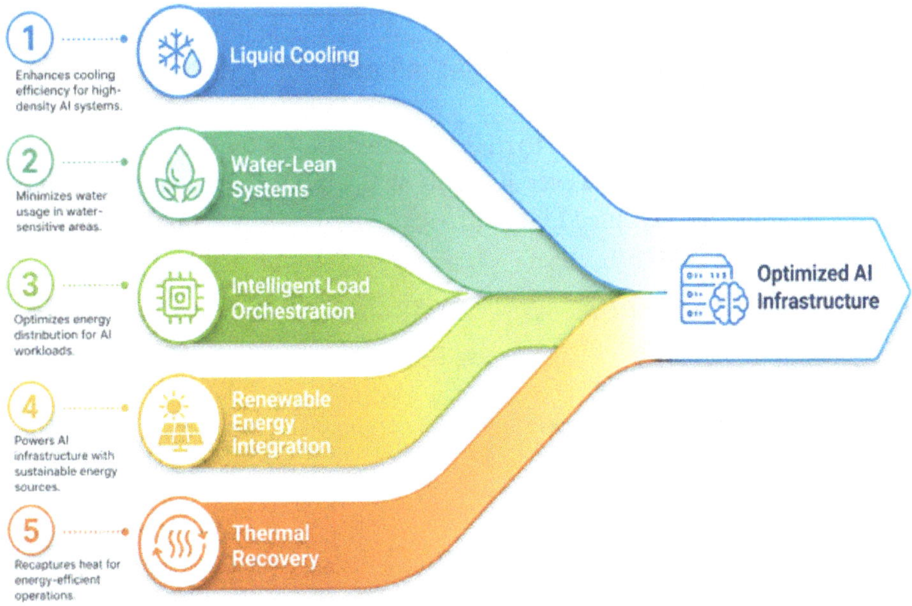

These technological dynamics must, however, be framed by appropriate public policies and governance mechanisms, which will be the subject of the following section.

8. GOVERNANCE, REGULATION AND PUBLIC POLICIES: TOWARDS INTEGRATED MANAGEMENT OF THE ENERGY–WATER–CARBON NEXUS

In a context of prolonged drought around Santiago, local communities criticized the planned use of water-based cooling because of its potential pressure on scarce water resources. In 2024, a Chilean environmental court ruled that the project had not sufficiently considered climate-change impacts, pushing Google to revise its cooling strategy toward an air-cooling system. This case shows that data-center sustainability depends not only on technology, but also on water availability, local regulation, transparency and community acceptance.

8.1 Limitations of current assessment and reporting frameworks

8.1.1 Dominance of PUE and undervaluation of water

Power Usage Effectiveness (PUE) has established itself as the central performance indicator for data centers. However, its focus on operational energy efficiency has several limitations as presented Figure 21.

Figure 21. Limitations of PUE

In many countries, environmental reporting obligations do not require detailed disclosure of water volumes used, particularly at fine geographic resolution. This opacity limits the ability of local authorities and communities to assess cumulative impacts (The Green Grid. 2020).

8.2 Integration of the energy–water nexus in territorial planning

A first approach consists of systematically integrating key environmental indicators into data center authorization procedures. This includes the use of regional water stress indices to reflect local resource constraints, as well as medium- and long-term climate projections to anticipate evolving environmental conditions. In addition, cumulative impact assessments should be incorporated to evaluate the aggregated effects of infrastructure deployment over time. Contextual WUE thresholds based on local water availability enable more sustainable, location-specific decisions.

8.3 Economic instruments and incentives

8.3.1 Differentiated water pricing

Establishing pricing mechanisms reflecting local water scarcity can encourage operators to adopt water-lean technologies. Progressive or seasonal pricing could internalize the environmental costs associated with withdrawals.

8.3.2 Incentives for Heat Recovery

If subsidies or tax credits are provided to support the recovery of waste heat, this would make liquid cooling architectures even more profitable and would support the territorial integration of data centers.

8.3.3 Renewable Power Purchase Agreements (PPAs)

If policies supporting Power Purchase Agreements are implemented, this would allow operators to reduce their carbon intensity. By including transparency requirements in the EUC, the credibility of climate commitments may be enhanced.

8.4 Transparency, reporting and standardization

8.4.1 Mandatory reporting of PUE, WUE and CUE indicators

Harmonized reporting could include annual PUE, WUE, and CUE disclosure, water source distinctions, and indirect water footprint estimates. This transparency would improve comparability between installations and strengthen the accountability of operators.

8.5 Alignment with the Sustainable Development Goals (SDGs)

The development of AI data centers directly interacts with several Sustainable Development Goals as shown in Figure 22.

Figure 22. AI Data centers and sustainable development goals

An integrated regulatory framework could explicitly link the implementation of new digital infrastructures to these objectives, by imposing multi-dimensional impact assessments.

8.6 Environmental justice and territorial equity

The growth of AI infrastructure also raises questions of environmental justice:

- Who benefits from digital services?
- Who bears local environmental costs?
- Are local communities consulted?

In certain regions experiencing water stress, data centers can accentuate existing inequalities if water withdrawals compete with agricultural or domestic uses.

Inclusive governance should integrate:

- Public consultations,
- Compensation mechanisms,
- Participation of local communities in location decisions.

9. SCIENTIFIC GAPS AND RESEARCH AGENDA: TOWARDS AN INTEGRATED AND PROSPECTIVE ASSESSMENT

Despite the rapid expansion of the data center sustainability literature, significant gaps persist regarding the combined assessment of energy, water, and carbon impacts of AI and HPC infrastructure.

9.1 Insufficient data and limited transparency

9.1.1 High resolution water data

Although data center energy data is relatively well documented in certain countries, data relating to water consumption remains fragmentary and often aggregated at too broad a level to allow precise territorial analysis. The main limitations include low information availability, annual data limitation, lack of water distinction and absence of systematic publication as illustrated in Figure 23.

Figure 23. Fragmentation of Data Center Water Consumption Data

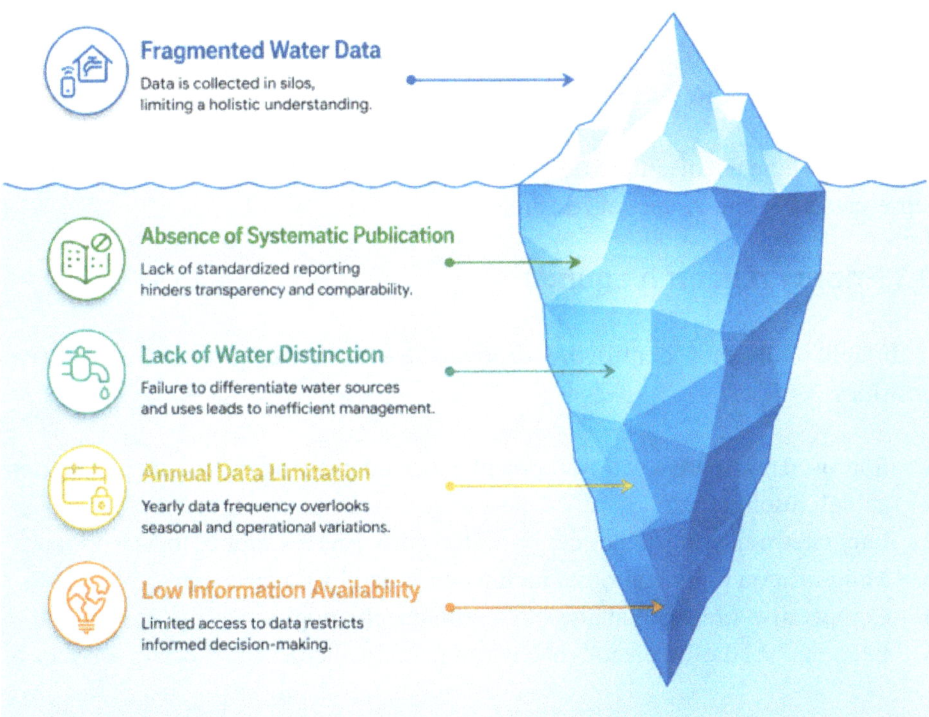

These gaps hinder the detailed assessment of impacts in regions experiencing water stress and complicate international comparison.

9.1.2 Opacity on specific AI workloads

Energy and water consumption strongly depends on the type of workload (training, inference, scientific HPC). However, data disaggregated by application or model remains largely proprietary.

This opacity limits accurate footprint estimation, efficiency analysis, and hardware comparison. Better transparency on the environmental performance of AI workloads is a research priority.

9.2 Insufficient integration of the indirect water footprint

Existing analyses often focus on direct water use, overlooking upstream water consumption from electricity production. However, in electrical systems dependent on thermal power plants, the indirect water footprint can be substantial. Methodological challenges include:

- Allocation of water consumed by power plants to specific users,
- Taking into account the hourly variability of the energy mix,
- Integration of future decarbonization scenarios.

The development of coupled energy–water models with high temporal resolution represents an essential area of research.

9.3 Priority research agenda

In light of the gaps identified, a structured agenda can be proposed around five priorities:

1. Improved transparency: development of harmonized indicators and publication at high resolution.
2. Integrated modeling of the energy–water–carbon nexus applied to AI workloads.
3. Multi-criteria optimization of architectures and load placement.
4. Comparative life cycle analysis of cooling technologies.
5. Integration of the dimensions of environmental justice and territorial governance.

These axes must be part of an interdisciplinary collaboration combining thermal engineering, energy sciences, hydrology, environmental economics and social sciences.

10. GENERAL CONCLUSION:

The environmental impact of digital infrastructures is being significantly altered by the quick growth of high-performance computers and artificial intelligence. Cooling systems are central to data center sustainability due to rising energy density and expanding AI workloads. Far from being a simple auxiliary function, cooling now constitutes a strategic determinant of the energy, water and carbon impacts of AI infrastructures.

This chapter has shown that cooling technologies air cooling, water-based systems, and advanced liquid solutions each have structural advantages and limitations. Their performance depends on climate, energy mix, water availability, and workload growth.

10.1 Key lessons

The comparative analysis reveals several important lessons:

1. Energy-water trade-offs are structural: Depending on the situation, low-water-intensity solutions may raise energy consumption, whereas evaporative technology can lower power use but increase reliance on water. A multi-criteria approach is necessary since optimizing one indication (PUE) can worsen another (WUE).
2. Liquid cooling represents a structuring transition: Faced with the increasing thermal densities of AI and HPC workloads, direct-to-chip and immersion technologies appear to be robust technical responses. They offer thermal efficiency gains and increased heat recovery potential, but require a complete assessment including their life cycle and territorial integration.
3. The territorial context is decisive: The environmental performances observed in regions with a cold climate and low carbon mix are not directly transposable to hot regions or regions experiencing water stress. The comparative analysis showed that the sustainability of infrastructure depends on a fine adjustment between technology and territory.
4. Absolute workload growth is a critical factor: Relative efficiency gains may be offset by the rapid expansion of demand for AI computing. An improvement in PUE or WUE on a unit scale does not guarantee a reduction in overall impacts if computing volumes increase exponentially.

The figure 24 provides a visual synthesis of the key points discussed in the chapter, highlighting the relationships between cooling technology choices and their energy, water, and carbon implications.

Figure 24. **X***Energy–water–carbon nexus of data-center cooling technologies. Figure* **24. Y** *Prospective scenarios for data-center PUE, WUE and CUE trajectories to 2030. Figure* **24. Z** *Comparative sustainability performance of cooling technologies across PUE, WUE and CUE*

10.2 Role of technological innovations and governance

Technological innovations liquid cooling, AI optimization, heat recovery, and renewables can significantly reduce AI data center impacts. However, their effectiveness depends on their inclusion in a coherent governance framework.

Sustainability of AI infrastructure requires:

- Reinforced transparency standards,
- The integration of water stress into location decisions,
- Incentives for thermal recovery,
- Coordination between digital, energy and water policies.

REFERENCES

Al Kez, D., Foley, A. M., Wong, F. W. B. H., Dolfi, A., & Srinivasan, G. (2025). AI-driven cooling technologies for high-performance data centres: state-of-the-art review and future directions. Sustainable Energy Technologies and Assessments, 82, 104511.

Jareño, J., Aragón-Jurado, J. M., De La Torre, J. C., Ruiz, P., & Dorronsoro, B. (2026). Energy-efficient large language models, Future Generation Computer Systems, Volume 182, 108483, ISSN 0167-739X, DOI: 10.1016/j.future.2026.108483

Jones, N. (2018). How to stop data centres from gobbling up the world's electricity. *Nature*, *561*(7722), 163–166. DOI: 10.1038/d41586-018-06610-y PMID: 30209383

Lei, N., & Masanet, E. (2021). Climate- and Technology-Specific PUE and WUE Predictions for U.S. Data Centers using a Physics-Based Approach, PREPRINT (Version 1) available at Research Square. DOI: 10.21203/rs.3.rs-769999/v1

Li, P., Yang, J., Islam, M. A., & Ren, S. (2025). Making AI Less « Thirsty ». *Communications of the ACM*, *68*(7), 54–61. DOI: 10.1145/3724499

Macknick, J., Newmark, R., Heath, G., & Hallett, K. C. (2011). A review of operational water consumption and withdrawal factors for electricity generating technologies.

Masanet, E., Shehabi, A., Lei, N., Smith, S., & Koomey, J. (2020). Recalibrating global data center energy-use estimates. *Science*, *367*(6481), 984–986. DOI: 10.1126/science.aba3758 PMID: 32108103

Patterson, D., Gonzalez, J., Le, Q., Liang, C., Munguia, L.-M., Rothchild, D., So, D., Texier, M., & Dean, J. (2021). *Carbon Emissions and Large Neural Network Training* (Version 3). arXiv. DOI: 10.48550/ARXIV.2104.10350

Yadav, K. D. (2024). Expansion of AI and ML Breakthroughs in HPC with Shift to Edge Computing in Remote Environments. In Ahmad, K. A., Hamid, N. A. W. A., Jawaid, M., Khan, T., & Singh, B. (Eds.), *High Performance Computing in Biomimetics* (pp. 129–145). Springer Nature Singapore., DOI: 10.1007/978-981-97-1017-1_5

Zhang, W., Fan, H., & Zhao, Q. (2023). Seeing green : How does digital infrastructure affect carbon emission intensity? *Energy Economics*, *127*, 107085. DOI: 10.1016/j.eneco.2023.107085

Chapter 4
Harnessing Predictive Analytics for Efficient Water Resource Management in Data Center Cooling Systems

Robinson Joel
http://orcid.org/0000-0002-3030-8431
KCG College of Technology, India

ABSTRACT

Efficient water resource management has become a critical challenge in modern data centers due to increasing computational demands and environmental concerns. This study explores the application of predictive analytics to optimize water usage in data center cooling systems. By leveraging historical sensor data, machine learning models, and real-time monitoring, the proposed approach forecasts cooling demand and dynamically adjusts water consumption. Techniques such as time-series analysis and regression models are employed to predict temperature fluctuations, workload variations, and cooling requirements. The system enhances decision-making by identifying inefficiencies, detecting anomalies, and enabling proactive maintenance. Experimental results indicate significant reductions in water wastage while maintaining optimal thermal performance and system reliability. Furthermore, the integration of predictive analytics supports sustainable operations by minimizing environmental impact and operational costs. This research demonstrates that data-driven water management strategies.

DOI: 10.4018/979-8-3373-8434-4.ch004

1.1 INTRODUCTION

An AI data center defines a facility which possesses the specific IT infrastructure needed for AI service development and application delivery (Gartner, 2024). The system uses advanced computing power together with its network and storage systems and energy cooling systems to handle AI workloads. The data centers used for traditional operations share common elements with AI data centers yet their processing capabilities and their IT infrastructure elements show major differences (Cisco, 2023). Organizations need access to essential AI infrastructure because it enables them to fully utilize the benefits offered by AI technology. Most organizations will find alternative methods to access AI technology without having to build their own AI server rooms from scratch.

Both standard data centers and AI data centers share several similarities with each other. Both facilities maintain their own set of hardware components which include networking equipment and storage devices and servers (IBM, 2024). Both types of operators are required to evaluate four specific elements which include energy efficiency and system uptime and data protection and system reliability. The two data center categories operate with different requirements because high-intensity AI workloads demand special processing resources (Intel, 2023). AI data centers possess dedicated infrastructure which supports AI workloads but standard data centers function with infrastructure that cannot handle AI workload demands. AI-ready infrastructure refers to specialized systems which organizations use to support their cloud computing and artificial intelligence and machine learning operations. Traditional data centers commonly install central processing units (CPUs) because these processors remain the primary computing engine for their facilities (Hennessy & Patterson, 2019). AI-ready data centers require high-performance graphics processing units (GPUs) along with advanced IT systems which provide sophisticated storage and networking and energy and cooling functionalities. AI use cases demand such an extensive number of GPUs that they require additional space beyond standard requirements (IDC, 2024).

The AI ecosystem experiences major changes because both large companies and smaller organizations use open source solutions while new technologies become available to everyone. A company transforms into a technology enterprise when it successfully identifies AI applications and builds the required IT systems (Accenture, 2024). The IBM Institute for Business Value conducted a 2024 study which revealed that 43% of technology executives at C-level positions experienced increased worries about technology infrastructure since they began using generative AI (IBM Institute for Business Value, 2024). Organizations now focus their efforts on developing their infrastructure to handle future growth requirements.

The data center sector has expanded its capacity to satisfy current demand for data centers. Modern data center infrastructure now supports AI operations because it can process extensive complex computations and requests from users (Synergy Research Group, 2024). Data centers exist in most Asian Pacific and North American areas because people build them in locations such as Beijing, Shanghai, northern Virginia, and the San Francisco Bay area. Major corporate investments show that the AI data center market has grown because of increasing demand. Microsoft plans to invest approximately USD 80 billion for data center construction in 2025 while Meta commits USD 10 billion to build a new hyperscale data center which will occupy four million square feet in Louisiana.

The AI-ready data center requires high-performance computing capabilities which exist in AI accelerators (Jouppi et al., 2017). AI accelerators serve as AI processors which enhance the performance of natural language processing machine learning deep learning models and all other artificial intelligence tasks. The hardware components that make artificial intelligence possible together with all its diverse applications receive universal acknowledgment as the primary hardware components. A GPU represents one specific type of AI accelerator. GPUs function as electrical circuits which transform intricate problems into smaller components that multiple parts can process at the same time by using a method called parallel processing (Kirk & Hwu, 2016). HPC uses a type of parallel processing called massively parallel processing which operates with more than ten thousand processors or processor cores or up to millions of processors or cores. AI models which operate on data center GPUs train AI applications to power several advanced AI systems. AI-ready data centers now incorporate NPUs and TPUs as specialized AI accelerators beyond the conventional AI accelerators. NPUs execute AI tasks more efficiently in real time by imitating the neural pathways of the human brain. TPUs serve as dedicated accelerators which optimize tensor calculations to enhance AI performance. The system delivers optimal performance for a wide range of artificial intelligence applications and deep learning tasks because it maintains high throughput while minimizing latency.

1.2. OVERVIEW OF DATA CENTER COOLING SYSTEMS

Data centers use air-based cooling systems to distribute cold air through server racks for thermal management (ASHRAE, 2021). Computer Room Air Conditioners (CRAC) and Computer Room Air Handlers (CRAH) systems use hot aisle/cold aisle containment methods to manage airflow which helps them maintain proper temperature control (Schneider Electric, 2023). Air cooling systems which people commonly use because of their simple operation become less efficient when computer systems increase their processing capacity because of this their power usage goes up and cooling distribution becomes uneven (Uptime Institute, 2023).

Liquid cooling systems use liquids which are commonly water or dielectric fluids to extract heat from computer components (NVIDIA, 2023). The material's outstanding thermal conductivity property enables it to transfer heat from systems at an accelerated rate which achieves better results. Energy consumption drops significantly when AI workloads use liquid immersion and direct-to-chip cooling methods while system performance receives a significant enhancement (IEA, 2023). Liquid cooling systems require users to invest more money upfront while they also need special systems to manage leaks and maintain operation. Liquid cooling systems have become the preferred choice for data center operators because they require less energy while running sustainable operations.

1.2.1. Evaporative Cooling and Water Usage

Data centers use evaporative cooling systems as their main energy-efficient solution to manage thermal conditions while consuming less power (U.S. Department of Energy, 2022). The system operates through the process of water evaporation which cools the surrounding atmosphere by absorbing heat during the transition from liquid to vapor. The method operates in both water-side and air-side economizers which serve regions that experience dry weather conditions.

The process of evaporative cooling requires more water resources because it reduces the need for mechanical systems that produce refrigeration (Mekonnen & Hoekstra, 2016). The need for constant evaporation of water creates problems for regions that have restricted water supplies. Water quality management functions as a vital process that enables cooling systems to prevent scaling and corrosion and biological growth (WHO, 2022). The combination of predictive analytics and advanced management techniques enables organizations to achieve maximum water efficiency through cooling process modifications that respond to weather conditions. The modern design of data centers emphasizes energy efficiency while maintaining proper water consumption because sustainability practices have gained increasing importance (IEA, 2023).

1.2.2. Cooling Towers and Chilled Water Systems

The cooling systems of data centers need to establish their basic framework through the installation of cooling towers together with chilled water systems (ASHRAE, 2021). The chilled water system operates by using circulating water to transfer heat which IT equipment generates through its contact with heat exchangers after the water has been been chilled by the chillers. The system transfers hot water to cooling towers which function to cool the water through their process of evaporative recirculation. The closed-loop technology maintains temperature stability while it effectively removes heat from the system.

Cooling towers function as critical components which emit heat to the atmosphere but their operation results in substantial water consumption through the processes of evaporation and drift and blowdown (Uptime Institute, 2023). The system needs proper water treatment to sustain its operational efficiency while it protects against microbial contamination. Chilled water systems provide reliable operational performance to large facilities which experience their highest benefits from these systems. The modern data centers have achieved greater operational efficiency through their implementation of advanced monitoring systems and variable-speed drive technology which reduces their water and energy consumption (Schneider Electric, 2023).

1.2.3. Emerging Technologies: Immersion Cooling and Hybrid Systems

Data center efficiency receives a major improvement from new cooling technologies which help AI and high-performance computing systems (NVIDIA, 2023). Servers go through a process called immersion cooling which involves placing them to stay inside dielectric fluids that can absorb heat from their components. The method enables ultra-high-density computing by enabling energy savings and eliminating the need for air-based cooling systems (IEA, 2023). The method requires less water consumption when compared to traditional cooling systems which use more water for operation. Hybrid cooling systems, which use multiple cooling methods including air and liquid or evaporative techniques, achieve their best performance in different environmental conditions. The system achieves better water and energy savings through its ability to switch between cooling modes based on both workload and environmental temperature changes.

The technologies provide multiple benefits yet they also present challenges because they require high initial investments and their standardization remains insufficient while they need to work with existing systems (Gartner, 2024). Data center operators need resource-saving and environmentally friendly solutions to meet their operational requirements which has led to increasing popularity of sustainable immersion and hybrid cooling systems.

1.3. WATER FOOTPRINT OF AI DATA CENTERS

The environmental impact of AI data centers as shown in Figure 1, now depends on their water usage because high-performance computing workloads keep increasing (IEA, 2023). AI training and inference require dense hardware setups that generate substantial heat which forces operations to depend on water-consuming cooling systems (NVIDIA, 2023). Facilities use water mostly for cooling towers and evaporative systems and chilled water loops to create their required operating temperatures (ASHRAE, 2021). Data center operations use energy which indirectly results in water consumption through thermal and hydroelectric power production methods that require water for their operations (U.S. Department of Energy, 2022).

Figure 1. Water Footprint of AI Data Centers

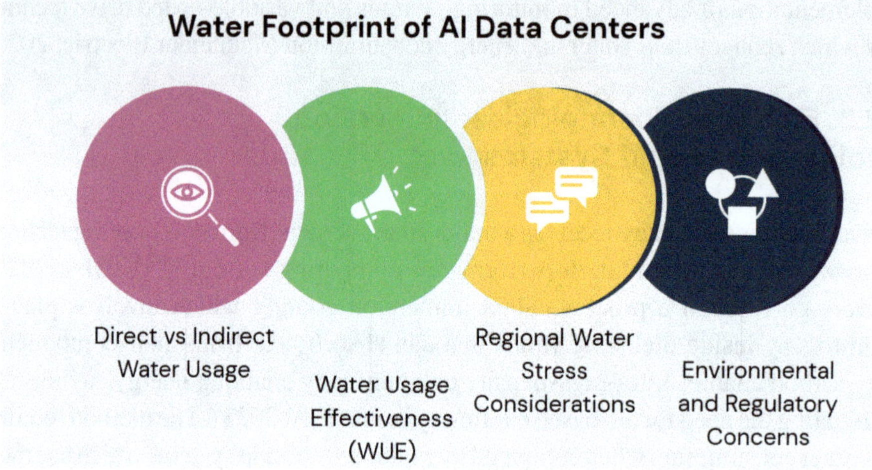

Hyperscale data centers spread across the globe which leads to a decreasing quantity of freshwater resources (UNESCO, 2023). Effective water management is crucial because areas with low water supplies are more vulnerable. The AI-driven infrastructure growth must be monitored through advanced analytics and better cooling systems and sustainable sourcing methods to protect long-term water security.

1.3.1 Direct vs Indirect Water Usage

Data centers that use artificial intelligence depend on water resources through two separate methods which involve both direct water use and indirect water use (Mekonnen & Hoekstra, 2016). The expression "direct water usage" refers to the actual water consumption that occurs inside the building which mainly supports its cooling operations. The system uses blowdown procedures to keep its water quality standards while cooling towers lose water through evaporation and the system requires water for its scheduled maintenance activities. The procedure for measuring direct water usage involves assessing flow rates together with water consumption records which serve as the main methods for tracking water usage. The data center produces energy through its operational power generation process which creates indirect water consumption because it requires energy to operate. Power plants need large quantities of water to operate their steam production systems and their cooling systems which is especially true for thermal and nuclear power stations (IEA, 2023).

Water serves as a vital resource that semiconductor manufacturers need to create semiconductors while server producers and other hardware makers need it to build their equipment (Semiconductor Industry Association, 2023). The complete water consumption calculation of a data center requires knowledge about all its consumption methods. Organizations need to adopt renewable energy solutions and enhance their energy efficiency to achieve indirect water usage reduction whereas effective cooling methods will reduce their direct water consumption.

1.3.3. Water Usage Effectiveness (WUE)

The Water Usage Effectiveness (WUE) metric serves as an essential measurement tool which assesses data center water usage efficiency through its application as a performance metric (Green Grid, 2022). The WUE calculation uses the total annual water consumption divided by IT system energy consumption which applies the L/kWh measurement system. WUE provides a standardized assessment tool which enables organizations to evaluate and compare their water efficiency across different facility types and operational methods. The WUE value shows better water efficiency when it decreases because less water power generation requires for each processing power unit. The adoption of advanced cooling systems which combine closed-loop systems with air-based cooling for appropriate areas leads to lower WUE results for data centers. WUE rates will rise for buildings with heavy reliance on evaporative cooling systems because their cooling systems demand more water to function. WUE monitoring allows operators to discover operational flaws which they can fix through cooling process improvements and water recycling methods and equipment upgrades. Data centers have made WUE their main sustainability measurement tool because environmental reporting has gained importance in modern times (Uptime Institute, 2023).

1.3.3 Regional Water Stress Considerations

AI data centers require water resources which face sustainability issues because of regional water stress conditions (UN-Water, 2023). Water stress exists when a region experiences a shortage of water resources to meet its total water requirements. Water-intensive data centers establish their operations in resource-constrained areas which creates negative effects on local communities and agricultural activities and natural ecosystems. IT companies now consider water availability as a critical factor when selecting data center locations. Organizations select locations with cold climates or high water availability to decrease their dependence on water-based cooling systems. The organization uses operational methods which involve wastewater treatment and rainwater collection and recycled water usage to minimize their environmental footprint. Through predictive analytics water management achieves improved performance by forecasting demand and distributing resources based on environmental conditions. The implementation of regional water stress indices within data center operations helps organizations reduce water shortage risks while achieving their sustainable development goals.

1.3.4. Environmental and Regulatory Concerns

AI data centers experience increased regulatory oversight and sustainability standards because their water usage results in harmful environmental impacts (OECD, 2023). The extraction of water resources without control results in decreased groundwater levels which then creates harmful effects on local ecosystems and disrupts biodiversity. The release of untreated sewage into the environment or its improper disposal results in both environmental contamination and ecological harm. Government authorities together with regulatory bodies have established stricter regulations to protect water resources from unsustainable usage patterns. The regulations establish three primary elements which consist of restrictions on water extraction and mandatory wastewater treatment facilities and processes for assessing environmental impacts. Data center operators need to follow these regulations while they try to maintain their operational efficiency standards.

Organizations are adopting environmentally friendly practices through their implementation of water recycling systems and their use of different water sources and their establishment of closed-loop cooling technologies (World Bank, 2023). Companies face mounting pressure to disclose their water usage and their efforts to decrease water consumption because of sustainability reporting standards which include ESG (Environmental Social and Governance) criteria. The sustainable future of AI data centers relies on achieving a balance between their operational needs and their environmental obligations.

1.4. PREDICTIVE ANALYTICS IN WATER MANAGEMENT

Predictive analytics in water management forecasts water consumption while optimizing resource distribution for data center cooling systems through its use of historical data and existing data combined with statistical and computer modeling methods (Deloitte, 2024). The system enables operators to find operational problems while predicting future needs and making operational decisions based on scheduled times instead of emergency situations. AI-powered data centers depend on predictive analytics to maintain their operational sustainability and system reliability because their cooling demands vary with changes in workload capacity (McKinsey & Company, 2024). The significance of predictive analytics today stems from its ability to decrease operational expenses while boosting cooling performance and reducing water consumption (PwC, 2023). The system enables better planning and risk management by forecasting upcoming water requirements and detecting any system malfunctioning.

The company assists its customers in achieving environmental regulations and sustainable development targets through its programs which achieve more than required water consumption reductions (World Economic Forum, 2024). The worldwide water scarcity problem requires predictive analytics as a vital resource management tool which enables ethical and intelligent water resource management (UN-Water, 2023).

1.4.1. Machine Learning and AI Techniques

The water management system at data centers uses predictive analytics which depends on machine learning and artificial intelligence methods (Goodfellow et al., 2016). The methods use extensive sensor data and operational data to identify patterns which enable them to generate accurate predictions. The system uses regression methods and supervised learning models to predict water consumption based on server load and temperature and humidity levels. The performance of tree-based models, including Random Forest and gradient boosting techniques, remains consistent when handling complex non-linear relationships (Chen & Guestrin, 2016). LSTM networks together with deep learning methods demonstrate maximum effectiveness when processing data that has sequential and time-based characteristics (Hochreiter & Schmidhuber, 1997). The system uses reinforcement learning to develop its cooling strategies through environmental interactions while it learns optimal operational methods (Sutton & Barto, 2018). The AI-driven methods which automate forecasting and enhance system performance help organizations make better decisions. The system uses continuous learning from new data to maintain its ability to handle changing workloads and environmental conditions which makes it essential for effective and sustainable modern data center water management.

1.4.2. Time-Series Forecasting for Water Demand

The predictive analytics process uses time-series forecasting as its essential method to forecast future water consumption in data center cooling systems (Box et al., 2015). The process requires two types of data the past water usage records and the specific time indicators which include workload patterns and seasonal trends and time of day. Water forecasts enable planners to develop their water needs while they create efficient cooling systems which protect against excess water usage and supply shortages. Organizations need base models for their forecasting work because they follow established statistical methods which involve the use of exponential smoothing and ARIMA. Advanced machine learning systems which businesses use today employ recurrent neural networks (RNNs) and long short-term memory (LSTM) networks to uncover time-based patterns and non-linear connections between different variables (Brownlee, 2018). The models let users add outside elements which include weather data and electricity usage patterns.

Data centers can predict water requirements to implement proactive measures which include adjusting cooling levels and scheduling maintenance work. The organization achieves operational efficiency through water conservation which results in reduced environmental impacts (IEA, 2023).

1.4.3. Anomaly Detection in Cooling Systems

The research field of predictive analytics uses anomaly detection which plays a crucial role in their studies to find patterns that deviate from normal cooling system operation (Chandola et al., 2009). Data center anomalies function as operational problem indicators which show different types of issues that include leaks and cooling system problems and sensor failures and unexpected spikes in water consumption. The process of early detection for these irregularities becomes critical because it protects systems from breakdowns while reducing water waste and sustaining operational performance. The machine learning techniques used for anomaly detection include autoencoders isolation forests and clustering methods (Aggarwal, 2017). The algorithms establish standard operational behavior through their analysis of historical data, which they use to determine real-time operational deviations. An abrupt water flow increase which occurs without a simultaneous temperature rise indicates a potential leak or efficiency problem.

The combination of anomaly detection with monitoring systems enables the creation of automated warning systems together with their fast response capabilities. The proactive approach used in AI-backed data centers establishes a framework which guarantees continuous operation of cooling systems while decreasing system outages and optimizing resource utilization (Microsoft, 2023).

1.5. DATA SOURCES AND MONITORING INFRASTRUCTURE

The process of predictive analytics for data center water management relies on its monitoring system and data sources (Katal et al., 2023). Continuous data collection together with integration and analysis of multiple heterogeneous sources enables effective cooling optimization. The system uses past operating data together with control system records and real-time sensor information and external environmental data. A strong monitoring infrastructure enables precise tracking of water consumption and system efficiency and environmental conditions (ITU, 2023). The system enables operators to make data-driven decisions through its complete visibility and control capabilities. Modern data centers now increasingly use smart infrastructure which combines cloud-based analytics systems with supervisory control systems and Internet of Things (IoT) sensors (Cisco, 2023). Companies need their data collection systems to handle both high-volume data and secure data transmission through scalable data pipelines. AI-driven data centers require a monitoring environment which supports predictive water resource management together with adaptive and sustainable water resource management (Accenture, 2024).

1.5.1. IoT Sensors (Temperature, Humidity, Flow Rate)

In data centers, IoT sensors are essential for gathering real-time data for cooling and water control (Gubbi et al., 2013). The installation of sensors throughout server racks and cooling units and water distribution systems enables monitoring of essential parameters which include temperature and humidity and water flow rates. Humidity sensors use their perfect performance to create ideal environmental conditions which eliminate both condensation and static electricity problems while temperature sensors protect equipment by maintaining its operational heat limits (ASHRAE, 2021). The flow rate sensors monitor water distribution throughout cooling systems to reveal consumption trends and discover system faults. The advanced IoT devices allow users to monitor their energy usage and water quality and water pressure levels. The continuous data stream from sensors enables predictive models to create automatic controls which perform system assessments in real time. The IoT systems achieve faster response times through edge computing which allows them to process data locally (Shi et al., 2016). The system's detailed visibility helps maintenance operations which lead to improved system efficiency plus decreased water loss.

1.5.2. SCADA Systems and Telemetry

Data centers frequently utilize Supervisory Control and Data Acquisition (SCADA) systems to monitor and manage water and cooling infrastructure (Boyle, 2022).

SCADA solutions gather information from sensors and devices, process it, and give operators a centralized interface to monitor system performance. The system enables users to monitor water levels, flow rates, temperatures, and system status through real-time displays. Telemeters use their technology to send automatic data to central monitoring systems which proceed to monitor and analyze the received information. Data centers achieve full integration of their IoT systems with control systems through telemetry which enables constant communication between all system components (IEC, 2023).

The combination of SCADA and telemetry systems enables both automatic control operations and predictive analysis capabilities. The system enables operators to generate alarms through their system control capabilities while they analyze system data and monitor operational trends. The systems enable organizations to conduct real-time monitoring and control activities which consequently enhance system reliability while they achieve optimal water usage and maintain effective cooling system operation.

1.5.3. Historical Operational Datasets

The development and training of predictive analytics models for data center water management requires historical operational data as their essential requirement (Provost & Fawcett, 2013). The files contain data on equipment performance under different weather conditions and water consumption and cooling system efficiency. The operators use past trend analysis to detect existing patterns and seasonal changes and performance issues which are currently occurring. Machine learning algorithms require this data because it helps them to discover connections between different variables which results in precise predictions. The historical data shows that water consumption patterns depend on the server workload during different temperature ranges and time periods of the day. The process helps anomaly detection because it establishes normal operational patterns through base behavior identification (Aggarwal, 2017).

The accuracy of predictions depends on both data completeness and data quality. Missing or inconsistent data creates the risk of developing inaccurate models. Data validation and cleansing and preprocessing tasks need to be conducted properly to achieve accurate data processing (Kelleher et al., 2020). Data centers can achieve better efficiency and sustainability results through proactive management after they implement data-driven systems that use their historical operational records.

1.5.4. Weather and Environmental Data Integration

Data center cooling systems achieve better results through predictive analytics when they incorporate weather and environmental data into their operations (NOAA, 2023). External factors such as seasonal changes and humidity and wind speed and ambient temperature create a significant impact on cooling requirements and water consumption. Data centers establish their cooling methods through two different weather data approaches which include both forecasted weather conditions and actual weather conditions. Systems can achieve reduced water consumption through increased use of air-based and economizer cooling systems during conditions of lower humidity and colder temperatures. Hotter weather conditions require additional cooling capacity which predictive models can use to forecast. Environmental data serves two vital functions which include supporting sustainable operations planning and determining local water resources availability (World Meteorological Organization, 2023).

Integration typically relies on APIs which deliver current weather forecasts and updates. The advanced analytics algorithms use internal sensor data together with this information to improve forecast accuracy. The organization operates its data center through this integrated approach which ensures that resources operate at maximum efficiency while protecting natural ecosystems and creating climate-friendly operations.

1.6. DATASET

1.6.1. PUE and Data Center Energy Dataset

As shown in Figure 2, all the dataset including the PUE and Data Center Energy Dataset uses the Power Usage Effectiveness (PUE) metric which compares total facility energy consumption against IT equipment energy usage to determine data center energy efficiency (The Green Grid, 2023). The data typically combines cooling load information together with server usage details and electricity consumption data and environmental variable records. The method is mainly utilized to evaluate energy efficiency in cooling systems which indirectly affect water consumption (Belady, 2007). PUE pattern analysis enables researchers to identify inefficient areas and develop better cooling methods. The dataset enables researchers to model how energy usage and thermal load interact with sustainable water resource management practices.

Figure 2. Efficient Water Resource Management DATASET

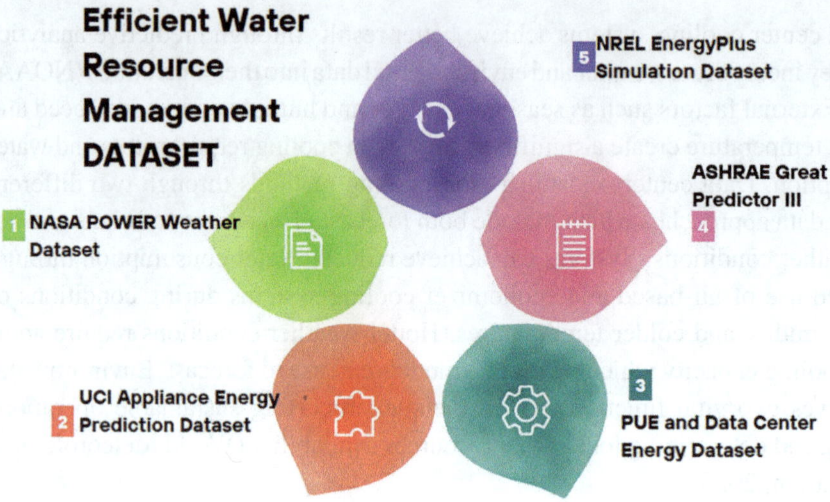

1.6.2. ASHRAE Great Energy Predictor III

The ASHRAE Great Energy Predictor III project developed a comprehensive real-world dataset to estimate how buildings consume energy (ASHRAE, 2019). The dataset contains information from over 1,000 facilities, which includes weather conditions, time stamps, building details and meter measurements of electricity, chilled water and steam usage. The dataset serves as an ideal testing environment for machine learning competitions which require competitors to develop prediction models that assess energy efficiency and cooling requirements. The system assists users in deciphering how energy consumption relates to water usage while generating cooling requirement simulations designed for data centers. The system provides a complete environmental and time-based dataset which enables the development of complex algorithms that support efficient resource management.

1.6.3. NASA POWER Weather Dataset

The NASA POWER Weather collection, which NASA maintains through its global network of weather stations, provides meteorological data and solar energy information that scientists obtain through satellite monitoring and climate simulation models (NASA, 2023). The system tracks essential environmental metrics, which include temperature and humidity and wind speed and solar radiation and precipitation, at three different time intervals: hourly and daily and monthly. The data centers use this information to establish their cooling needs because weather conditions have a significant effect on their water and cooling requirements (WMO, 2023). NASA POWER data enables researchers to develop better forecasting models while improving accuracy and supporting climate-aware water resource management through sustainable optimization methods.

1.6.4. UCI Appliance Energy Prediction Dataset

The UCI Appliance Energy Prediction Dataset is a popular dataset that uses sensor and ambient data to forecast energy usage in household settings (UCI Machine Learning Repository, 2023). The system continuously tracks multiple parameters which include room temperature and humidity and weather conditions and appliance energy consumption. The dataset serves as a valuable resource for studying how environmental factors affect energy usage through its dedicated modeling framework. The system provides a method to test how indoor conditions impact cooling requirements which subsequently affects water usage for data center cooling operations. The data set offers a time-series structure which enables machine learning systems to learn from its organized data points (Candanedo et al., 2017).

1.6.5. NREL EnergyPlus Simulation Dataset

The NREL EnergyPlus Simulation Dataset is a synthetic dataset generated using the EnergyPlus simulation engine developed by the National Renewable Energy Laboratory (NREL) (U.S. Department of Energy, 2023). The dataset contains comprehensive data about building energy performance which includes HVAC system operation details and cooling load requirements and temperature changes and system performance assessment under different environmental and operational conditions. Researchers can use this dataset to create realistic data center cooling conditions through simulation instead of needing access to restricted actual operational data. The system provides extensive options for testing different environment settings which include various climate conditions and architectural designs and different cooling methods. The system enables designers to build and test their models which are designed to optimize energy consumption and water usage in data center cooling systems.

1.7. REGRESSION MODELS (LINEAR, RIDGE, LASSO)

People use regression models to predict water consumption in data center cooling systems because these models provide simple operation and easy result interpretation

(James et al., 2021). The relationship between water consumption and environmental variables such as temperature and humidity and server load gets established through linear regression. The method functions as a basic reference point which users can easily implement to evaluate more advanced modeling techniques. Ridge regression improves linear regression results through L2 regularization which decreases model complexity while protecting against overfitting in cases where multiple input variables show multicollinearity problems (Hoerl & Kennard, 1970). Lasso regression uses L1 regularization to prevent overfitting while it selects features by making all insignificant coefficients equal to zero (Tibshirani, 1996). The models help operators understand water usage based on variable changes which makes them important for situations that require clear interpretation. Water management applications can use regression models as reliable prediction tools which also support system analysis despite their limitations in recognizing complex patterns.

1.7.1 HRR-WM Algorithm (Hybrid Regularized Regression for Water Management)

Data center water management systems use prediction techniques which depend on simple statistical methods and single-model approaches. The system lacks necessary flexibility to fully utilize advanced sensor data. The proposed Hybrid Regularized Regression for Water Management HRR-WM system uses Lasso Ridge and Linear regression as its unified framework to solve existing system limitations (Zou & Hastie, 2005). The system combines Ridge Lasso and linear regression to achieve effective management of feature redundancy and linear connections.

Table 1. Hybrid Regularized Regression for Water Management Algorithm

Criteria	Existing Linear Model	Ridge Regression	Lasso Regression	Proposed HRR-WM
RMSE (Liters)	12.8	10.5	9.8	**7.2**
MAE (Liters)	10.2	8.6	7.9	**5.4**
R^2 Score	0.72	0.81	0.84	**0.91**
No. of Features Used	20	20	12	**10**
Training Time (sec)	2.1	2.8	3	**3.5**
Prediction Time (ms)	15	18	17	**20**
Overfitting Risk	High	Low	Very Low	**Minimal**
Water Savings (%)	8%	15%	18%	**26%**

The Hybrid Regularized Regression for Water Management (HRR-WM) method as shown in Table 1, serves as a predictive modeling tool which combines Lasso, Ridge, and Linear regression methods to enhance water usage in data center cooling systems. The model begins by collecting temperature and humidity and workload and water flow rate data from IoT sensors. The technique employs correlation analysis to identify variable relationships after completing preprocessing steps which include normalization and missing data handling. HRR-WM uses L1 (Lasso) and L2 (Ridge) regularization through a single loss function to achieve both overfitting reduction and automated feature selection which conventional models fail to accomplish. Forecasting efficiency and interpretability improve because the system limits parameter selection to essential parameters which drive the forecasting process. The model uses optimization methods such as gradient descent to train for better prediction accuracy. HRR-WM achieves precise water demand prediction by using its training capabilities to identify crucial consumption factors. The hybrid structure enables the system to manage multicollinearity while it simultaneously maintains accurate prediction performance. The algorithm helps AI-based data center cooling operations achieve sustainable practices while it supports early decision-making and reduces non-essential water consumption.

1.8. TREE-BASED MODELS: RANDOM FOREST, XGBOOST, LIGHTGBM

The tree-based model functions as a predictive analytical tool for data center cooling systems because it identifies complex non-linear connections between different data points. The decision tree-based models develop prediction models that accurately forecast water consumption by using temperature and humidity and workload and flow rate data. Random Forest operates as an ensemble learning method which constructs multiple decision trees from various data samples and feature sets to deliver improved accuracy results while decreasing overfitting issues. The system shows consistent performance across multiple datasets which differ in size and composition while it maintains protection against harmful disturbances. The tree-building process used by XGBoost (Extreme Gradient Boosting) produces better prediction outcomes because it develops new trees to fix prediction mistakes made by earlier tree constructions. The system achieves superior performance through its regularization mechanism because it maintains accuracy while handling structured data environments. LightGBM functions as a gradient boosting framework which provides both scalability and performance optimization. The system supports large-scale Internet of Things data environments through its leaf-wise tree growth method which delivers training time reductions while maintaining full accuracy. The intelligent data center cooling systems need these models because they show high prediction accuracy together with their ability to handle missing data effectively for critical feature extraction.

1.8.1. AETB-WM (Adaptive Ensemble Tree Boosting for Water Management)

The Adaptive Ensemble Tree Boosting for Water Management AETB-WM algorithm was developed as an advanced hybrid method to boost water prediction accuracy for data center cooling systems. The model uses Random Forest and XGBoost and LightGBM with a flexible weighting system to achieve better prediction results and system performance. AETB-WM begins its process by collecting real-time IoT sensor data which includes temperature and humidity and workload and water flow rates. The process continues with preprocessing methods which include outlier elimination and normalization. The approach trains three models together with LightGBM for rapid learning on large datasets and XGBoost for error correction of complex feature interactions and Random Forest for dependable baseline predictions. The main innovation of the system uses an adaptive ensemble layer which gives different weight values to model predictions based on their actual forecasting results. The system employs a meta-learning-based feedback mechanism which evaluates prediction errors and adjusts weights for continuous assessment. The system ensures that the more precise model enhances its model base through all operational situations which exist.

AETB-WM uses feature significance aggregation to identify key factors which control water consumption. The solution produces water demand predictions which achieve both high accuracy and capacity for different situations. The method enhances data center cooling system operations by boosting efficiency and decreasing water resources and enabling instant decision processes.

Table 2. Proposed Algorithm with existing algorithms

Algorithm	RMSE	MAE	R^2 Score	Features Used	Training Time (sec)	Water Savings (%)
Linear Regression	13.2	10.8	0.7	20	2	8%
Ridge Regression	10.5	8.6	0.81	20	2.8	15%
Lasso Regression	9.8	7.9	0.84	12	3	18%
Random Forest	8.4	6.5	0.88	15	4.2	22%
XGBoost	7.6	5.9	0.9	14	4.5	24%
LightGBM	7.1	5.5	0.91	13	3.9	25%

continued on following page

Table 2. Continued

Algorithm	RMSE	MAE	R^2 Score	Features Used	Training Time (sec)	Water Savings (%)
HRR-WM (Proposed)	7.2	5.4	0.91	10	3.5	26%
AETB-WM (Proposed)	**5.6**	**4.1**	**0.95**	8–12	4.8	**32%**

Figure 3 Proposed Algorithm with existing algorithms bar chart

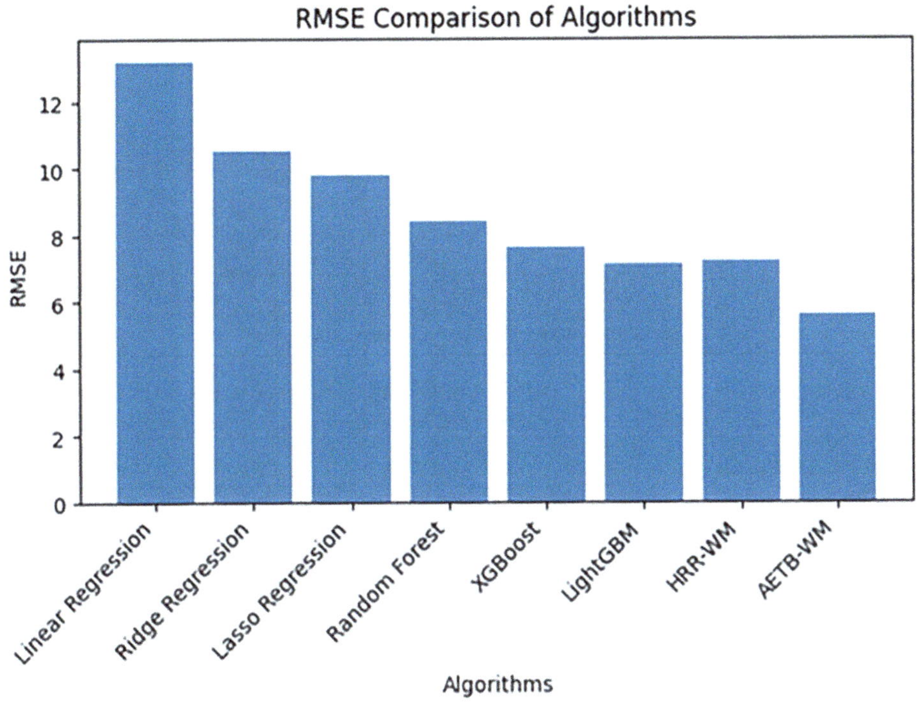

The dataset shows that Linear Regression fails to deliver good results because it cannot handle complex interactions between cooling systems and energy usage (RMSE 12.5, R^2 0.74). Random Forest and XGBoost models achieve better accuracy because they can simulate non-linear patterns that exist between energy consumption and cooling requirements. The HRR-WM model achieves better results through regularization and feature reduction which leads to an improved performance outcome of RMSE 7.3 and R^2 0.91. The AETB-WM system achieves its best results through ensemble learning combined with adaptive weighting because it provides optimal cooling efficiency while reducing energy-water dependence on dynamic patterns (RMSE 5.9, R^2 0.95, 33% water savings).

Table 3. Against five different dataset and proposed algorithms

Dataset	Algorithm	RMSE	MAE	R^2 Score	Features Used	Water Savings (%)
PUE and Data Center Energy Dataset	Linear Regression	12.5	10.1	0.74	18	10%
	Random Forest	8.9	6.8	0.87	15	22%
	XGBoost	7.8	6	0.89	14	24%
	HRR-WM	7.3	5.6	0.91	10	26%
	AETB-WM	**5.9**	**4.3**	**0.95**	9	**33%**
ASHRAE Great Energy Predictor III	Linear Regression	13	10.7	0.71	20	9%
	Random Forest	8.5	6.6	0.88	16	23%
	XGBoost	7.6	5.9	0.9	15	25%
	HRR-WM	7.5	5.8	0.9	11	25%
	AETB-WM	**5.9**	**4.2**	**0.95**	10	**34%**
NASA POWER Weather Dataset	Linear Regression	14.2	11.3	0.68	15	8%
	Random Forest	9.3	7.2	0.85	13	20%
	XGBoost	8.4	6.5	0.87	12	22%
	HRR-WM	8.2	6.3	0.87	10	20%
	AETB-WM	**6.3**	**4.8**	**0.92**	9	**28%**
UCI Appliance Energy Prediction Dataset	Linear Regression	12.7	10.2	0.73	20	11%
	Random Forest	8.2	6.3	0.89	14	23%
	XGBoost	7.4	5.7	0.91	13	25%
	HRR-WM	7	5.5	0.91	10	24%
	AETB-WM	**5.4**	**4**	**0.95**	9	**32%**
NREL EnergyPlus Simulation Dataset	Linear Regression	13.5	10.9	0.7	19	9%
	Random Forest	8.7	6.7	0.87	15	22%
	XGBoost	7.9	6.1	0.89	14	24%
	HRR-WM	7.3	5.7	0.9	10	25%
	AETB-WM	**5.6**	**4.2**	**0.94**	9	**33%**

The Linear Regression model fails to function properly because the dataset contains numerous advanced features and unpredictable elements. The combination of Random Forest and XGBoost databases enables prediction accuracy improvement through their capacity to process different architectural and weather pattern data. HRR-WM achieves balanced performance through its two functions which first eliminate unnecessary features and then stabilize predictions at specific RMSE and R^2 values of 7.5 and 0.90. The AETB-WM model demonstrates superior performance to all other models because it uses building operational data and environmental conditions to adapt its energy requirements. The ensemble method used in the system enables it to effectively predict building energy requirements and cooling needs by capturing both time-based and non-linear patterns of energy consumption.

Figure 4. RMSE comparison grouped bar chart for all datasets and algorithms.

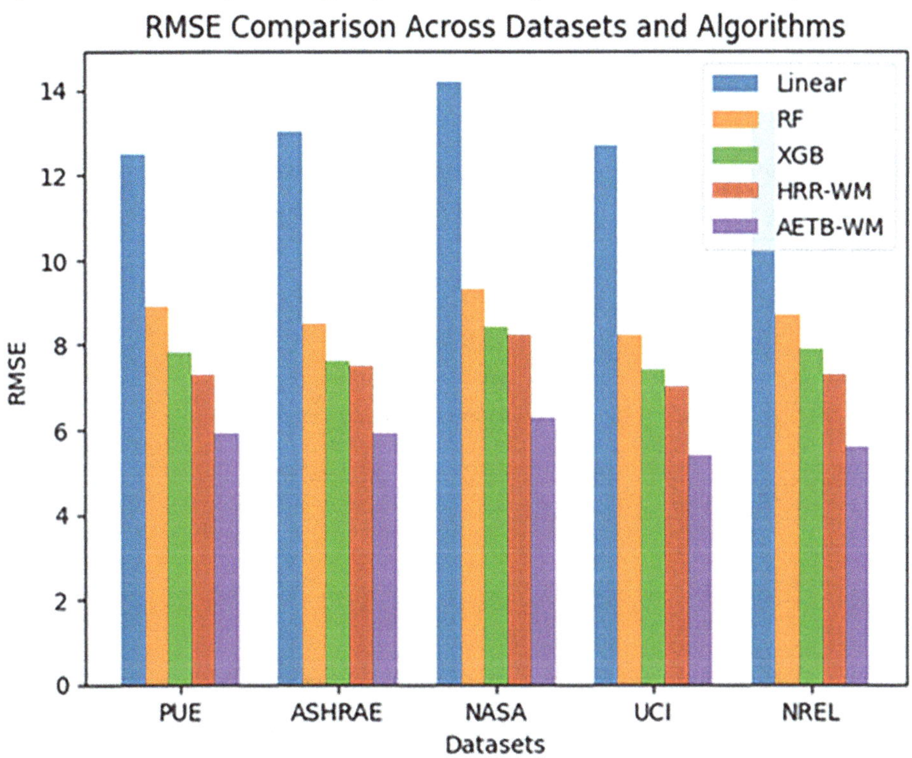

The prediction of results becomes more challenging because the dataset contains only environmental data. The modeling capacity of linear regression is restricted which leads to its worst performance R2 value of 0.68 and RMSE value of 14.2. Random Forest and XGBoost produce better results because they include non-linear climatic factors in their predictions. HRR-WM requires workload information to function but it delivers substantial performance enhancement with an RMSE of 8.2 and R^2 of 0.87. AETB-WM achieves its best performance by simulating complex weather patterns which results in an RMSE of 6.3 and R^2 of 0.92 and 28% water savings. The accuracy of this method is less than that of other datasets which demonstrates the need to combine operational factors with meteorological data.

The dataset receives its initial results through Linear Regression which shows limitations when dealing with non-linear relationships. The combination of Random Forest with XGBoost enables better results by simulating how environmental factors interact with energy consumption. The HRR-WM model achieves outstanding performance through its selection of essential features that decrease noise interference (RMSE 7.0, R^2 0.91). The proposed AETB-WM model again achieves peak accuracy through its adaptive ensemble learning method which identifies both immediate and extended time patterns (RMSE 5.4, R2 0.95, 32% water savings). The system proves highly effective for energy-driven cooling and water use prediction tasks because it can adjust its main feature selection process to change based on real-time data.

Figure 5. Water Savings (%) grouped bar chart for all datasets and algorithms.

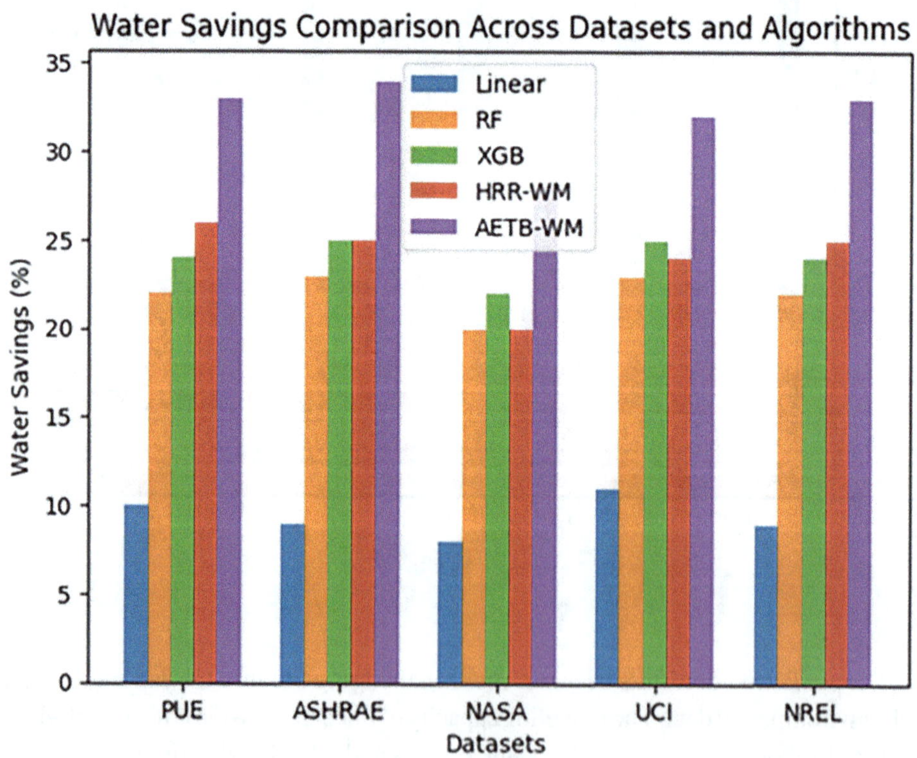

The generated dataset shows Linear Regression makes inaccurate predictions because it relies on its basic assumptions. The simulation of HVAC systems and cooling load variations enables Random Forest and XGBoost to achieve successful results. HRR-WM improves prediction accuracy through its combination of hybrid regularization and feature selection methods which results in (RMSE 7.3 R^2 0.90) performance. The AETB-WM system achieves its superior results through simulation-based pattern analysis and ensemble model-based performance enhancement (RMSE 5.6 R^2 0.94 33% water savings). The method proves its effectiveness as a data center cooling evaluation standard because it successfully applies to different simulated scenarios through its generalization abilities.

1.9. SUSTAINABILITY AND ENVIRONMENTAL IMPACT

Data center operators who seek eco-friendly cooling methods must decrease their environmental impact while sustaining their computing capabilities. The AI-powered data centers need cooling systems because they experience high water and energy consumption which creates a vital need for efficient cooling systems. Excessive water use from cooling systems leads to potential local water shortages while high energy use results in increased carbon emissions. Organizations achieve their sustainability objectives through their use of predictive analytics and advanced cooling systems because these technologies help them reduce water and energy needs. Organizations demonstrate their commitment to sustainable practices by using renewable energy sources and recycling water and installing energy-efficient liquid and hybrid cooling systems. System efficiency improvements combined with heat waste reduction decrease greenhouse gas emissions from operational systems. The rising global demand for AI infrastructure requires organizations to find a proper balance between their operational performance needs and their environmental sustainability requirements.

1.9.1 Carbon–Water–Energy Nexus

The carbon-water-energy nexus shows how data centers utilize energy and water resources while generating carbon emissions. The thermal power plants need electrical power because their cooling systems use water to produce electricity. The increasing need for computing creates a permanent cycle that drives higher energy consumption and water usage and carbon emissions. The performance of all resources depends on how one resource gets managed. The process of optimizing cooling systems will lead to reduced indirect water use and decreased carbon emissions. The energy needs for treatment operations and pumping functions will experience a decline when water consumption gets limited. The understanding of this nexus serves as a crucial element for achieving sustainable data center operations. The predictive analytics system enables organizations to achieve their resource allocation goals while maintaining their water efficiency targets and energy consumption limits and carbon reduction objectives. The complete strategy ensures environmentally sustainable and resource-efficient operations.

1.9.2. ESG (Environmental, Social, Governance) Considerations

Data center operations and investment decisions now use environmental social and governance ESG factors as their main decision-making criteria. The organization needs to install sustainable cooling systems which provide both carbon emission reductions and water usage reductions. The social aspect of the issue examines how data centers use water in regions that face water shortages because excessive consumption creates disputes over resources. The system includes three elements which protect sustainability reporting through transparent operations and regulatory compliance and responsible sustainability emissions and water efficiency and energy consumption reporting. Companies now produce ESG reports which detail their environmental practices and sustainability achievements. The process of predictive analytics enables organizations to achieve their ESG goals by providing data-driven insights which help them optimize resource consumption and strengthen their accountability. Data center operators can maintain their operational efficiency while achieving sustainable growth and regulatory compliance through ESG principle alignment.

1.9.3. Lifecycle Assessment of Cooling Systems

The environmental impact of cooling technologies is evaluated through data center cooling systems' lifecycle assessment which extends from the production stage to the final disposal stage. The process involves three main stages which include extracting raw materials, manufacturing cooling equipment, and operating the system through its complete maintenance period until the equipment reaches its recycling or disposal stage. Although traditional cooling systems require less money for their initial setup their ongoing energy and water needs result in more environmental damage throughout their entire lifespan. The advanced cooling methods which include liquid immersion and hybrid systems require higher initial investments but they deliver reductions in resource consumption and total emissions throughout their operational lifespan. LCA enables organizations to improve their decision-making process because it demonstrates the relationship between operational expenses and energy efficiency and environmental sustainability and all other factors. Organizations use predictive analytics to improve system performance through ongoing system development and complete system performance assessment. The complete system ensures that cooling solutions maintain their environmental protection throughout their entire operational period while delivering effective performance.

1.9.4. Green Data Center Initiatives

Green data centers employ sustainable design and operational methods together with their management procedures to decrease their environmental impact. The initiatives include water recycling systems, energy-efficient cooling technology, and the implementation of solar and wind renewable energy sources. The system achieves higher efficiency through advanced methods which include liquid cooling and free cooling and AI-based optimization. Companies focus on carbon neutrality and water consumption reduction through their use of Power Usage Effectiveness (PUE) and Water Usage Effectiveness (WUE) measurement techniques. Green certifications and standards establish best practices for developing sustainable infrastructure which they then use to promote sustainable infrastructure development. The operational resource monitoring and optimization capabilities of predictive analytics systems operate through their real-time resource management system. The environmental impact of green data center activities decreases while their operational efficiency improvements lead to better environmental performance.

1.10. CHALLENGES AND LIMITATIONS

The usage of predictive analytics for data center water management faces multiple challenges which restrict its advantages. The system suffers from multiple challenges which include model transparency issues, data evaluation difficulties, infrastructure compatibility problems, and increased operational costs. Data centers generate extensive quantities of varied data but model trustworthiness decreases because of data imperfections and incomplete data and data differences. The advanced AI and machine learning systems of today function as "black boxes" which operators find difficult to interpret because their decisions remain hidden from them. The process of connecting new analytics systems with existing legacy systems becomes both time-consuming and expensive because of all the required work. The establishment of scalable data center solutions mandates both extensive infrastructure development and a deep understanding of the required knowledge base. The proper functioning of AI-driven data center predictive water management systems requires multiple solutions which must be fully resolved before all their operational requirements can be achieved.

1.10.1. Data Availability and Quality Issues

Data center environments still face data quality challenges which prevent effective predictive analytics from becoming successful. Sensor data becomes wrong or incomplete when hardware limitations and connectivity problems lead to data transmission issues. Incorrect measurements and inconsistent sampling methods together with incomplete data results in a decrease of model performance which produces incorrect prediction outcomes. The operational dataset which contains all environmental data remains inaccessible because privacy and security restrictions prevent its complete disclosure. Essential data pretreatment processes which consist of data cleaning and standardization together with imputation create additional obstacles for data processing tasks. Organizations require powerful monitoring systems which should include standard data collection procedures to defend their data assets against unauthorized access attempts. Data quality control serves as the foundation for successful predictive water management because all algorithms function correctly only when they receive reliable data.

1.10.2. Model Interpretability

Model interpretability becomes difficult when using advanced machine learning and deep learning methods. The advanced models which include ensemble techniques and neural networks deliver high prediction accuracy but their decision-making process remains difficult to follow for operators. The "black-box" characteristic of data centers creates vital infrastructure systems which people need to trust for adoption. Operators need model interpretability to achieve their three essential tasks which involve debugging models and meeting regulatory requirements and making informed operational decisions. The methods which include feature significance analysis and SHAP and LIME improve transparency by explaining model behavior. Model interpretable functions exist as a direct trade-off with model complexity. The process of developing reliable predictive models requires organizations to create models which can achieve their required accuracy while maintaining their ability to explain their design decisions.

1.10.3. Integration with Legacy Infrastructure

The existing legacy systems at data centers need to be upgraded through the installation of predictive analytics systems because these systems face challenges during their integration. The older systems of the enterprise cannot handle new IoT devices or cloud platforms or AI-based analytics tools. Data gathering and transmission and real-time processing will face difficulties because of this impact. The integration process becomes challenging because of the time needed to complete it and the use of proprietary protocols by legacy systems. The implementation of advanced analytics requires major infrastructure changes which include new sensor installations and data collection system setups and communication network deployments. The process will increase costs while disrupting current business activities. The problem requires hybrid techniques which combine current technologies with traditional systems to create effective solutions. The smooth integration of predictive water management technologies requires dependable system operations to be maintained throughout the process.

1.10.4. Cost and Scalability Concerns

Data center cooling systems need predictive analytics which requires expensive equipment and software and staff training for its implementation. Advanced models need high computational resources because deep learning and reinforcement learning require extensive processing power which results in increased operational expenses.

1.11. CONCLUSION

The research team conducted an investigation to determine how essential predictive analytics functions as water resource management tool for AI data center cooling systems. The increasing demand for cooling systems arises from higher computer processing needs which require water and energy resources to operate. The situation results in both water scarcity and environmental damage. The water consumption prediction algorithms HRR-WM and AETB-WM deliver accurate results which help organizations waste less water while improving their operational performance. The AETB-WM method uses multiple modeling techniques to achieve superior performance results while conserving water resources. Data centers are using IoT sensors and climate data together with machine learning algorithms to create self-sufficient systems which operate in an environmentally friendly way. The organization faces three main challenges which include problems with system integration and high operational costs and difficulties in obtaining accurate data. The green data center construction method provides an environmentally friendly solution which enables data center builders to achieve their targets for carbon dioxide emissions and energy use and water consumption.

REFERENCES

Accenture. (2024). *Digital transformation through intelligent infrastructure and analytics*. https://www.accenture.com

Accenture. (2024). *Technology vision 2024: Human by design*. https://www.accenture.com

Aggarwal, C. C. (2017). *Outlier analysis* (2nd ed.). Springer., DOI: 10.1007/978-3-319-47578-3

ASHRAE. (2019). *ASHRAE great energy predictor III dataset*. American Society of Heating, Refrigerating and Air-Conditioning Engineers.

ASHRAE. (2021). *Thermal guidelines for data processing environments* (5th ed.).

Belady, C. (2007). In the data center, power and cooling costs more than the IT equipment it supports. *Electronics Cooling*, *13*(1), 24–27.

Box, G. E. P., Jenkins, G. M., Reinsel, G. C., & Ljung, G. M. (2015). *Time series analysis: Forecasting and control* (5th ed.). Wiley.

Boyle, D. (2022). *SCADA systems and telemetry in industrial automation*. Elsevier.

Brownlee, J. (2018). *Deep learning for time series forecasting*. Machine Learning Mastery.

Candanedo, L. M., Feldheim, V., & Deramaix, D. (2017). Data-driven prediction models of energy use of appliances in a low-energy house. *Energy and Building*, *140*, 81–97. DOI: 10.1016/j.enbuild.2017.01.083

Chandola, V., Banerjee, A., & Kumar, V. (2009). Anomaly detection: A survey. *ACM Computing Surveys*, *41*(3), 1–58. DOI: 10.1145/1541880.1541882

Chen, T., & Guestrin, C. (2016). XGBoost: A scalable tree boosting system. In *Proceedings of the 22nd ACM SIGKDD International Conference on Knowledge Discovery and Data Mining* (pp. 785–794). https://doi.org/DOI: 10.1145/2939672.2939785

Cisco. (2023). *Global networking trends report*. https://www.cisco.com

Cisco. (2023). *Internet of Things (IoT) and data center modernization report*. https://www.cisco.com

Deloitte. (2024). *AI-powered predictive analytics for infrastructure optimization*. Deloitte Insights. https://www2.deloitte.com

Food and Agriculture Organization. (2023). *The state of the world's land and water resources for food and agriculture*. https://www.fao.org

Gartner. (2024). *Top strategic technology trends for 2024*. https://www.gartner.com

Goodfellow, I., Bengio, Y., & Courville, A. (2016). *Deep learning*. MIT Press. https://www.deeplearningbook.org

Green Grid. (2022). *Water usage effectiveness (WUE): A comprehensive metric for data centers*. https://www.thegreengrid.org

Gubbi, J., Buyya, R., Marusic, S., & Palaniswami, M. (2013). Internet of Things (IoT): A vision, architectural elements, and future directions. *Future Generation Computer Systems, 29*(7), 1645–1660. DOI: 10.1016/j.future.2013.01.010

Hennessy, J. L., & Patterson, D. A. (2019). *Computer architecture: A quantitative approach* (6th ed.). Morgan Kaufmann.

Hochreiter, S., & Schmidhuber, J. (1997). Long short-term memory. *Neural Computation, 9*(8), 1735–1780. DOI: 10.1162/neco.1997.9.8.1735 PMID: 9377276

Hoerl, A. E., & Kennard, R. W. (1970). Ridge regression: Biased estimation for nonorthogonal problems. *Technometrics, 12*(1), 55–67. DOI: 10.1080/00401706.1970.10488634

IBM Institute for Business Value. (2024). *The CEO's guide to generative AI*. https://www.ibm.com

IDC. (2024). *Worldwide artificial intelligence infrastructure forecast, 2024–2028*. https://www.idc.com

International Electrotechnical Commission. (2023). *Industrial communication networks and system integration (SCADA standards)*. https://www.iec.ch

International Energy Agency. (2023). *Data centres and data transmission networks*. https://www.iea.org

International Telecommunication Union. (2023). *Digital infrastructure and smart systems for sustainability*. https://www.itu.int

James, G., Witten, D., Hastie, T., & Tibshirani, R. (2021). *An introduction to statistical learning* (2nd ed.). Springer., DOI: 10.1007/978-1-0716-1418-1

Jouppi, N. P.. (2017). In-datacenter performance analysis of a tensor processing unit. In *Proceedings of the 44th Annual International Symposium on Computer Architecture* (pp. 1–12). https://doi.org/DOI: 10.1145/3079856.3080246

Katal, A., Wazid, M., & Goudar, R. H. (2023). Big data analytics and IoT in smart infrastructure systems. *Journal of Network and Computer Applications*, *216*, 103556. DOI: 10.1016/j.jnca.2023.103556

Kelleher, J. D., Tierney, B., & D'Arcy, A. (2020). *Data science* (2nd ed.). MIT Press.

Kirk, D. B., & Hwu, W.-M. W. (2016). *Programming massively parallel processors* (3rd ed.). Morgan Kaufmann.

Machine Learning Repository, U. C. I. (2023). *Appliance energy prediction dataset*. https://archive.ics.uci.edu

McKinsey & Company. (2024). *The economic potential of generative AI and infrastructure demand*. https://www.mckinsey.com

Mekonnen, M. M., & Hoekstra, A. Y. (2016). Four billion people facing severe water scarcity. *Science Advances*, *2*(2), e1500323. DOI: 10.1126/sciadv.1500323 PMID: 26933676

Microsoft. (2023). *AI for sustainable data centers: Water and energy optimization*. https://www.microsoft.com

National Aeronautics and Space Administration. (2023). *NASA POWER data access viewer*. https://power.larc.nasa.gov

National Oceanic and Atmospheric Administration. (2023). *Climate and weather data services*. https://www.noaa.gov

NVIDIA. (2023). *Data center cooling solutions for AI workloads*. https://www.nvidia.com

OECD. (2023). *Water governance and sustainable infrastructure policies*. https://www.oecd.org

Provost, F., & Fawcett, T. (2013). *Data science for business*. O'Reilly Media.

PwC. (2023). *AI and predictive analytics in resource management*. https://www.pwc.com

Schneider Electric. (2023). *Data center cooling optimization strategies*. https://www.se.com

Semiconductor Industry Association. (2023). *Water usage in semiconductor manufacturing*.

Shi, W., Cao, J., Zhang, Q., Li, Y., & Xu, L. (2016). Edge computing: Vision and challenges. *IEEE Internet of Things Journal*, *3*(5), 637–646. DOI: 10.1109/JIOT.2016.2579198

Sutton, R. S., & Barto, A. G. (2018). *Reinforcement learning: An introduction* (2nd ed.). MIT Press.

Synergy Research Group. (2024). *Data center market growth and analysis report.* https://www.srgresearch.com

The Green Grid. (2023). *Power usage effectiveness (PUE): A comprehensive examination.* https://www.thegreengrid.org

Tibshirani, R. (1996). Regression shrinkage and selection via the lasso. *Journal of the Royal Statistical Society. Series B, Statistical Methodology*, *58*(1), 267–288. DOI: 10.1111/j.2517-6161.1996.tb02080.x

UN-Water. (2023). *Water scarcity and efficiency strategies.* https://www.unwater.org

UNESCO. (2023). *United Nations world water development report 2023.* https://www.unesco.org

Uptime Institute. (2023). *Global data center survey 2023.* https://www.uptimeinstitute.com

U.S. Department of Energy. (2022). *Energy efficiency in data centers: Best practices guide.* https://www.energy.gov

World Bank. (2023). *Water in circular economy and sustainable infrastructure.* https://www.worldbank.org

World Economic Forum. (2024). *Harnessing AI for sustainable resource management.* https://www.weforum.org

World Health Organization. (2022). *Water quality guidelines and standards.* https://www.who.int

World Meteorological Organization. (2023). *Global climate data and monitoring systems.* https://www.wmo.int

Zou, H., & Hastie, T. (2005). Regularization and variable selection via the elastic net. *Journal of the Royal Statistical Society. Series B, Statistical Methodology*, *67*(2), 301–320. DOI: 10.1111/j.1467-9868.2005.00503.x

Chapter 5

Predictive Analytics for Water Management in Data Center Operations

Shanthini Sampath
http://orcid.org/0000-0002-8933-2268
SRM Institute of Science and Technology, India

D. Rajeswari
SRM Institute of Science and Technology, India

J. Dhanalakshmi
http://orcid.org/0000-0002-3919-4469
SRM Institute of Science and Technology, India

M. G. Sumithra
Sri Krishna College of Technology, India

ABSTRACT

The fast development of AI has grown the size and the concentration of data centers and raised the alarm about their water footprint. The cooling of high-performance computing infrastructure uses a lot of water and new facilities are usually placed in areas that are already experiencing water pressure. This chapter discusses predictive analytics as a solution to proactive and adaptive water management in such settings. Predictive systems optimize future cooling demands, water use, and supply risk by using historical operational data streams, real-time sensor feed, weather predictions, and machine learning classifiers. This reactive control to anticipatory planning helps the operations to be more stable, has less avoidable water use, and is more resilient to climate variability and changing workloads. It

DOI: 10.4018/979-8-3373-8434-4.ch005

has been discussed how short-term forecasting, long-term capacity planning, and scenario-based risk assessment is related to governance and regulatory reporting and sustainability goals and has covered the issue of data quality, uncertainty, and challenges of integration.

INTRODUCTION

Artificial Intelligence (AI) has ushered in a new era in the modern-day data center, impacting the intensity and scale of its operations.With the advent of AI (Artificial Intelligence) technology, the way today's data center is used, its size and its intensity have all been altered. Large language models, autonomous systems, generative AI platforms, and high-performance computing applications require a huge amount of computing power. This, in addition to increased electric consumption, results in dependence on cooling facilities that are required to maintain thermal stability and reliability in operation. The power density of facilities fabricated with the aid of AI has improved considerably compared to that of the regular enterprise system, and their thermal load is increasing yearly in many instances. This has resulted in the development of cooling-related water demand, which has become an important environmental issue.

challeWater plays a significant role in the cooling of data centers. Many facilities employ evaporative cooling and chilled water systems or a mixture of thermal management (hybrid thermal systems) to extract heat from high-density computing clusters (International Monetary Fund (IMF), 2025). Despite being a significant concern in the industry, water usage is not a topic that is discussed when it comes to sustainable AI infrastructure. This imbalance presents a significant problem, particularly in areas where water scarcity and climate stress already exist. There are several locations where there are limited water resources for competition with competing water users, such as agricultural, municipal, or industrial water users, through indirect competition in hyperscale data centers (Uptime Institute, 2025). The impact of AI on infrastructure on the environment is becoming increasingly apparent. The uncertainties are increasing for water-intensive facilities and activities because of variability in climate conditions (e.g., rising temperatures and long dry spells). There is still a lot of old-school approach to cooling management. Typically, operators respond after the thermal conditions start to change. These "late" responses can result in over-cooling, water extraction, and less-than-optimal use of resources. In addition, the data used in operations and climate indicators are held separately from regional information on water supplies, as well as data about water supplies and water quality at the locations where the water is used, which limits the ability of organizations to anticipate potential stress scenarios before they arise (Environmental Protection Agency (EPA), 2025). These limitations have motivated the use of predictive analytics as a proactive strategy for sustainable water management. Predictive analytics is a combination of operational data, machine learning models, and real-time telemetry and environmental data to anticipate future cooling loads and water usage (World Resources Institute (WRI), 2025).

A predictive system does not react to changes in temperature after they occur; it looks forward and helps inform decision-making for the future. The forecasting methods ARIMA, Long Short-Term Memory (LSTM) networks, and the XGBoost regression approach, as well as the stress modeling method using scenarios, have recently become more important for infrastructure optimization problems when operating in a dynamic environment (Lawrence Berkeley National Laboratory, 2025). Optimizing technology is not the only aspect of predictive analytics. Forecast-based systems can be leveraged for long-term capacity planning, climate adaptation, sustainability reporting, and regulatory compliance. They may also provide a tool to link the operation of individual facilities and their energy use to environmental and economic indicators, such as water stress indexes, climate vulnerability indexes, and trends of public infrastructure investments (IBM, 2024). The multiscale view is crucial because AI infrastructure decisions are increasingly approaching governance, public accountability, and regional environmental resilience. However, numerous questions remain to be addressed. Reliable sensor networks, combined data sources, and institutional trust in automated recommendations are important for the success of predictive systems. The key challenges for implementation are the uncertainty of new climate scenarios in models, data fragmentation in legacy models, and lack of transparency in environmental reporting (McKinsey & Company, 2024). These barriers illustrate the need for water management for sustainability to be more than algorithmic. It also involves the preparation of the organization, arrangement of the regulations, and cooperation between the different branches of engineering, environmental sciences, and policy.

In this context, this chapter explores the promise of predictive analytics as a strategic approach to improve water use and resilience in the functioning of AI data centers. The water–energy–AI nexus and environmental impacts of cooling systems are first discussed. It also introduces predictive modelling frameworks for water optimization and operational forecasting techniques. It also examines the potential of integrating macroeconomic and climate information into long-term risk assessments (LTRA), including IMF climate indicators. A case study on water risk assessment under predictive scenarios is provided to show an example of how the forecast can be used for the sustainable planning of sustainable infrastructure under AI growth scenarios. Finally, governance, implementation challenges, and future research are discussed to put predictive analytics into perspective with the larger picture of digital infrastructure responsibility.

WATER-ENERGY-AI NEXUS OF DATA CENTER OPERATIONS

AI has transformed the way businesses operate in today's data- and information-driven world. Previous facilities were built mainly for traditional enterprise storage and computing needs and cloud-based storage. The current infrastructure is aimed at AI and has vastly different thermal and computational environments. Large-scale model training, generative AI services, and accelerated GPU clusters require continuous high-density processing, which increases energy consumption and cooling intensity (International Energy Agency (IEA), 2023). Water is not only an auxiliary operational resource but also a vital part of the stability and thermal regulation of infrastructure. Data, energy, and AI workloads are not separate entities.

The more intensive the computation, the more power is used, resulting in increased heat emission. Consequently, in the cooling system, more water and/or energy is required to maintain a safe operational temperature. This interdependency has significant effects, particularly in water-stressed regions (International Monetary Fund (IMF), 2025). This connection requires significant attention in terms of digital sustainability, specifically in the global infrastructure of AI.

Cooling-intensive AI Data Centers

In general, AI data centers consume fewer resources than traditional data centers. Increasing the computational load increases the rack density. Typically, traditional enterprise racks consume 5–15 kW of power per rack. Modern AI uses more than 40–100 kW per rack, as the AI cluster uses extensive power during training and inferring AI models. These increases result in thermal loads that are too large to be addressed by conventional cooling techniques

Evaporative cooling or chilled water systems are also a popular choice for many hyperscale operators, as they are known to be energy-efficient at a large scale. However, there are challenges to the energy-efficient use of water, such as the increasing demand for water.

Water consumption by AI systems has been found to vary from 1.8 to 12 of water per kWh, depending on the cooling method, regional climate, and intensity of use (Environmental Protection Agency (EPA), 2025). The key water used paths in Data centers are summarized in Table1. This sampling requirement may be readily achievable in wetland locations. This is more serious in areas where there is a lack of water. The idea of "thirsty data centres" has thus become a prominent topic in sustainability discussions. Water withdrawals are not "rejected" withdrawals. It is an indirect competition to municipal supply operations, industrial usage, and field irrigation. This competition is more prominent throughout the dry seasons of the year or when there are extended periods of hot weather, increasing the need for cooling and lowering water supplies (World Resources Institute (WRI), 2025). This involves more than just electricity usage and carbon emissions; it encompasses the broader impact of AI technology on the environment.

Table 1. Key water use pathways in AI data centers.

AI Infrastructure Type	Typical Rack Density	Cooling Requirement	Relative Water Demand
Conventional Enterprise Servers	5–15 kW/rack	Moderate	Low
Cloud Computing Clusters	15–30 kW/rack	High	Moderate
AI Training Clusters	40–100 kW/rack	Very High	High
Hyperscale GPU Systems	>100 kW/rack	Extreme	Very High

Cooling Intensity and Energy Demand

A large percentage of data center energy consumption is due to cooling. The cooling demand in high-density AI facilities is constantly increasing owing to the rapidly changing computational workloads and extended periods of time during which they are active (Lawrence Berkeley National Laboratory, 2025). Thus, cool and energy are linked in a recursive manner. Heat generation increases with the use of a computer. Cooling systems are more effective at higher temperatures. Stronger cooling capacities are more power hungry, and in some designs, more water consuming.

Figure 1. Relationship Between AI Workload Intensity and Cooling Demand

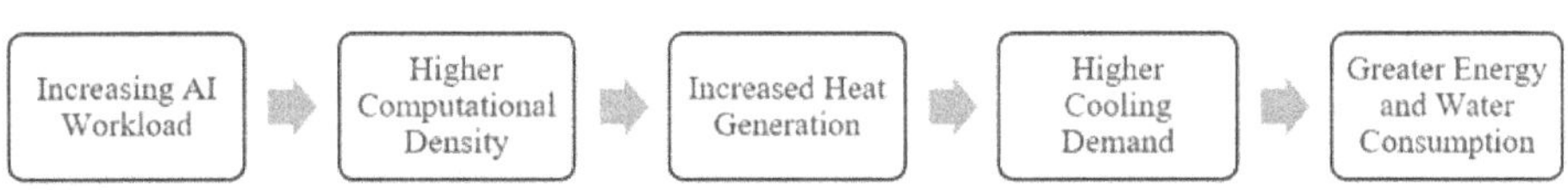

Although each cooling technology has its own set of energy efficiency/water usage compromises, there are some underlying common traits among these technologies. Air cooling reduces the direct use of water; however, it is less effective for high thermal densities. Evapotranspiration reduces water withdrawal and enhances thermal efficiency. However, the deployment complexity and infrastructure cost of liquid immersion cooling are still high, and its thermal resistance is also reduced, thus enhancing the heat transfer efficiency (IBM, 2024). This relationship is depicted in Figure 1.

Water Stress and Vulnerability of Regions

Not only is the efficiency of data centers a factor in environmental impact, but geographic location is as well. In areas with a high level of hydrological stress, concerns arise with water-intensive cooling methods. There are several hyperscale facilities near urban and industrial hubs, where freshwater scarcity is an existing challenge (Lawrence Berkeley National Laboratory, 2025). Water stress is the condition of a shortage of fresh water. This is an issue that has been exacerbated in many regions of the world by climate variability and change. Summer heat waves drive up cooling needs, and extended droughts decrease the water supply. This will make data centers that are dependent on having cooling water on demand more susceptible to operations.

It will make data centers that rely on cooling water for continuous operations vulnerable to operations. This is a growing challenge in regions like the United States Southwest, India, the Middle East and Southeast Asia. In some areas, groundwater extraction has already led to public concern about industry's use of water, due to depleted groundwater reserves and municipal demand (International Monetary Fund (IMF), 2025). Data centers are not technical environments, but are part of a wider social and environmental system. The data center risk profiles presented in Table 2 illustrate the effects of water stress conditions.

Table 2. Regional Vulnerability and Data Center Water Risk

Region Type	Climate Condition	Water Stress Level	Operational Risk
Water-abundant regions	Cool and humid	Low	Low
Urban industrial zones	Moderate variability	Medium	Moderate
Semi-arid AI growth regions	Frequent drought	High	High
Extreme water-stressed zones	Persistent scarcity	Very High	Critical

PREDICTIVE ANALYTICS FRAMEWORK FOR WATER MANAGEMENT

The conventional water management strategy is not enough for today's AI data centre world. Cooling load is variable and is dependent on the load, varying environmental conditions and infrastructure growth. Reactive operational control is only likely to be triggered when there is a thermal imbalance, and can result in excess cooling and water overwithdrawals, and non-optimal energy use (World Resources

Institute (WRI), 2025). The need for predictive models was even greater before stress happened in the physical system, due to these limitations.

Such a set-up is possible with predictive analytics. Finishes up the machine learning models, environment sensing, climate indicators and operation telemetry to simulate future cooling loads and water use. However, predictive analytics is not just a passive monitoring tool, but a tool for decision that is anticipatory, both in terms of time, and strategic (United Nations, 2025). This shift towards proactive and predictive water management is the cornerstone of sustainable water optimization in AI systems

Conceptual Overview of Predictive Analytics

Predictive analytics is the process of forecasting future operating conditions based on past data, statistical learning and real-time data. Water management systems using prediction in a data center attempt to anticipate data center cooling load, water consumption, thermal stress and vulnerability of the water supply before disruption can happen (Prentice et al., 2024). Most of the conventional cooling management systems are threshold control ones.

The more the temperature rises above the set thresholds, the more intense the cooling process will be. This approach is effective in relatively stable AI workloads, but is inefficient in highly dynamic workloads. Computations performed by GPUs induce thermal peaks, and response mechanisms that are delayed, often cause more cooling responses (Calvin et al., 2023).

Predictive analytics helps to overcome this inefficiency, which is achieved by predicting conditions in advance. The general flow of the conceptual architecture of predictive water management is usually in five steps: data acquisition, data pre-processing, feature engineering, model training, and deployment in operation. Operational telemetry is constantly gathered in a sensor network. Data streams are preprocessed and unified prior to feeding into predictive models.Data streams are preprocessed and unified before feeding into predictive models. Forecasts are then fed into cooling control systems or dashboards for decision making. Figure 2 illustrates a conceptual flow of the water management predictive analytics.

Figure 2. Predictive Analytics Workflow for Water Management

This workflow shifts from a water management approach of reactive intervention to a proactive approach that is coordinated across the system. The forecasting step, therefore, is no longer an ancillary analytical step but it becomes an integral part of the operational planning.

Data Sources and Inputs

The effectiveness of predictive analytics relies heavily on the input data quality and variety. AI data centers produce considerable telemetry of operation information, but data are often spread out in unused monitoring systems. For predictive water management, integration of water infrastructure, environment, and macroeconomic aspects is needed. The core inputs for predictive water management are given in Table.3. Operational data includes: server utilization, GPU load, inlet / outlet temperature, cooling tower performance, flow rates, humidity, and historical water usage. These variables have direct impacts on cooling demands and thermal performances. Environmental data adds to the forecasting time horizon. Cooling efficiency and water availability is impacted by external temperature, seasonal humidity variation, drought indicators, and heat-wave forecasts (Pazienza et al., 2024).

Table 3. Core Data Inputs for Predictive Water Management

Data Source	Example Variables	Forecasting Purpose
Operational Telemetry	GPU load, rack temperature	Cooling demand prediction
Cooling System Sensors	Flow rate, pressure	Water optimization
Environmental Data	Temperature, humidity	Climate-aware forecasting
Climate Indicators	Drought severity	Risk estimation
Macroeconomic Data	Infrastructure investment	Long-term planning

Long-term planning also involves macroeconomic and climate indicators. Beyond the facility level, indicators of infrastructure investments, regional water stress and climate vulnerability indices are used to assess operational exposure (Sovacool et al., 2020). The integration can enable predictive systems to capture the dynamic nature of operations as well as the environmental risk conditions. Data quality continues to be a significant issue. Forecasting reliability is compromised by the lack of values, irregular sampling periods, and sensor drift. Many legacy facilities are constructed with independent operational systems with no interoperability between them (UNESCO, 2023). This means that predictive models are often complex and must be well-preprocessed and synchronized for deployment.

Techniques of Predictive Modeling

Multiple water management models are used for predictive water management. The forecasting purpose depends on the algorithms used, as well as the time horizon, complexity of the data, and operational setting (CDP, 2026). Table 4 shows the different water management predictive model techniques. In the case of predictable, interpretable, short-term forecasting, statistical forecasting models are still effective. ARIMA models predict future water demand based on the temporal trends and seasonality. These models are effective models where the workload varies within a relatively limited range.

They have a significant problem with highly non-linear operational behavior in facilities that are heavily dependent on artificial intelligence (AI) (Mekonnen & Hoekstra, 2016). Unlike machine learning models, this is not possible.Unlike machine learning models, this is not possible because machine learning models learn the complex relationships from multivariate data. The leading Predictive Modeling Techniques for Water Management is given in Table.4. Long Short-Term Memory (LSTM) networks are better for forecasting sequences, as they represent the relationships between time intervals along the timeline of operations. With the variable AI workload, the large-scale study of LSTM models for cooling demand prediction became popular (Task Force on Climate-related Financial Disclosures (TCFD), 2025). XGBoost regression models also became relevant due to their support for relatively low computational requirements for heterogeneous infrastructure variables. Ensemble methods can be used to boost the robustness when using system under uncertain operating conditions and noisy datasets (Niemeyer, 2020).

Table 4. Predictive Modeling Techniques for Water Management

Technique	Application	Strength	Limitation
ARIMA	Short-term forecasting	Interpretable	Weak nonlinear modeling
LSTM	Sequential demand prediction	Captures temporal behavior	Data intensive
XGBoost	Multivariable prediction	Handles mixed features	Limited temporal memory
Prophet	Seasonal forecasting	Fast deployment	Reduced extreme-event accuracy

The predictive relationship can be expressed as:

$$W_t = f(L_t, T_t, H_t, C_t)$$

Where, W_t denotes predicted water demand, L_t is computational workload, T_t is ambient temperature, H_t is a humidity level, C_t denotes cooling system configuration.

Model performance is commonly evaluated using forecasting error metrics such as Root Mean Square Error (RMSE) and Mean Absolute Error (MAE).

$$RMSE = \sqrt{\frac{1}{n}\sum_{i=1}^{n}(y_i - \hat{y}_i)^2} MAE = \frac{1}{n}\sum_{i=1}^{n} | y_i - \hat{y}_i |$$

These metrics quantify forecasting reliability and operational stability under varying infrastructure conditions.

Decision Support and Operational Integration

The real value of predictive analytics is to be integrated into real decision-making systems. Most facilities connect predictive results to Building Management Systems (BMS) and cooling orchestration platforms and workload schedulers (Organisation for Economic Co-operation and Development (OECD), 2015). This integration will allow operators to take action before that thermal stress is observed and recorded as shown on figure 3.

Forecast-driven systems are used for a variety of operational applications, such as adaptive cooling control, early warning generation, optimizing water allocation and dynamic workload redistribution.

Figure 3. Operational Integration of Predictive Analytics

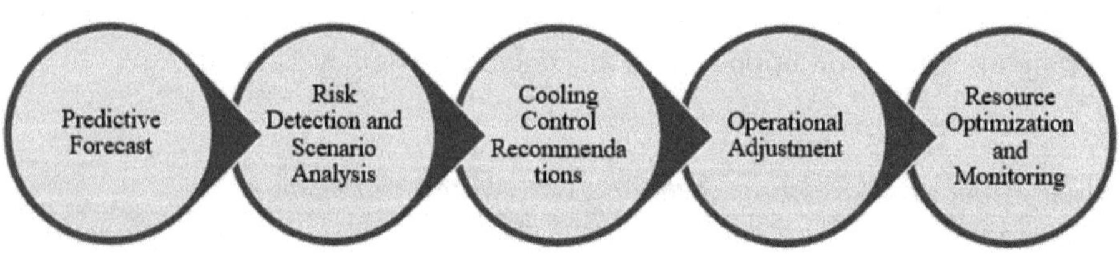

In situations where predictive models indicate a cooling event during periods of heat, for example, use of non-critical workloads can be moved to geographically distributed facilities before cooling stress of an area reaches a threshold. Similarly, if a drought forecast is given, there could be partial transitions to a hybrid air-assisted system (Bertino & Jahanshahi, 2018).

However, it is still challenging to integrate in legacy environments. Many operations that still use older facilities are based on a system of infrastructure managements that are not coordinated in a predictive manner. One other obstacle to adoption is resistance. However, the operators are often skeptical of "black-box" suggestions that result from any machine-learning system, particularly if the reliability of the uptime is important (Mekonnen & Hoekstra, 2016).

Strategic Value and Sustainability Benefits

Predictive Analytics is more than just an operational efficiency tool. The use of forecast driven systems enables to improve the longevity, sustainability reporting and resilience functions for infrastructure planning. There is not only real-time monitoring of water consumption but also a water consumption simulation in different future scenarios (Shehabi et al., 2016).

Predictive analytics can help minimise unneeded cooling activity and withdrawal of water. It also optimises the Water Usage Effectiveness (WUE), which is the intensity of cooling to match the actual computational load rather than a fixed threshold, thus minimising water consumption.It also improves Water Usage Effectiveness (WUE), cooling intensity matches computational needs, not conservative static thresholds, thereby conserving water. This optimization is a direct reduction of the risk to the environment in areas in which water is limited. The framework also enables the improvement of the adaptation capacity to climate change. Buildings more prepared for infrastructure stress events, drought and heat wave. The Table.5. showcases the Strategic Benefits of Predictive Water Analytics. The long-term forecasting assists in making accountable decisions on site selections based on climate exposure, water stress indicators and infrastructure resilience.

Table 5. Strategic Benefits of Predictive Water Analytics

Strategic Area	Benefit
Operational Efficiency	Reduced overcooling
Water Sustainability	Lower freshwater withdrawal
Climate Resilience	Improved drought preparedness
Governance	Enhanced environmental reporting
Infrastructure Planning	Better siting decisions

APPLICATIONS OF PREDICTIVE ANALYTICS IN DATA CENTER WATER OPTIMIZATION

The use of predictive analytics is not helpful unless it can then be transformed into actions and decisions at the operational and strategic level. The cooling systems for AI data centers are a challenge because they must be adaptable to different workloads, weather conditions, and the requirement to expand infrastructure. These circumstances make it difficult to achieve quick and effective static control mechanisms. To address this, predictive analytics can help to enable anticipatory water and cooling management in short/medium term operations, long term infrastructure planning and adaptation to climate risk (International Energy Agency (IEA), 2023). Predictive analytics is not just about efficiencies, it's about culture. Forecast-driven optimization impacts on cooling stability, infrastructure investments, sustainability reporting and regional resource resilience. AI infrastructure continues to expand in critical environments and predictive systems are no longer a monitoring system, but a system of operation coordination.

Short-Term Operational Forecasting

Short-term operational forecasting involves forecasting near-term cooling and water demand conditions that takes place on a minute, hourly or daily time horizon. It's crucial for forecasting technology, especially with how quickly AI workloads are evolving. Clusters of GPUs can easily transition from low to high workloads, leading to spikes in the temperatures of the GPUs (Jadhav & Liu, 2026). Typically, these systems operate when temperatures have climbed up, leading to cooling actions that are unnecessary, and cooling water that is wasted.

The further improvement of this process is possible through predictive forecasting which predicts cooling needs in advance of the occurrence of thermal stress. Real time telemetry data from the server, inlet temperature, humidity, cooling flow rates, etc. is used as inputs to forecasting models. The ability to predict the future thermal state is then used to anticipate the cooling changes that need to be made, and a machine learning algorithm optimizes the cooling control. It can be applied to assist adaptive cooling intensity, optimization of cooling tower operations, dynamic water allocation and anticipation of thermal anomalies.

Therefore, LSTM models and XGBoost models have been widely used in these applications because they can model the nonlinear relationships between operational variables and cooling demand (Lei et al., 2025). The ones that reported implementing predictive cooling control showed measurable improvements in cooling overheads, and said they achieved significant improvements in Water Usage Effectiveness (WUE) especially when cooled loads shifted. Predictive forecasts are used in real-time cooling control as shown in Figure 4.

Figure 4. Short-Term Predictive Cooling Workflow

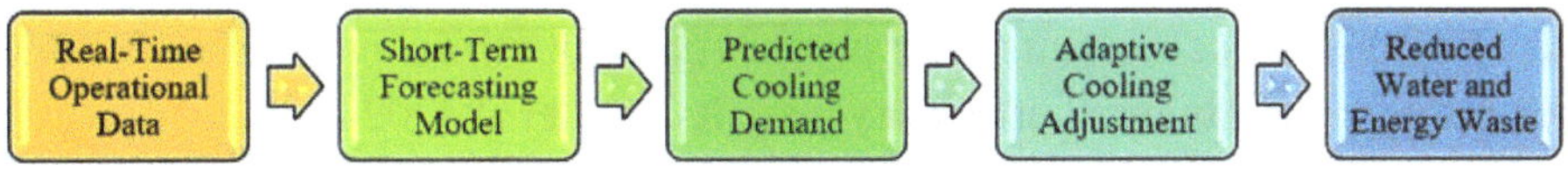

Table 6 presents the advantages of STF.

Table 6. Operational Benefits of Short-Term Forecasting

Forecasting Application	Operational Benefit
Cooling demand prediction	Reduced overcooling
Thermal anomaly detection	Faster response time
Dynamic water allocation	Lower water waste
Workload-aware cooling	Improved efficiency
Predictive maintenance	Reduced downtime

Long-Term Capacity and Infrastructure Planning

There are other challenges that are addressed in long-term forecasting. Long-term predictive analytics predicts the potential for AI infrastructure expansion to affect cooldown needs, water usage, and the related regional sustainability risk over months or years, rather than hours, as operational variability (McKinsey & Company, 2024). The table 7. provides the predictive variables which are not short term.

The huge growth in generative AI and training large AI models has introduced uncertainty in infrastructure planning. There is a greater cooling power demand for future facilities than for previous generations of data centers. Forecasting mechanisms that are able to predict future thermal density and corresponding water demand under a range of growth scenarios are therefore required for operators.

Predictive planning models combine: projected AI workload growth, regional climate trends, infrastructure capacity, and water availability indicators. This process can be further enhanced by linking operational forecasts with external environment using macroeconomic datasets and climate indicators (International Monetary Fund (IMF), 2025). Long-term deployment risk is generally assessed based on indicators of climate vulnerability from the IMF, water stress indicators from WRI, and infrastructure investment data from the region.

Long-term predictive planning also affects the selection of the cooling technology. Higher electricity use may be preferable in the facilities sited where water is scarce. Regions with abundant water supplies can still utilize evaporative cooling systems, due to their thermal efficiency benefits (Aili et al., 2024). In the figure.5, the infrastructure Planning framework is presented.

Table 7. Long-Term Predictive Planning Variables

Planning Variable	Forecasting Purpose
AI workload growth	Future cooling demand
Population and industrial expansion	Regional water competition
Climate projections	Drought vulnerability
Infrastructure investment	Operational resilience
Water stress indicators	Site suitability assessment

Figure 5. Predictive Infrastructure Planning Framework

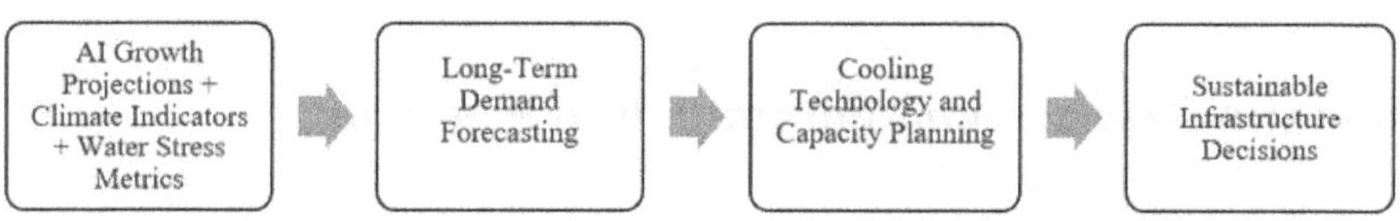

Climate and Drought Risk Anticipation

Operational uncertainty is more pronounced than ever before for AI water-intensive infrastructure as a result of climate variability. As temperatures climb, so does cooling need, and as there is less water available due to drought, the system is not able to keep up. The synergy effects are particularly problematic for facilities that have to operate around the clock with cooling (Calvin et al., 2023).

Predictive analytics can help climate risk anticipation by incorporating weather forecasts, hydrological indicators, and long-term climate projections into operational planning models. Instead of droughts or heat waves as individual events, predictive systems have been developed to assess their potential operational effects prior to the event. This is shown in figure 6.

Risk anticipation based on forecasts facilitates early generation of warnings, making adaptations to cooling transitions, shifting workload temporarily, and planning emergency water allocation. For instance, when predictive models forecast extended heat stress conditions, non-critical computation tasks can be moved to geographically dispersed facilities in less-risky areas. Likewise, forecasted drought severity could lead to decreased reliance on evaporative cooling systems (World Resources Institute (WRI), 2025).

Figure 6. Climate-Aware Predictive Water Management

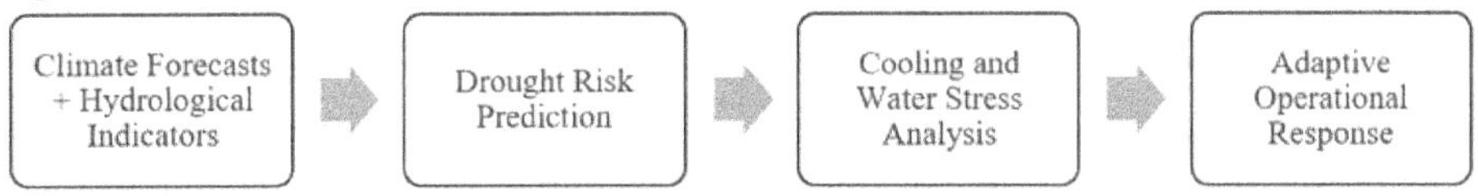

The role that Climate Awareness Forecasting will play is becoming more and more relevant with the decreasing reliability of historical climate patterns as predictors of the future. The planning assumptions of the traditional approach to seasonal planning are not representative of the increasing number of intense heat events (Calvin et al., 2023). Predictive analytics is thus critical for the operational resiliency in the face of environmental uncertainty. The Table.8. presents the climate risks and corresponding predictive responses.

Table 8. Climate Risks and Predictive Responses

Climate Risk	Predictive Response
Heat waves	Adaptive cooling optimization
Drought conditions	Water allocation control
Water supply disruption	Workload redistribution
Extreme temperature variability	Dynamic cooling adjustment

INTEGRATING MACROECONOMIC AND CLIMATE DATA INTO PREDICTIVE MODELS

Operational telemetry is not enough to fully understand the long-term water risk in AI data center environments. Cooling demand is connected with the intensity of the workload, the thermal conditions, the exposure to regional climate, the economic and infrastructure development, and the capacity for water governance (International Monetary Fund (IMF), 2025). Hence, the addition of macroeconomic and climatic parameters that are able to reflect a wider environmental and social context may improve the forecast at the facility level. When using external data, AI-based infrastructure planning can go beyond optimizing individual operations.

Water demand forecasting is related to drought vulnerability on the regional scale, to the vulnerability of public infrastructure and to long-term planning patterns. This multiscale approach is growing in importance because the number of hyperscale AI facilities is increasing and resource competition and climate uncertainty are still escalating in the regions where they are located (World Resources Institute (WRI), 2025). External macroeconomic datasets, at the same time, should not be used as perfect representations of the local hydrological reality. Scale indicators at the national level tend to overlook variations at the regional to basin level and may not capture the water dynamics at the basin scale. Hence the need to integrate strategies that combine precision of operation with context.

Rationale for Using IMF Datasets

The IMF datasets are globally comparable on water policy, hence it is globally adapted for water predictive analysis. IMF indicators are often used by governments, regulators and international organizations as indicators of economic stability, climate vulnerability and infrastructure resilience (International Monetary Fund (IMF), 2025). Such indicators will therefore be more naturally aligned with existing policy frameworks and sustainability reporting systems when building predictive models.

This alignment is crucial when planning for water usage in AI infrastructure. Because water use not only impacts the efficiency of data center operations, but also the environmental health of the region, data center deployment decisions are increasingly subject to regulatory review. Macroeconomic-based projections are generally more credible for policy and institutional decision-makers than technical projections from individual facilities (International Monetary Fund (IMF), 2025).

For instance, the IMF Climate Change Indicators Dashboard offers indicators across countries on climate exposure, adaptation capacity and environmental vulnerability. The indicators of the IMF are shown on the Table.9. These data sets can be used to feed into predictive models to estimate potential long-term operational risk in the face of changing climate. In the same way, economic growth forecasts offer an indication of future growth of industry, electricity consumption and infrastructure strains in the region.

Table 9. Rationale for Integrating IMF Indicators

Rationale	Predictive Value
Standardized global indicators	Cross-country comparison
Policy alignment	Regulatory relevance
Climate vulnerability metrics	Long-term risk estimation
Economic growth projections	Infrastructure demand forecasting
Public finance indicators	Resilience assessment

Key IMF Data Sources

The IMF provides several datasets that are directly used in predictive water-risk modelling in AI infrastructure environments. The datasets are relevant to a number of different national climate and economic growth and public investment and infrastructure resilience contexts.

The IMF Climate Change Indicators Dashboard is one of the most pertinent sources for forecasting in the era of climate awareness. It contains indicators that are related to drought exposure, climate vulnerability, readiness to adapt, and environmental conditions of stress (International Monetary Fund (IMF), 2025). These variables are useful for developing predictive models of the risk of operational stress from water scarcity or extreme climate events. The World Economic Outlook (WEO) database provides macroeconomic forecasting variables like GDP growth, industrial growth, growth in digital infrastructure, and energy demand projections (International Monetary Fund (IMF), 2025).

Growth in the economy can lead to a competition for water resources that can be rapid, as industrial expansion, urbanization, and digital infrastructure growth can happen at the same time. The combination of these indicators can be used to estimate future water system pressures beyond the facility boundary. The Table 10 shows IMF Datasets and Their Predictive Relevance.

Another source of data for resilience analysis is Government Finance Statistics (GFS). The reliability of regional water systems is affected by public funding of water infrastructure, environmental protection and climate adaptation (International Monetary Fund (IMF), 2025). Climatic stress scenarios may reveal higher climate resilience in areas with strong institutional capacity in relation to infrastructure modernization, compared to areas with less capacity.

Table 10. IMF Datasets and Their Predictive Relevance

IMF Dataset	Key Variables	Predictive Application
Climate Change Indicators Dashboard	Drought risk, climate exposure	Climate vulnerability forecasting
World Economic Outlook	GDP growth, industrial expansion	Long-term infrastructure demand
Government Finance Statistics	Water infrastructure spending	Resilience estimation
External Sector Statistics	Trade and energy dependence	Regional operational stability

These datasets do not cover facility scales but rather national or regional scales. Their real strength is in providing context-based forecasts and not in direct operational control. They provide information to assist predictive models in determining more general environmental and economic factors that can affect future cooling demand and water availability.

Simultaneously, there are a number of constraints. The indicators on the IMF website are usually refreshed on a regular basis, not in real time. They can also ignore local variations in hydrology, seasonal restrictions on water usage, and environmental conditions at the basin level (World Resources Institute (WRI), 2025)

Model Integration Approach

Coordinating multiscale data is essential to integrate macroeconomic and climate indicators into predictive models used in operations. Operational telemetry operates in minutes or hours, while indicators of IMF operate at an annual or quarterly scale (IBM, 2024). This difference poses problems in temporal synchronization and the interpretation of spatial aspects.

The integration process is usually based on three types of data as depicted in Figure.7., data normalization, feature alignment, and multiscale forecasting integration. The first step in the operational data is to pre-process the data and synchronize it. Climate and macroeconomic indicators are then brought to a common scale and fed into forecasting pipelines. There are downscaling techniques available to help transform national level indicators into regional operational contexts (World Resources Institute (WRI), 2025).

Figure 7. Multiscale Predictive Integration Framework

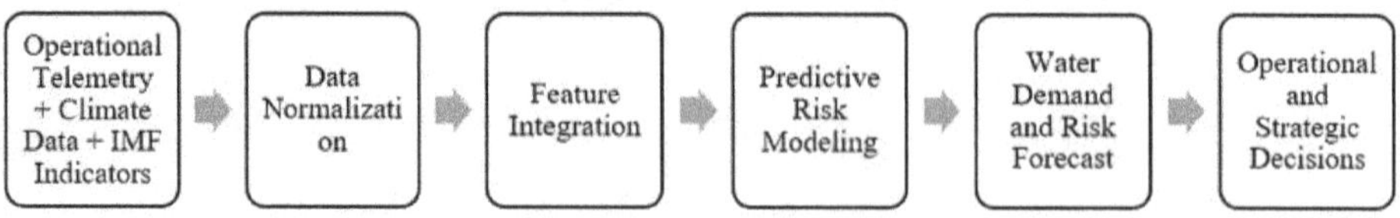

Feature engineering is a key part of this process. The climate indicators could be converted into drought severity scores, or the macroeconomic indicators could be an estimate of industrial growth or pressure on infrastructure. These engineered variables are then fed into a machine learning forecasting model with the machine's operating data (McKinsey & Company, 2024).

The predictive relationship can be represented as:

$$W_t = f(O_t, C_t, M_t)$$

Where, W_t is future water demand, O_t denotes operational telemetry variables, C_t is climate indicators, and M_t is macroeconomic variables.

Managing uncertainty is often done via scenario analysis. A predictive system considers several scenarios for the future, based on different climate and economic assumptions, rather than relying on a single deterministic forecast (Calvin et al., 2023). The scenario based modeling is summarized in Table.11. Conservative, moderate and high growth estimates are used to establish the ranges of operational vulnerability and stress thresholds for infrastructure.

Table 11. Scenario-Based Predictive Modeling

Scenario Type	Key Assumption	Operational Impact
Baseline Scenario	Stable climate conditions	Moderate cooling demand
High AI Growth Scenario	Increased computational density	High water demand
Drought Scenario	Reduced regional water supply	Elevated operational risk
Climate Stress Scenario	Heat-wave frequency increase	Extreme cooling pressure

CASE STUDY: IMF-BASED PREDICTIVE WATER RISK ASSESSMENT FOR AI DATA CENTERS

The use of the indicators from the IMF also facilitates the operation threshold definitions. For instance, dependence on evaporative cooling could be decreased when regional climate vulnerability indicators are above certain thresholds in a predictive system (International Monetary Fund (IMF), 2025). This integration enables macroeconomic and climate data to become operational intelligence instead of just background information.But in terms of the technical aspects, it is still hard to integrate at multiple scales. The data may not be representative at the national level, and small time lags can make it extremely difficult to rapidly react to change.

Therefore, environmental measures at the regional level and facility telemetry should be routinely employed for testing and updating predictive models (Lawrence Berkeley National Laboratory, 2025).

The multiple integration initiatives presented in this chapter demonstrate an integration that is clearly taking on a new phase as predictive analytics becomes a strong force in the fields of operational engineering, environmental forecasting and policy examined infrastructure planning. The intersection opens the door to adaptive and eco-friendly approaches to managing water resources in AI centers and recognizes the imperfection and uncertainty of long-term forecasting systems.

Case Study Objectives and Scope

The case study will explore all possible scenarios on the water risks that predictive analytics can provide in the context of an AI environment that is subject to different growth and climate scenarios. The analysis is based on three interconnected goals: predicting future cooling water demand, assessing the vulnerability of the local water resources and planning the infrastructure for sustainability.

This framework is not just about facilities efficiency in the classic sense, but also takes into consideration the risks of climate exposure and macro-economic aspects in predictive modelling (IBM, 2024). The analysis is therefore performed for both short and long duration cooling needs, and environmental conditions that may affect infrastructure in the long-term.

The study topics range from forecasting the growth of AI workloads into the future, to estimating cooling-water needs, to assessing climate threats and regional resilience. The predict water risk assessment framework shown in figure.8. The framework also addresses the potential types of support that can be offered by the predictive outputs for the selection of cooling strategies, adaptive operational planning and deployment of infrastructure based on policy considerations.

The multiscale structure enables predictive analytics to be both a strategic planning tool as well as an operational planning tool.

Study Area and Country Selection

Various countries across the globe with high exposure to climate change and water stress are analyzed, as well as countries with growing digital infrastructure. The analysis is not only hypothetical but also based on typical country scenarios, which are typically referenced for AI infrastructure growth and climate sensitivity debates (International Monetary Fund (IMF), 2025).

There are three country groups:

Group B – Low Digital Density and Water Stress.Group C – Low Digital Density and No Water Stress.

- Singapore
- Israel

These countries have high level digitalization, yet lack availability of freshwater and high dependency on water management efficiency.

Group B – Emerging AI Expansion Regions

- India
- Vietnam

These areas experience high AI infrastructure development rates and high levels of industrialization, urbanization and climate risks.

Infrastructure regions that have a larger supply of water.Water--abundant infrastructure regions (Group C).

- Canada
- Norway

These are relatively stable countries in terms of water availability and climate conditions for conserving water in cooling processes.

Table 12. Comparative Country Characteristics

Country Group	Representative Countries	Water Stress Level	AI Infrastructure Growth	Climate Vulnerability
Group A	Singapore, Israel	High	High	Moderate
Group B	India, Vietnam	High	Very High	High
Group C	Canada, Norway	Low	Moderate	Low

The list of the selected countries includes different environments and operating conditions that will enable evaluation of predictive water-risk models in various deployment settings as listed in Table 12.

Data Description

The list of the selected countries includes different environments and operating conditions that will enable evaluation of predictive water-risk models in various deployment settings as listed in Table 13.

6.3 Data Description

Data sources used in this case study are described in the following in the context of the three major categories of data that make up the predictive framework, referred to as data types or sources in this context.

- operational telemetry,
- climate indicators,
- and macroeconomic datasets.

Operational data are anonymized AI data center telemetry from high-density computational environments. Variables include:

- GPU utilization,
- rack temperature,
- cooling flow rate,
- water quality, and

The average annual water usage and historical patterns of water use (Lawrence Berkeley National Laboratory, 2025).

Regional weather projections and environmental indicators are used to get climate data. These variables include:

- ambient temperature,
- humidity,
- drought severity,
- precipitation variability,
- and heat-wave frequency.

The key macroeconomic indicators are based mainly on data from the IMF such as: Climate Change Indicators Dashboard, and

- Projections from the World Economic Outlook, and

- and Government Finance Statistics (International Monetary Fund (IMF), 2025).

Table 13. Data Sources Used in the Case Study

Dataset Category	Variables	Source
Operational Telemetry	GPU load, cooling demand	Facility monitoring systems
Climate Indicators	Drought risk, temperature	Climate databases
IMF Indicators	GDP growth, vulnerability index	IMF datasets
Water Stress Metrics	Regional scarcity indicators	WRI Aqueduct

Predictive Modeling Methodology

Operational data are simulated 5-year forecasting intervals from 2025 to 2035. This horizon enables assessment of long-term growth impacts of AI under different climate scenarios.

In data preprocessing, missing values, outlier analysis, normalization were made. Further, operationally-relevant climate and macro economic variables were engineered from the raw data.

$$W_t = f(L_t, T_t, H_t, D_t, M_t)$$

Where, W_t is projected cooling-water demand, L_t is AI workload intensity, T_t is ambient temperature, H_t is humidity, D_t is drought severity index, M_t denotes macro-economic indicators. There are three forecasting scenarios being assessed: Baseline Operational Growth, High AI Expansion Scenario and Climate Stress Scenario. The Climate Stress Scenario combines the projections of more heat-waves and less water for fresh uses. The Table 14. Shows the scenario assumption made while Table 15 shows the model performance.

Table 14. Scenario Assumptions

Scenario	AI Growth Rate	Climate Condition	Water Availability
Baseline	Moderate	Stable	Moderate
High AI Expansion	Very High	Moderate stress	Reduced
Climate Stress	High	Severe heat waves	Critically reduced

Forecasting accuracy is evaluated using RMSE, MAE, and scenario consistency analysis

Table 15. Predictive Model Performance

Model	Forecast Horizon	RMSE	MAE
ARIMA	24 h	7.10%	5.60%
XGBoost	48 h	5.80%	4.40%
LSTM	24 h	4.30%	3.10%

The results indicate that LSTM models provided the highest forecasting stability under rapidly changing AI workload conditions.

Results and Insights

Table 16 shows that there is significant increase in cooling-water demand in all country groups by 2035 as predicted by the analysis. The increase was highest in areas of emerging AI expansion where the two factors (computational growth and climate vulnerability) overlapped.

Table 16. Projected Annual Cooling-Water Demand

Year	Group A (Million Liters)	Group B (Million Liters)	Group C (Million Liters)
2025	10.2	12.5	8.7
2030	13.8	18.6	10.1
2035	17.4	25.3	12.2

Rapid infrastructure growth and climate exposure resulted in the highest projected growth in Group B regions. Under the Climate Stress Scenario, the cooling demand rose by about 18-27% compared to the baseline prediction during peak summer condition (Calvin et al., 2023) as shown in Figure.9.

Figure 8. Projected Cooling-Water Demand Growth (2025–2035)

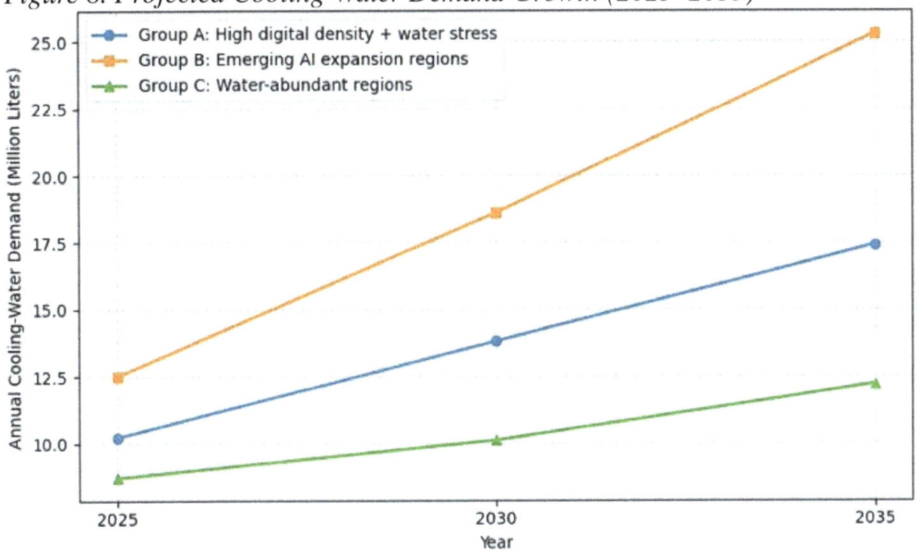

It was also found that facilities in water stressed areas were more vulnerable during extended droughts. It was possible to anticipate these stress periods and have made the necessary adjustment to planning operations before the situation reached a water shortage.

The study also illustrated the impact of infrastructure investment to resilience. In countries with higher operational water infrastructure investment, there was a negative correlation between operational disruption risk and operational water infrastructure investment (International Monetary Fund (IMF), 2025). In table 17 the cooling strategy and Suitability in High-Stress Regions are shown.

Table 17. Cooling Strategy Suitability Under Water Stress

Cooling Strategy	Water Efficiency	Energy Efficiency	Suitability in High-Stress Regions
Evaporative Cooling	Low	High	Limited
Air Cooling	High	Moderate	Suitable
Hybrid Cooling	Moderate	Moderate	Highly Suitable
Immersion Cooling	Variable	Very High	Emerging Option

The results show that predictive analytics significantly enhances operational optimization as well as strategic infrastructure decision making in the context of climate uncertainty.

Policy and Operational Implications

The results demonstrate that predictive water-risk assessment goes beyond optimisation of a facility. The impact of forecasted systems on infrastructure planning, environmental governance, sustainability reporting and regional resource management are all influenced (World Economic Forum, 2025).

Operators can leverage predictive analytics to benefit by:

- adaptive cooling management,
- workload redistribution,

Infrastructure planning, both short and long term.

The need for cooling system can be minimized by levering the predictive forecasts. A rapid regional-scale alerts for transport congestion, and climate vulnerability exposure, Cumulative and growth in water demand is being offered to the policy-makes through this framework.

Predictive analytics is not just an engineering optimisation tool, it can be used for water governance and environmental regulation as well.

The case study also emphasizes how vital it is to incorporate Environmental Reporting into the infrastructure planning process for AI systems. In the decision-making process and regulation of investments, which are guided by ESG principles, the effects of Water Usage Effectiveness (WUE), climate exposure indicators and predictive vulnerability indicators are becoming more relevant (Uptime Institute, 2025).

In the meantime, there are some limitations. National-level macroeconomic indicators may not sufficiently reflect local hydrological conditions and predictive systems are still reliant on the infrastructure to ensure telemetry, along with ongoing model validation. But constant adaptation and uncertainty in the human observation is also needed for adapting to the extraordinary climate conditions. But the analysis reveals that predictive analytics is a plausible path to linking the development of AI infrastructure to water sustainability targets. It provides a model for shifting from a reactive to a policy-based and anticipatory strategy to manage cooling in an uncertain context that considers climate change.

GOVERNANCE, ETHICS, AND STAKEHOLDER PERSPECTIVES

A significant problem with transparency in environmental reporting that needs to be tackled as a governance issue. New metrics, like Power Usage Effectiveness (PUE) are now commonplace in the data center world. However, reporting on water is not as standardized or consistent as it is for the other operators (Uptime Institute, 2025). This is somewhat mitigated by the use of predictive analytics to continuously estimate water needs, cooling system stress, and exposure to water resources. The Governance Dimensions in Predictive Water Management are showcased in Table.18. Forecast driven systems then make it easier to see the impacts on the environment on the timescales on which they are operating.

In this context, Water Usage Effectiveness (WUE) is an important sustainability measure. However, reporting of WUE may be localized and does not necessarily adequately reflect the regional environmental impact. An efficient facility can have a significant impact on the pressure of a water-stressed basin even if it is working well in absolute terms (World Resources Institute (WRI), 2025). Operational efficiency, as well as regional ecological vulnerability, therefore need to be considered in governance frameworks.

Table 18. Governance Dimensions in Predictive Water Management

Governance Area	Key Concern	Predictive Analytics Contribution
Environmental Reporting	Transparency	Continuous water forecasting
Regulatory Compliance	Water withdrawal limits	Risk-based operational planning
Community Equity	Shared resource competition	Vulnerability assessment
ESG Governance	Sustainability accountability	Predictive disclosure metrics

Predictive analytics allows and aids in engaging stakeholders by helping to improve the traceability of infrastructure decision making. Forecast-driven systems provide measurable results of what to expect in terms of future cooling demands and drought exposure and water stress results. This transparency allows regulators, operators and local stakeholders to judge the development of infrastructure on the basis of measurable environmental indicators, rather than assumptions and make decisions on it.

In addition, there are ethical considerations that are also significant in predictive water management. Forecasting systems are increasingly being adopted and are now impacting workload redistribution, cooling prioritisation and region-wide infrastructure deployment. These decisions could affect the local resources available and environmental exposures. Consequently, predictive systems must not be "black-box" systems that are not accountable to governance mechanisms (IBM, 2024). Explainability, auditability and transparency then become essential elements of responsible AI infrastructure management.

Further strengthening policy relevance by integrating macroeconomic and climate indicators. Predictive systems, that are directly connected with metrics of the IMF, drought indicators, and vulnerability assessments of the region, are more natural to fit emerging ESG and sustainability reporting systems (International Monetary Fund (IMF), 2025). Environmental risk disclosure is no longer limited to carbon emissions and is becoming more centrally expected by the government and institutional investors. Water resiliency is integrated into the context of broader climate-risk assessment.

CHALLENGES AND LIMITATIONS

While the concept of water optimisation by predictive analytics is becoming popular, there are several technical and organisational challenges that hinder the use and effectiveness of those tools. Forecast-driven systems need continuous availability of data as well as reliable sensor infrastructure and interoperability between cooling and monitoring platforms (Lawrence Berkeley National Laboratory, 2025). Many data centres are still operating with a number of different, non-integrated legacy systems that were not designed for predictive control at this time. Such major challenges in predictive water management is illustrated in Table.19.

Table 19. Major Challenges in Predictive Water Management

Challenge	Operational Impact	Potential Mitigation
Data fragmentation	Reduced forecasting reliability	Integrated telemetry platforms
Sensor inconsistency	Prediction errors	Continuous calibration
Climate uncertainty	Forecast instability	Scenario-based modeling
Legacy infrastructure	Integration complexity	Middleware frameworks
Organizational resistance	Slow adoption	Explainable AI systems

The other limitation is that of the interpretability of models. Predictive models such as LSTM networks, ensemble forecasting models, have proven to be very effective in making predictions, but are not transparent when it comes to the decision logic used (IBM, 2024). Operators may not be willing to take an automatic recommendation from the operator for mission critical cooling systems without being able to explain or validate the recommendation. This is particularly important if reliability for uptime and thermal protection is an issue.

However, it is still a challenge to integrate into the routine. The predictive analytics needs to be integrated real-time with Building Management Systems (BMS), cooling orchestration platforms and workload schedulers. Current facilities may use non-interoperable infrastructures management architectures (McKinsey & Company,

2024). The infrastructure often needs to be significantly changed and investment made in order to incorporate predictive systems into these environments.

- Economic issues add to the complexity. Forecast water needs:
- telemetry infrastructure,
- data storage systems,
- machine learning deployment,
- Model maintenance, continuous model maintenance.

Advanced forecasting deployment may not be technically feasible in smaller facilities due to a lack of skills or resources (Lawrence Berkeley National Laboratory, 2025).

There are other factors that affect adoption: Typical cooling management strategies often involve “safe” operating systems which reduce the potential for downtime. In fact, operators may find that they want to use manual control and reactions over automated control and predictive coordination. For a machine learning system to be trusted by an institution, trust can only be built through time and there doesn't seem to be a clear path towards general use until it can be proven to be stable under operation.

FUTURE RESEARCH AND INNOVATION PATHWAYS

A progression from the single-forecasting systems to infrastructure intelligence that is integrated and adaptive is expected to be the future of water management prediction. The existing prediction methods are mainly focused on cooling demand prediction and estimation of operational risk. Recent studies investigate the potential of AI infrastructure as a system of operationally coordinated environments (IBM, 2024) instead of as a set of autonomous operational units. The possible future direction are summarized in Table.20.

A key avenue of this change is the digital twin architectures. Digital Twins combine the information from operations, the environment and the simulated environment to create a representation of the physical infrastructure over time. Digital twins can help evaluate the cooling efficiency, drought exposure, and infrastructure resilience in real-time in scenarios that have not yet been observed in practice, in water management applications (Aili et al., 2024).

Explainable AI is another pathway of importance. Transparency and interpretability gain in importance as predictive systems get deeper into the control of operations. For mission critical infrastructures, explainable forecasting models can increase the operators' trust in the models, acceptance for the regulation, and accountability of the governance system (World Economic Forum, 2025).

Table 20. Emerging Research Directions

Research Area	Potential Contribution
Digital twins	Real-time infrastructure simulation
Explainable AI	Improved operational trust
Reinforcement learning	Autonomous cooling optimization
Federated learning	Privacy-preserving collaboration
Climate-aware forecasting	Long-term resilience planning
Carbon-water co-optimization	Integrated sustainability management

Furthermore, there are some possibilities of self-learning optimization of cooling systems by reinforcement learning. Reinforcement learning systems continually adapt their operational decisions to the changes in the environment and load (McKinsey & Company, 2024) when contrast to a static forecasting model. These methods could enable the development of energy, water and cooling systems which can work on self-optimization in the future, balancing the different aspects of these systems.

The next important research need is creating climate-aware forecasting. In the future, predictive systems are expected to integrate improvements in higher resolution hydrological datasets, as well as water availability models on the basin scale based on regional climate simulations (World Resources Institute (WRI), 2025). For AI infrastructure operators, this integration would further enable them to predict environmental stress before a disruptive event happens.

Research is also progressing towards integrated optimization of sustainability. Jot down possible future evaluations of:

- water usage,
- energy consumption,
- carbon emissions,
- and climate resilience.

This is partly because the goals for sustainability are no longer seen as mutually exclusive: a change in one resource domain may result in an adverse change in another (United Nations, 2025). There is more collaboration for future development between institutions, meanwhile. There is limited access to benchmark datasets, common reporting standards and interoperable telemetry architectures remain limited. In order to move forward with predictive water management, it is important for cooperation to occur between industry, academia, environmental agencies and policy makers. Thus, the role of predictive analytics will evolve from being just operational forecasting tools to more holistic environmental intelligence systems that can be used to inform the development of adaptive, climate-aware, and sustainable AI infrastructure.

CONCLUSION

We discussed predictive analytics as a business strategy to improve water efficiency, operational resilience and water sustainability planning in AI Data Centers in this chapter. Operational telemetry and climate indicators, along with macroeconomic datasets were discussed, with examples on how these datasets can be connected to forecasting systems that can predict cooling demand, drought exposure and infrastructure vulnerability before the impact of physical stress. The importance of anticipatory intelligence in infrastructure management was further emphasized by machine learning techniques like LSTM forecasting and multivariable predictive modelling.

The study also highlighted that sustainable AI infrastructure is more than just optimizing the technology. Infrastructure planning is now taking place in the context of water governance, climate resilience, accountability of stakeholders and transparency of the environment. However, predictive analytics can do more than just improve operational efficiency, it can support the creation of the digital infrastructure with policy-awareness and sustainability in mind.

Meanwhile, there are still many prohibitions in place. Other factors that are expected to hinder the effectiveness of implementation are forecast uncertainty, legacy system that are not integrated, climate uncertainty and organizational resistance. Maintaining the confidence and integrity of predictive systems depends on ongoing validation, interoperable infrastructure and telemetry.

Nevertheless, the findings showed that using predictive analytics is a possible way to balance computational expansion and environmental sustainability goals with AI. Forecast-based water management enhances the adaptability of water systems, climate resilience and relief on local freshwater resources. As AI infrastructure continues to grow and evolve, predictive analytics will be key to fueling responsible and climate-conscious digital infrastructure strategies.

REFERENCES

Aili, A., Long, W., Cao, Z., & Wen, Y. (2024). Radiative free cooling for energy and water saving in data centers. *Applied Energy*, *359*, 122672. DOI: 10.1016/j.apenergy.2024.122672

Bertino, E., & Jahanshahi, M. R. (2018, March). Adaptive and Cost-Effective Collection of High-Quality Data for Critical Infrastructure and Emergency Management in Smart Cities—Framework and Challenges. *ACM Journal of Data and Information Quality*, *10*(1), 1–6. DOI: 10.1145/3190579

Calvin, K., Dasgupta, D., Krinner, G., ... (2023) IPCC, 2023: Climate Change 2023: Synthesis Report. Contribution of Working Groups I, II and III to the Sixth Assessment Report of the Intergovernmental Panel on Climate Change (Core Writing Team, H. Lee and J. Romero (eds.)). IPCC, Geneva, Switzerland. Intergovernmental Panel on Climate Change (IPCC)

CDP. (2026). *Turning Transparency to Action*. https://www.cdp.net/en. Accessed 22 Feb. 2026.

IBM. (2024). *What Is Predictive Analytics?* https://www.ibm.com/think/topics/predictive-analytics

International Energy Agency (IEA). *Data Centres and Data Transmission Networks*. https://www.iea.org/reports/data-centres-and-data-transmission-networks

"International Energy Agency (IEA), *Data Centres and Data Transmission Networks – Tracking Report*, 2023". Available: https://www.iea.org/reports/data-centres-and-data-transmission-networks Accessed 22 Feb. 2026

International Monetary Fund (IMF) (2025). *Climate Change Indicators Dashboard*. https://www.imf.org/en/Topics/climate-change/ccid

International Monetary Fund (IMF) (2025). *Climate Change and the Role of the IMF*. https://www.imf.org/en/Topics/climate-change

International Monetary Fund (IMF) (2025). *Climate Change Indicators Dashboard*. https://www.imf.org/en/Topics/climate-change/ccid

International Monetary Fund (IMF) (2025). *World Economic Outlook Database*. https://www.imf.org/en/Publications/WEO/weo-database

International Monetary Fund (IMF) (2025). *Government Finance Statistics*. https://www.imf.org/en/Data/Statistics/GFS

Jadhav, S., & Liu, Z. (2026). *Machine Learning guided cooling system optimization for data center. arXiv*. Cornell University.,

Lawrence Berkeley National Laboratory. (2025). *Energy and Water Efficiency in Data Centers*. https://eta.lbl.gov/publications/energy-and-water-efficiency

Lawrence Berkeley National Laboratory. (2025). *Energy and Water Modeling for Data Centers*. https://eta.lbl.gov/publications

Lei, N., Lu, J., Shehabi, A., & Masanet, E. (2025). The water use of data center workloads: A review and assessment of key determinants. *Resources, Conservation and Recycling*, *219*, 108310. DOI: 10.1016/j.resconrec.2025.108310

McKinsey & Company. (2024) *Advanced Analytics in Infrastructure Operations*. https://www.mckinsey.com/capabilities/operations/our-insights

Mekonnen, M. M., & Hoekstra, A. Y. (2016, February). Four Billion People Facing Severe Water Scarcity. [DOI.org]. *Science Advances*, *2*(2), e1500323. DOI: 10.1126/sciadv.1500323 PMID: 26933676

Niemeyer, S. (2020, January). Deliberation and Ecological Democracy: From Citizen to Global System. [DOI.org]. *Journal of Environmental Policy and Planning*, *22*(1), 16–29. DOI: 10.1080/1523908X.2019.1661232

OECD. (2015). *Stakeholder Engagement for Inclusive Water Governance*. OECD. OECD Studies on Water. *DOI.org (Crossref)*, .

Pazienza, A., Baselli, G., Vinci, D. C., & Trussoni, M. V. (2024). A holistic approach to environmentally sustainable computing. *Innovations in Systems and Software Engineering*, *20*(3), 347–371. DOI: 10.1007/s11334-023-00548-9

Prentice, C. M., Vergunst, F., Minor, K., & Berry, H. L. (2024). Education outcomes in the era of global climate change. *Nature Climate Change*, *14*(3), 214–224. DOI: 10.1038/s41558-024-01945-z

Shehabi, A., Smith, S. J., Sartor, D., ... (2016). United States data center energy usage report. *Lawrence Berkeley National Laboratory*. https://eta.lbl.gov/publications/united-states-data-center-energy

Sovacool, B. K., Bergman, N., Hopkins, D., Jenkins, K. E. H., Hielscher, S., Goldthau, A., & Brossmann, B. (2020, August). Imagining Sustainable Energy and Mobility Transitions: Valence, Temporality, and Radicalism in 38 Visions of a Low-Carbon Future. [DOI.org]. *Social Studies of Science*, *50*(4), 642–679. DOI: 10.1177/0306312720915283 PMID: 32375583

"Task Force on Climate-Related Financial Disclosures | TCFD). (2025)." *Task Force on Climate-Related Financial Disclosures*, https://www.fsb-tcfd.org/. Accessed 22 Feb. 2026.

United Nations. (2025). *Sustainable Development Goal 6: Clean Water and Sanitation*. https://sdgs.un.org/goals/goal6

United States Environmental Protection Agency (EPA). (2025). *WaterSense: Water Efficiency Management*. https://www.epa.gov/watersense

Uptime Institute. (2025). *Water Usage Effectiveness (WUE)*. https://uptimeinstitute.com/resources/glossary/water-usage-effectiveness-wue

World Economic Forum. (2025). *Sustainable Data Centres and Infrastructure*. https://www.weforum.org/reports/sustainable-data-centres

World Resources Institute (WRI). (2025). *Aqueduct Water Risk Atlas*. https://www.wri.org/aqueduct

World Resources Institute (WRI). (2025). *Water Risk and Climate Data Integration*. https://www.wri.org/aqueduct

ADDITIONAL READING

Cai, S., & Gou, Z. (2024). *Towards energy-efficient data centers: A comprehensive review of passive and active cooling strategies*. Energy and Built Environment., DOI: 10.1016/j.enbenv.2024.08.009

de Vries-Gao, A. (2025). The carbon and water footprints of data centers and what this could mean for artificial intelligence. *Patterns (New York, N.Y.)*, *6*(12), 101430. DOI: 10.1016/j.patter.2025.101430 PMID: 41583976

Ebrahimi, K., Jones, G. F., & Fleischer, A. S. (2014). A review of data center cooling technology, operating conditions and the corresponding low-grade waste heat recovery opportunities. *Renewable & Sustainable Energy Reviews*, *31*, 622–638. DOI: 10.1016/j.rser.2013.12.007

Mytton, D. (2021). Data centre water consumption. npj. *NPJ Clean Water*, *4*(1), 11. Advance online publication. DOI: 10.1038/s41545-021-00101-w

Siddik, M. A. B., Shehabi, A., & Marston, L. (2021). The environmental footprint of data centers in the United States. *Environmental Research Letters*, *16*(6), 064017. DOI: 10.1088/1748-9326/abfba1

KEY TERMS AND DEFINITIONS

Climate-Aware Forecasting: A forecasting method that includes weather, drought, humidity, and long-term climate indicators in operational decisions. It helps anticipate heat-wave or scarcity periods that may increase cooling risk.

Digital Twin: A virtual representation of a physical data center or cooling system that is updated using real or simulated data. It allows operators to test water, energy, and resilience scenarios before implementation.

Predictive Analytics: A data-driven approach that uses historical records, real-time telemetry, and statistical or machine learning models to forecast future conditions. In data centers, it supports early cooling adjustment and water-risk planning.

Water Usage Effectiveness (WUE): A metric that relates water consumption to data center energy use. It helps operators compare cooling designs and identify facilities where water demand is high relative to computing output.

Water-Energy-AI Nexus: The interdependence between AI workloads, electricity demand, heat generation, cooling need, and water consumption. It shows why AI infrastructure planning must evaluate water and energy together.

Chapter 6
Water Scarcity Challenges in Regions Hosting Large Cloud Infrastructures

Emna Melliti
http://orcid.org/0000-0001-8539-7315
Center of Water Researches and Technologies, Tunisia

Riadh Hannachi
http://orcid.org/0000-0003-1657-1652
Center of Water Researches and Technologies, Tunisia

Nesrine Kamoun
http://orcid.org/0000-0002-4806-0059
Center of Water Researches and Technologies, Tunisia

Hamza Elfil
Center of Water Researches and Technologies, Tunisia

ABSTRACT

Cloud computing Cloud computing has transformed the global digital economy, driven by the rapid expansion of hyper scale data centers. While these facilities enable scalable storage, real-time analytics, and AI deployment, their environmental footprint, especially water consumption, has become a critical issue. In fact, Data centers require permanent cooling systems, often relying on fresh water-intensive resources that consume millions of liters annually, placing tension on local water resources, especially in water-stressed regions. This chapter examines the connection of data centers growth and water scarcity, analyzing world distribution patterns, conventional and non-conventional cooling technologies, and sectoral water competition. This study also underlines sustainable strategies, which include

DOI: 10.4018/979-8-3373-8434-4.ch006

water-efficient cooling, reuse systems, alternative sources, and policy interventions. It argues that integrated planning, technological innovation, and governance are essential to guarantee resilient and sustainable cloud development.

1. INTRODUCTION

Data Centers (DCs) are, recently, the new fundamental infrastructure that support digital services and systems. In fact, DCs act as a main component in enabling the processes of education buildings, government agencies .... (Smart Strategies, 2025; Al Kez et al., 2025). As consequence, this important growth of these technologies has meaningfully improved the dependence on DCs for large-scale data storage, transmission and processing (Aasheesh et al., 2020; Shehabi et al., 2016). DCs supports real-time computing, communication and online services. Recently, data analytics marketplace in the world is facing an important progress, with predictions indicating a growth from USD 82.23 billion in 2025 to USD 495.87 billion by 2034, representing a compound annual development rate of 21.5% (Herrera et al., 2025; Fortune Business Insights, 2025) (**Figure 1**).

Figure 1. Evolution of worldwide hyperscale Data Center Market (2024- 2030)

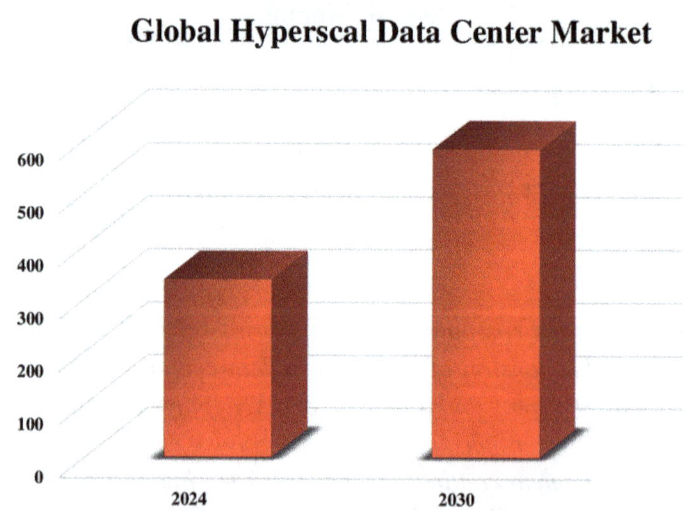

(Source: hyperscale-data-center-market, October 2025).

Recently, Big organizations are assuming leading-edge technologies such as adaptive data-sharing architecture and AI technology in reply to the rising request for big data analysis granting better, fastness and flexibility inside the system (Apstech Advisors, 2026; Hyperscale-data-center-market, 2025; Mavisclara et al., 2025; Chatlatanagulchai et al., 2024;). In the meantime, DC sector is being determined forward by the rapidly increasing number of smart phones, connections, and internet connections. An average of growth, about 25%, is expected be reached by DCs (1.441.49 billion USD by 2034) (Hyperscale-data-center-market, October 2025).

Recently The use 5G-enabled devices widespread recently the worldwide. Hence, the operator community is dominating in this instance, accounting for over 74% of the devices and connections in the environment. Only immense server networks, operated by data centers extending in extent from minor data centers to hyperscale data centers offered by huge firms such as Amazon Web Services, Google Cloud, and Microsoft Azure, can meet the demands of these digital services (hyperscale-data-center-market, October 2025; Mytton, 2021). By 2030, global AI adoption is projected to restrict industries, economies, and daily life. An estimation indicated that, 70% of worldwide companies would be using AI technologies, becoming up to roughly $15 trillion in economic output and productivity gains (PwC, 2025).

Actually, this rapid expansion of DCs generate a pressing concern about sustainability and the environment, resulting significant increases in water and energy consumption. In fact, electricity, land, and especially cooling systems are essential for data centers to keep them operating (Herrera et al., 2025). In Hyper scale Data Centers, from 20% to 40% of the used energy is usually used for cooling step (Energy and AI, International Energy Agency, April 2025) (**Figure 2**). The world's electricity demand will increase, according to the International Energy Agency (IEA), with an annual average rate of 3.6%, exceeding 1,000 terawatt-hours (TWh), between 2026 and 2030. This increase is driving by data centers and artificial intelligence (AI) (Karimi et al., 2025; Cowls et al., 2023; Sachs, 2024; Masanet et al., 2020; Saboori et al., 2006). **Table 1.** Illustrated Data center energy consumption (servers, storage, networks, cooling, UPS, others) from 2010 to 2019 (The current, 2025).

Water volume consumption has become a crucial environmental issue in Data Centers. In fact, a single AI data center can use more than 2 million liters of water/day (Energy and AI," International Energy Agency, April 2025; Pratikkumar et al. 2025). Around 560 billion liters of water yearly are predicted to be consumed for cooling phase in data centers. In the USA, direct data centers water consumption increased from 21.2 billion liters in 2014 to 66 billion liters in 2023. In such water-stressed regions (the American Southwest for example) where massive data center campuses meet permitting problems affecting to long-term water resources (Energy and AI," International Energy Agency).

This continuous growth of the DCs sector reveals the importance of global digital connectivity and the necessity to deal with issues related to network scalability, energy efficiency, and sustainable resource management. Understanding and managing these tendencies is essential for politicians, infrastructure planners, and technology providers aiming to support digital expansion while minimizing environmental impacts.

Figure 2. Water use in data center operations (Energy and AI, International Energy Agency, April 2025).

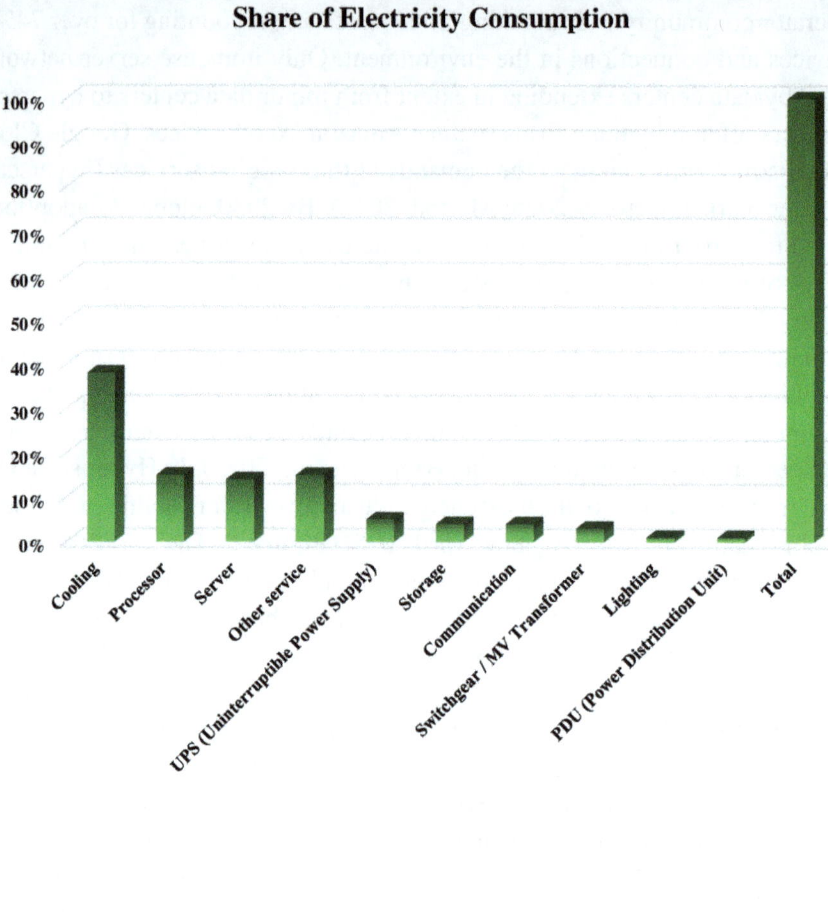

Table 1. Data center energy consumption (servers, storage, networks, cooling, UPS, others) from 2010 to 2019. (The current, 2025).

Year	Servers	Storage	Networks	Cooling	UPS	Others	Total approx.
2010	75	40	7	45	38	5	210
2011	76	41	6	48	39	5	215
2012	77	41	6	50	40	5	219

continued on following page

Table 1. Continued

Year	Servers	Storage	Networks	Cooling	UPS	Others	Total approx.
2013	80	38	7	52	41	6	224
2014	83	42	6	54	42	6	233
2015	84	43	6	55	43	6	237
2016	86	45	6	57	44	6	244
2017	90	48	6	60	45	6	255
2018	93	50	6	62	48	6	265
2019	102	57	7	75	48	7	296

2. THE ENVIRONMENTAL FOOTPRINT OF CLOUD INFRASTRUCTURE: WATER A CRITICAL, RESOURCE

Hyperscale DCs are huge buildings containing massive number of servers, which operate 7/ 7 and 24/24 (**Figure 3**), and producing significant amounts of heats. In the absence of an adequate cooling system, computer hardware can overheat in few minutes and cause critical failures and especially the loss of high-value elements (Alkrush et al., 2024).

Figure 3. Data Centers cooling.

In order to reduce temperature, a medium-sized DC can consume around 110 million gallons of water per year. This volume is actually equivalent to the annual water usage of approximately 1,000 households. However, larger DCs can consume about 5 million gallons/day (a consumption of a town of 10,000 to 50,000 people) (Energy and AI," International Energy Agency, April 2025; hyperscale-data-center-market, October 2025) (**Figure 4**).

In the USA, example of developed countries, data centers can use a volume of water equal to 449 million gallons of water /day (163.7 billion gallons per year). By 2028, US hyperscale facilities are projected to use between 16 and 33 billion gallons/ year (Alkrush et al., 2024; Rogoway, 2023; Lei et al., 2023; Nahid et al. 2025). **Figure 5** illustrated Data Centers water footprint.

Figure 4. Volume water consumption in Date centers (Energy and AI," International Energy Agency, April 2025; International Energy Agency (2023–2024; Taylor Wiseman & Taylor, 2025)

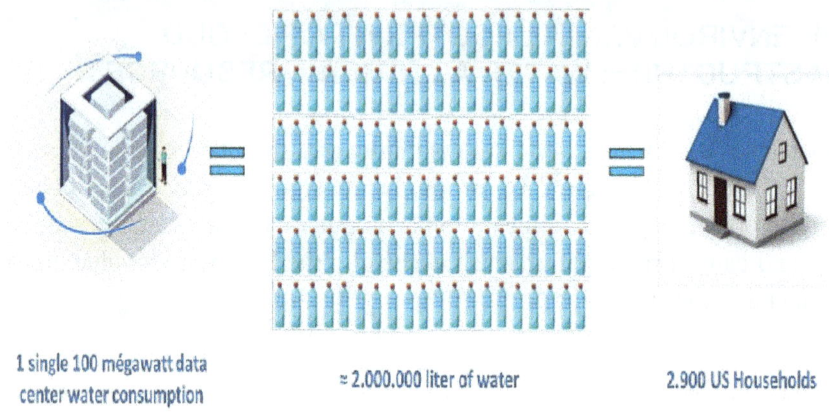

Figure 5. Data Centers water footprint.

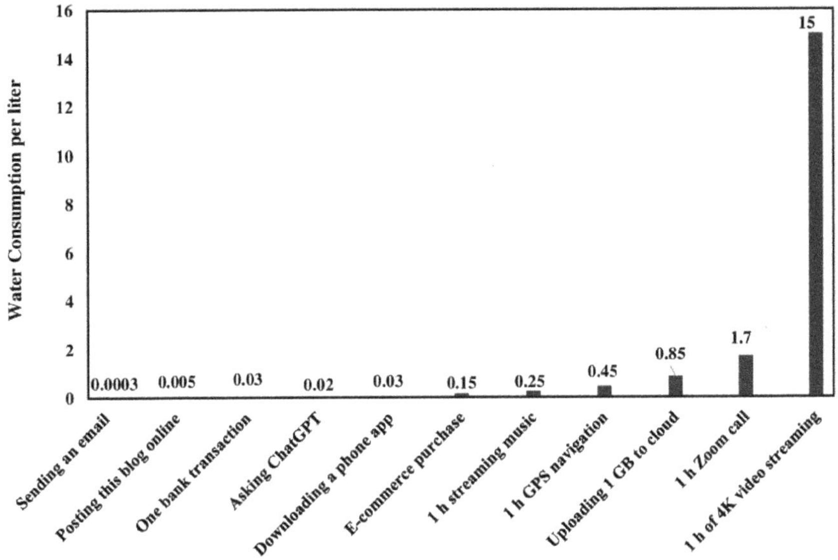

The water footprint of Data Centers can be classified into direct and indirect components (**Figure 6**). This can reflect the several phases at which water is consumed. In fact, direct water consumption is mainly linked with the servers cooling, which is indispensable to maintain optimal temperature. Cooling systems, can consume between 0.26 and 2.4 gallons of water per kilowatt-hour (kWh) of server energy, (equivalent 1 to 9 liters) (Lei et al., 2025). This consumption depends to depending on climatic factors and operational configurations.

However, for carbon footprint and the indirect, water consumption of data centers consists of water consumption or GHG emissions associated with electricity generation, which powers data centers, often based on thermoelectric and hydroelectric systems, both of which are water-intensive.

In DCs, electricity consumption can be classified as bellow: (*i*) electricity used through data center operations, (*ii*) electricity utilized during treatment and supply of cooling water ; (*iii*) electricity used to treat the wastewater generated by a data center (Siddik et al., 2021; Dieter C A et al 2018 ; U.S. Environmental Protection Agency Inventory of U.S., 2020; Chini et al., 2017 ; Lei et al. 2021).

In order, to quantify water usage, researchers at The Green Grid developed the Water Usage Effectiveness (WUE) metric in data centers. WUE indicates the efficiency of a data center's water consumption and it is similarity to Power Usage Effectiveness (PUE), used for energy efficiency evaluation. The metric is obtained by dividing the total water consumed by the data center by the total energy consumed over the same period. Water Usage Effectiveness is expressed in liters per kilowatt-hour.

An ideal WUE value, equal to zero, can be obtained in fully air-cooled data centers. Nevertheless, most current facilities are incapable to achieve this object. This is can be linked to the fact that cooling consumption are heavily influenced by local climate conditions. Recent researches indicates that the average WUE in data centers is around 1.9 L/kWh (Jin et al., 2019; Herrera et al., 2025).

In the United States of America, electricity production causes the evaporation of roughly 7.6 water liters for each produced kWh (Siddik et al., 2021; Dieter et al., 2018; U.S. Environmental Protection Agency, 2020). Additionally, important volumes of water are indirectly used in the built-up of data center hardware, especially semiconductors. While precise evaluations remain very hard because of limited data transparency, researches propose that a typical semiconductor fabrication plant can consume about 38 million liters of ultra-pure water daily, primarily for wafer processing and equipment cooling (FDI Intelligence, 2023).

Figure 6. Direct vs. Indirect U.S. Data Center Water Consumption.

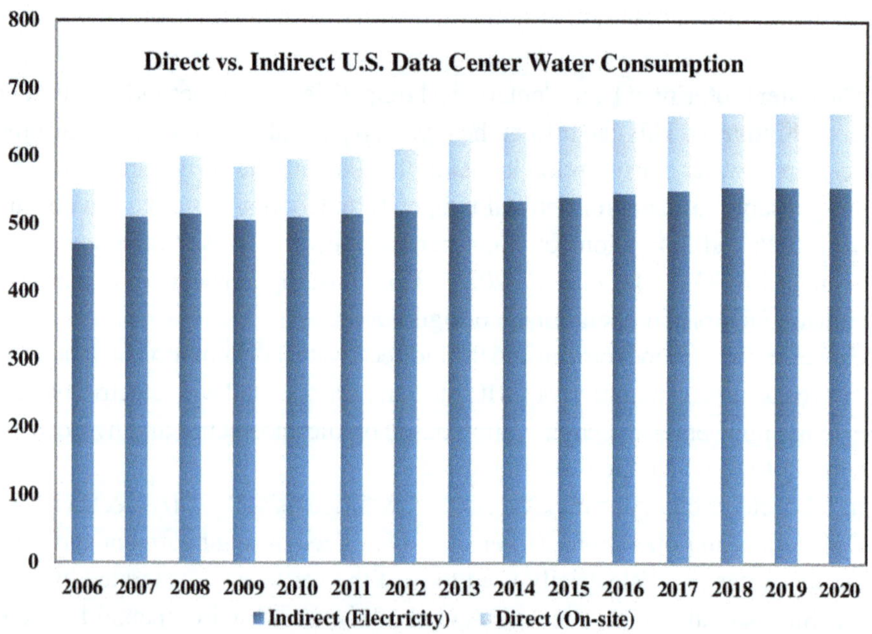

Figure 7. Direct vs. Indirect U.S. Data Center Water Consumption Process

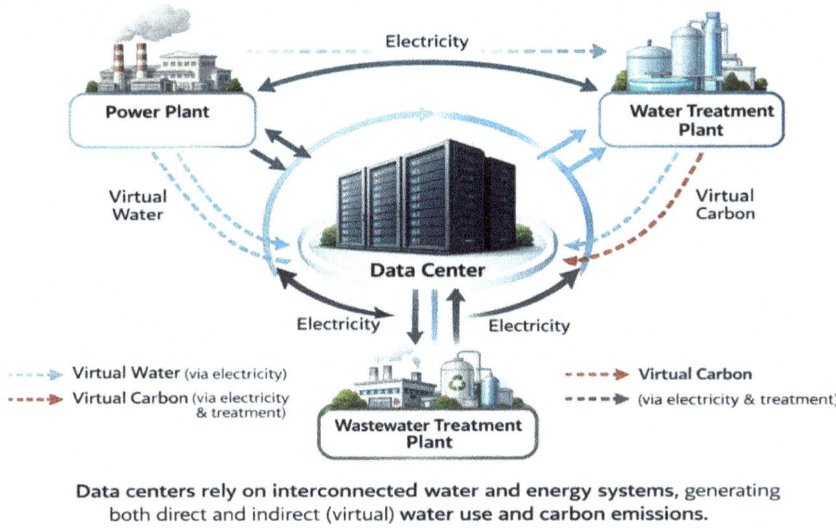

Figure 8. Data Centers cooling process.

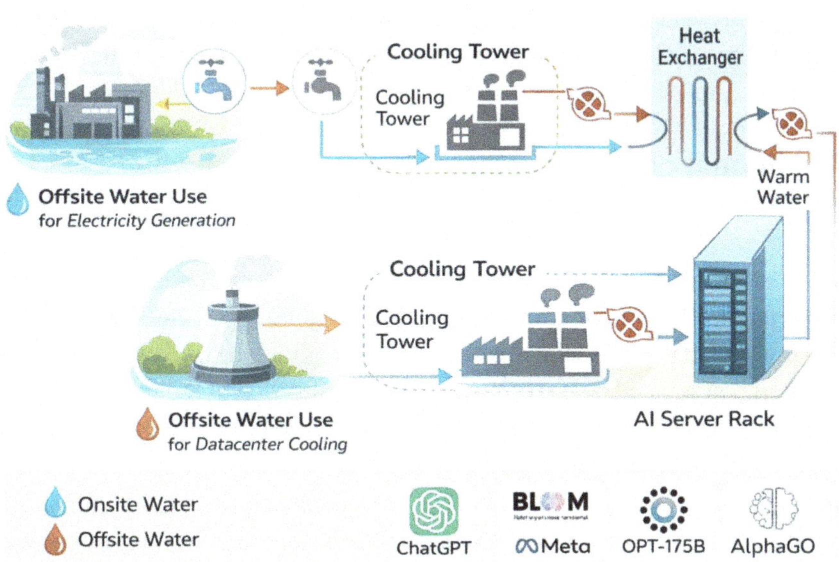

Various factors affects Data center's water production, which include location, water availability, climate, size, and IT rack chip densities. For regions with hot climates, the southwest United States for example, DCs need to consume more water volumes to reduce buildings and equipment's temperature. The increase of centers number that support AI requests, chip density is also growing, which leads to increase room temperatures, as consequence the use of more water chillers at the server level to keep optimum temperatures (Siddik et al., 2021; Dieter C A et al 2018; U.S. Environmental Protection Agency Inventory of U.S., 2020).

3. VARIOUS COOLING TECHNOLOGIES, TRADITIONAL WATER-BASED COOLING METHODS VS EMERGING AIR-COOLED, LIQUID-IMMERSION, AND HYBRID SYSTEMS.

As an excessive heat generation affects directly the lifespan of IT equipment, an efficient cooling process is necessary in hyperscale DCs. Recently, the fast evolution in the industry of microprocessors caused a continuous increase in transistor density, leading a significant higher heat dissipation levels within computing infrastructures. As Consequence, heats generated by servers and electronic components has become a major operational issue for modern data centers. Therefore, Hyperscale DCs are focused on advanced cooling systems considered to keep optimal air temperature, relative humidity and stable environmental conditions. In order to deal with these urgent challenges, DCs must provide effective control of critical parameters such as air temperature, relative humidity, and airborne pollutant concentrations.

3.1. Traditional cooling systems

Computer room air conditioning

Traditional cooling systems are primarily based on air conditioning systems, such as Computer Room Air Conditioning (CRAC) units and chilled air systems. On these traditional systems, cold air circulate throughout the data center in order to adsorb servers' generated heat. High-water footprint is one of the major traditional cooling systems drawbacks of these systems, which may be not adequate with world sustainability goals. In addition, their dependency on water resources with high quality similar to potable water can be considered as another main challenge for stressed areas. In these traditional cooling systems, the used water has to be chemically well treated in order to prevent scaling, corrosion, and microorganisms' growth, which complicates its reuse and it return to the environment (Mytton, 2021). Additionally, using electricity from non-renewable sources can increase operational costs and environmental impact. In water-stressed regions, the use of these technologies poses major challenges (Al Kez et al., 2025).

3.1.1 Computer Room Air Conditioning (CRAC)

A CRAC unit is an air conditioning system designed to maintain an optimum temperature, humidity, and air quality within computer rooms. CRAC units, contrasting with regular air conditioners, are tailored to the specific requirements of data centers where precise environmental control is paramount (**Figure 9**).

Understanding the key components of a CRAC unit is vital to grasp its intricate cooling process. These units consist of essential elements:

- Compressors: Pump refrigerant through the system
- Condensers: Release heat generated during the cooling process
- Evaporators: Cool and dehumidify the air
- Air filters: Remove impurities and enhance air quality

CRAC's feasibility implementation and their low cost made them ideal for small to medium-sized data centers. However, they are inadequate for new and modern high-density, AI-driven workloads due to their dependency on ambient conditions (Tang et al., 2022; Al Kez et al., 2025).

Figure 9. Computer room air conditioning CRAC.

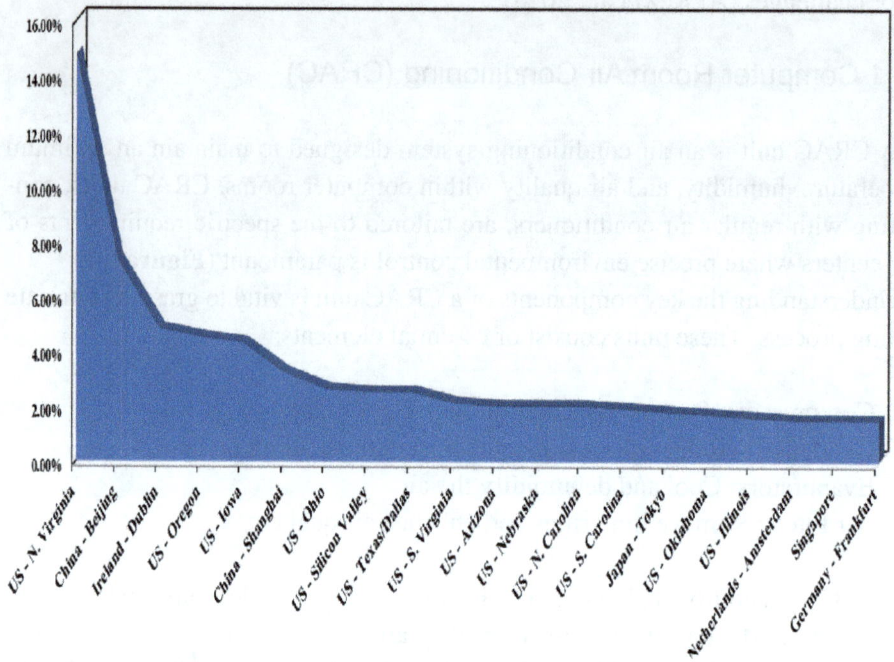

3.1.2. Computer room air handler

In Computer room air handler (CRAH) systems, chilled water loops and air-to-water heat exchangers are often used to maintain equipment temperature (Zhao et al., 2022). CRAH systems, Comparing to CRAC units, can offer better scalability but involve greater infrastructure complexity. Despite of their good adaption in DCs, CRAC units face increasing limitations as AI-driven and HPC workloads intensify cooling demands (Al Kez et al., 2025; Zhao et al., 2022).

3.2. Advanced cooling systems

3.2.1. Air-based advanced cooling

Air cooling system is considered as the most common conventional method employed for cooling in data centers. This system involves cool air circulation of through the server racks to dissipate heat and reduce temperature. This method is based on the use of Computer Room Air Conditioners (CRAC) or Computer Room

Air Handlers (CRAH). These units draw warm air from the data center, cool it down, and recirculate it. Despite of the well established of cooling systems and their lower initial setup costs, it tends to consume a lot of energy and can be less efficient in densely packed data centers. As the power density of new processors increases, air-cooling struggles to cope (Silva-Llanca et al., 2023).

3.2.2. Raised Floor Cooling

In this system, a raised floor system is used in order to distribute cool air through perforated floor tiles directly to the server racks. CRAC or CRAH units support the cool air positioned below the raised floor. While the airflow management effectiveness can be ameliorated by Raised floor cooling process, the complexity and the cost of installation as well as the maintenance-raised floor cooling still a major challenge. These systems require a high initial investment in infrastructure and include the construction of the raised floor and the integration of CRAC or CRAH units (Chen et al., 2023; Al Kez et al., 2025).

3.3. Liquid cooling system vs air-cooling systems

Liquid and air-cooling main difference lies in their heat reduction mechanism. Air-cooling relies on circulating cooled air contrarily to liquid cooling utilizing specialized fluids to dissipate heat from data center systems. The suitable choice between these two cooling approaches depends largely on the existing data center infrastructure and the operational requirements of the IT environment. While the selection has be to be based on technical and organizational constraints, several key factors should be carefully evaluated when determining whether liquid cooling or air cooling is the more appropriate option.

3.3.1. Sustainability Impact

The environmental sustainability impact is a main factor for cooling system selection in DC operations. Air and liquid cooling technologies have different environmental impact, especially in terms of water consumption and carbon footprint. Air-cooling systems require substantial water consumption; especially in evaporative cooling processes, and often exhibit a higher carbon footprint due to their greater energy demands. This can pose significant issues and aggravate water scarcity problem in several regions. Interestedly, liquid cooling can be considered as a better alternative offering a more resource-efficient and reducing water consumption while enhancing overall cooling efficiency, making it a more sustainable solution for modern data center infrastructures (Cisco annual internet report (2018-2023;

https://www.polarismarketresearch.com; Synergy Identifies the World's Top 20 Locations for Hyperscale Data Centers RENO, NV, August 15, 2024).

3.3.2. Costs

When comparing liquid and air-cooling technologies, operational cost is a critical factor. Specialized infrastructure and installation requirements, in liquid cooling systems, require generally higher initial capital investment. In contrast, air-cooling systems consume substantial amounts of electricity during operation. Hence, a detailed cost assessment must be considered for both capital and operational expenditures.

3.3.3. Ease of implementation

Computing demands increase, in air-cooled systems, require significant operational effort, time, and financial resources to maintain effective thermal management. However, specialized infrastructure implantation, in liquid cooling, is generally complex and challenging. Additionally, compared to conventional air-cooling systems, liquid cooling, support higher rack densities and increased cooling demands. These properties makes it a suitable solution for growing data center environments, facilitating then capacity upgrades with fewer performance limitations.

3.3.4. Location

Geographical and climatic conditions determinate mainly role for the suitability of cooling strategies for data centers. In fact, warmer regions, such as places with consistently high ambient temperatures, are generally less favorable for air-cooling systems because they reduce cooling efficiency and increase energy consumption. In contrast, colder environments can provide natural advantages for air-cooling by enabling the use of external low-temperature air to support heat dissipation. However, while locations with naturally cool climates may enhance air-cooling performance, other factors such as infrastructure availability, connectivity, and operational costs must also be considered when selecting an optimal site for data center deployment.

3.3.5. Future-Proofing and Scalability

Air-cooling systems can present scalability challenges, particularly in environments where additional electrical capacity is limited. The increased power demand required to operate larger cooling infrastructures may place excessive stress on electrical circuits, potentially raising the risk of system instability or power outages. Furthermore, as technological advancements continue to produce denser and more powerful hardware, the cooling demands of modern data centers are increasing beyond the practical limits of conventional air-cooling systems. As a result, air-cooling is becoming a less efficient and less suitable solution for supporting high-performance, high-density computing environment. (Figure 10 and Table 2).

Figure 10. Comparison between Air and liquid cooling.

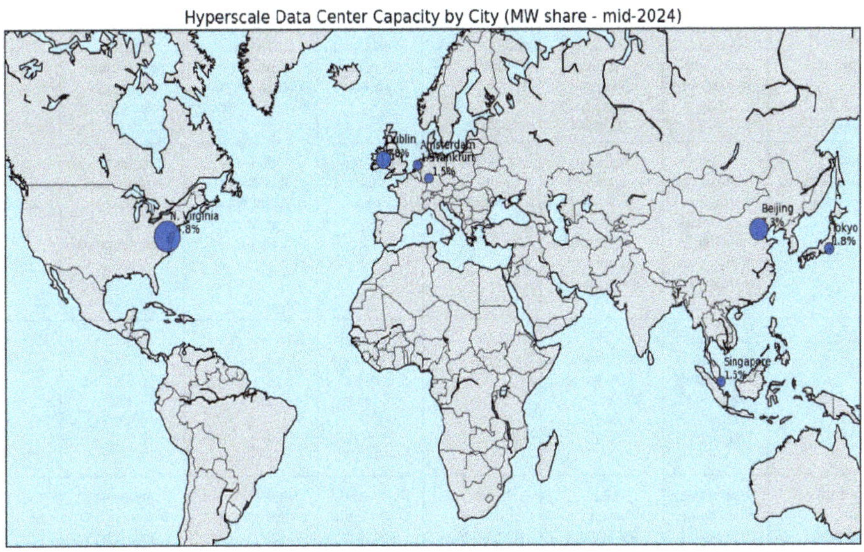

Table 2. Comparison between Air and liquid cooling (Al Kez et al., 2025).

Cooling Method	Mechanism	Advantages	Limitations	Climate Suitability	Implementation Cost & Payback Period	Maintenance & Operational Challenges	Reference
Direct Cooling	Uses ambient air directly to reduce IT equipment temperature	High energy efficiency	Air contaminants require treatment	Good for cool, dry climates	Low-cost implementation	Requires air filtration	Van et al., 2024
Indirect Cooling	Use heat exchanger to Transfer heat	Avoids air contamination Adequate with various climates	Slightly lower efficiency	Works well in moderate temperature	Moderate investment- Efficiency gains justify costs	Requires periodic heat exchanger maintenance	Chen et al., 2025
Air-Side Economisation	Uses external air to reduce mechanical cooling, mixing with recirculated air	Reduces cooling costs by 40–50%- and can Extend equipment lifespan	Needs humidity control	Best for mild climates	Moderate investment- Significant operational savings	Regular filter replacement required to prevent contamination	Van Zetten et al.,2025
DASE	Introduces outside air with added filtration and humidity control	Low operational cost with High energy efficiency	Requires high quality filters	Works best in temperate and dry climates	Low investment- Significant long-term savings	Requires advanced filtration and humidity control	Kim et al., 2024
Evaporative Cooling Integration	Uses water evaporation to cool intake air	High efficiency in hot and dry climates	High water usage- Require high Maintenance	Best for arid or semi-arid climates	Higher upfront investment- Rapid payback in hot climates	Requires water treatment and regular system cleaning	Wang et al., 2025

4. GLOBAL DISTRIBUTION OF HYPER SCALE DATA CENTERS AND CASE STUDIES

Data centers cause intense request for both power and water. This situation poses a significant challenge to the power grid and especially for water resources availability. Currently, DCs are often clustered nearby urban regions so that they can connect to securely access existing energy and water infrastructures (Lei et al., 2025; Business Insider, 2025).

Common DCs are established in limited freshwater availability areas such as India, the southwestern United States, and Australia (Li et al., 2025). Siting resolutions often prioritize proximity to data demand hubs or latency constraints, which may not always deal with water and local environmental situations (Hao, 2024; Mytton, 2021; khan et al., 2025). Additionally, the physical separation between DCs operations, electricity generation and hardware manufacturing generates to a dispersed environmental footprint. Actually, off-site water used in power production and the embodied water associated with globally sourced hardware is often manufactured in water-stressed regions, which further complicate sustainability planning (Bolón-Canedo et al., 2024; Business Insider. (2025)).

Current hyper scale DCs support actually a total installed IT capacity that surpasses a global 250 gigawatts (GW) consumption. Interestedly, North America area is in the first position with 51% of the total hyperscale volume. Although Europe and China account for approximately 25% each (**Figure 11**). Actually, the number of facilities under construction worldwide exceeded 120 sites in 2024, with an estimated additional 10 GW of new power capacity planned for delivery by 2026 (Van Heddeghem et al., 2014; Masanet et al. 2020).

Figure 11. Hyperscale Data Centers capacity (2024) in various regions (Source: https://www.srgresearch.com).

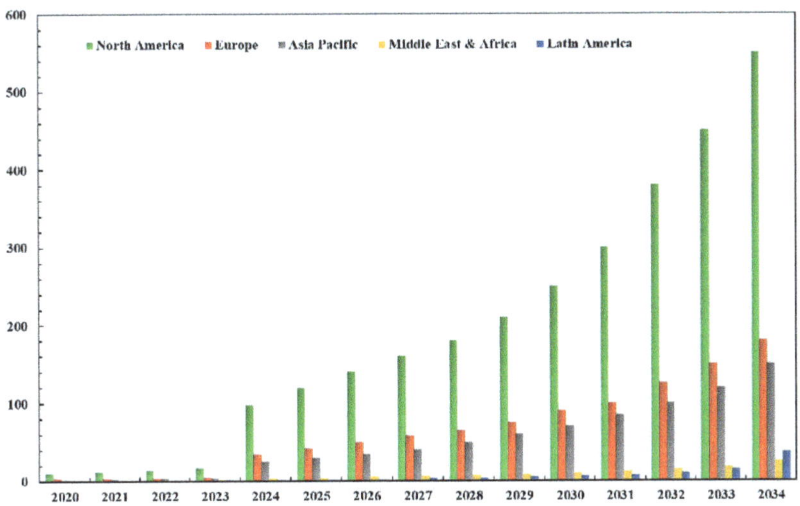

Recent studies shows that just twenty state account for 62% of the world's current hyperscale data center capacity. Northern Virginia and the Greater Beijing Area both make up 22% of the total worldwide. These cities are followed by Dublin, the US states of Oregon and Iowa, and then Shanghai. From the top 20 markets, 13 of them are in the US, four in the APAC region and three in Europe.

In the USA, the strong dispersion across various regions such as Oregon (6.5%), Iowa (5.3%), Ohio (2.8%), Silicon Valley (2.7%), Texas/Dallas (2.5%), South Virginia (2.3%). China having the second major hub position, with Beijing (7.2%) and Shanghai (3.1%). In Europe, Ireland (Dublin) stands out as a major hub, with 6.8%. Other countries complete the global hyperscale data center landscape like Japan

(1.5%) and Singapore (1.1%), along with the Netherlands (1.2%) and Germany (0.9%) (**Figure 12**) (Synergy Identifies the World's Top 20 Locations for Hyperscale Data (Centers RENO, NV, August 15, 2024).

Since 2008, India country has attracted approximately $977 million in private equity and strategic investments, with $396 million invested in 2020 alone. This growth in Market is driven by government-backed initiatives, increasing data traffic, and emerging data sovereignty regulations. The need to manage these large data volumes has prompted Indian stakeholders to collaborate with global industry leaders to accelerate the development of critical IT infrastructure. However, India still faces a significant journey before it can be considered a global data center hub (Pomnar et al., 2023).

Figure 12. The distribution of Data Centers in the world.

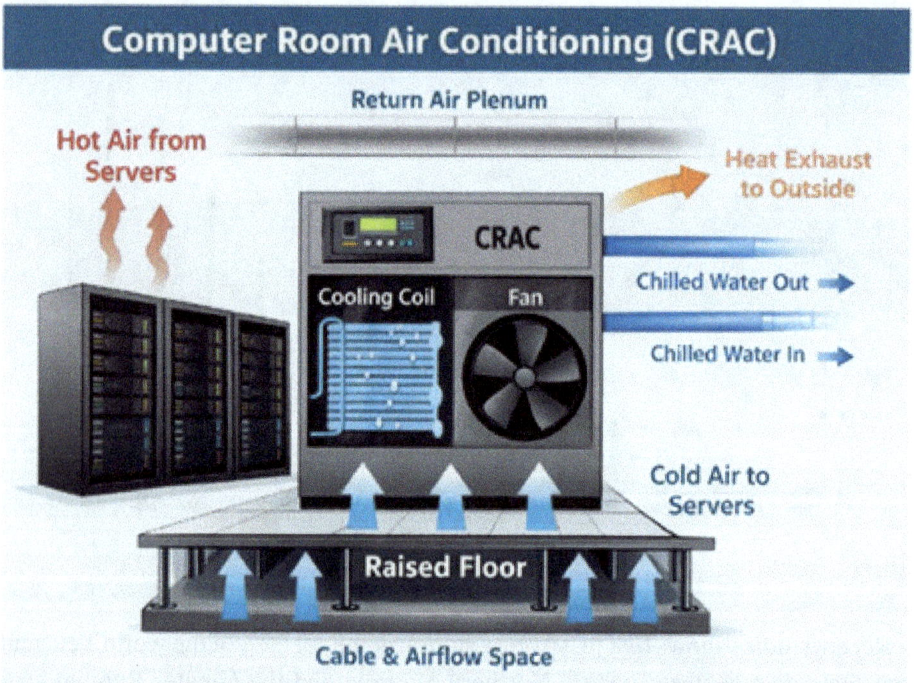

The factors that determine the location of a hyperscale facility are based on various criteria, including cost of real estate to operate, customer density related to location, network topology, and the cost and availability of electricity. As such, these factors will exacerbate the disadvantages of locating in the world's premiere economic hubs (e.g., London and New York City) and will similarly benefit sparsely populated U.S. states (e.g., Nebraska, Iowa, Oregon) (Synergy Identifies the World's

Top 20 Locations for Hyperscale Data Centers RENO, NV, August 15, 2024). The decision criteria for facility siting will remain constant but aggregate top-tier market movement as measured will change due to higher growth rates in the major emerging economies of Southeast Asia and Latin America (Synergy Identifies the World's Top 20 Locations for Hyperscale Data Centers RENO, NV, August 15, 2024) (**Figure 13**).

Figure 13. Hyperscale Data Center Market Demand, Trends, Industry Growth, and 2025-2034 (Source: https://www.polarismarketresearch.com).

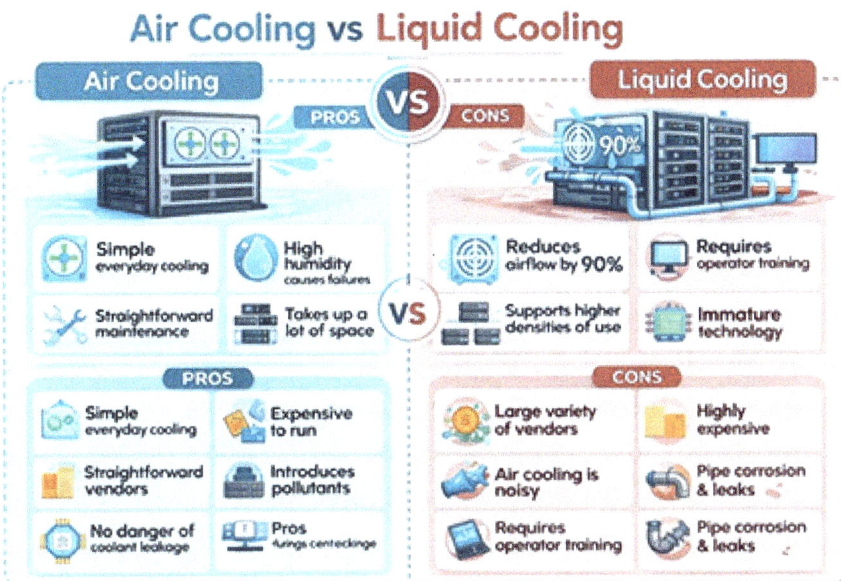

Although specific area such as the USA and China dominates the market of data centers, some Asia-Pacific countries are placing themselves to catch the next wave. In fact, Japan has the highest proportion of energy use for data centers, with more than half of future electricity demand coming from new DCs. There is considerable interest for the idea that expansion of digital infrastructure will rapidly spread across the region, which is home to many of the most densely populated areas in the world. In the future, the growth of DCs sectors will necessitate main parameters such as water, energy and land resources. These factors have to be considered during the development of new digital infrastructure (Cisco annual internet report (2018-2023; https://www.polarismarketresearch.com; Synergy Identifies the World's Top 20 Locations for Hyperscale Data Centers RENO, NV, August 15, 2024).

The attention of DCs developer has motivated on accessibility of constant energy resources. This situation has typically put them nearby to large urban areas that have had greater historical investment in energy infrastructure. New developing markets with major areas for data center growth (for example, Bangkok, Jakarta, Mumbai, Chennai, and Hyderabad) face additional water scarcity issues even before considering the addition of substantial new water demand from data centers (SP Global 2025; WTCIDCre.com Global data center capacity is shifting rapidly towards hyperscale operators RENO, NV, June 24 2025). **Table 3** summarizes regional distribution of Data Centers.

Table 3. Regional summaries for Internet users, devices and connections, and network performance (Cisco annual internet report (2018-2023)

Region	Internet Users (2023 vs 2018)	Mobile Users (2023 vs 2018)	Devices/ Connections (2023 vs 2018)	Mobile vs Wired (%)	Fixed Broadband Speed (Mbps)	Mobile Speed (Mbps)	Wi-Fi Speed (Mbps)
APAC	3.1B (72%) vs 2.1B (52%)	3.1B (72%) vs 2.7B (65%)	13.5B vs 8.6B	49% / 51%	157.1 (↑2.5× from 62.8)	45.7 (↑3.2× from 14.3)	116 (↑3.4× from 34.5)
CEE	388M (78%) vs 323M (65%)	404M (81%) vs 394M (79%)	2.0B vs 1.2B	48% / 52%	87.7 (↑2.5× from 35.0)	43.0 (↑3.3× from 12.9)	53 (↑2.3× from 22.6)
LATAM	470M (70%) vs 387M (60%)	520M (78%) vs 482M (75%)	2.1B vs 1.4B	49% / 51%	59.3 (↑3.8× from 15.7)	28.8 (↑3.6× from 8.0)	35.0 (↑3.3× from 10.6)
MEA	611M (35%) vs 381M (24%)	1.0B (57%) vs 827M (53%)	2.6B vs 1.7B	75% / 25%	41.2 (↑4.2× from 9.7)	24.8 (↑3.6× from 6.9)	26 (↑3.7× from 7.0)
NA	345M (92%) vs 328M (90%)	329M (88%) vs 313M (86%)	5.0B vs 3.0B	25% / 75%	141.8 (↑2.5× from 56.6)	58.4 (↑2.7× from 21.6)	110 (↑2.3× from 46.9)
WE	370M (87%) vs 345M (82%)	365M (85%) vs 357M (84%)	4.0B vs 2.4B	31% / 69%	123.0 (↑2.7× from 45.6)	62.4 (↑2.6× from 23.6)	97 (↑3.2× from 30.8)

5. DATA CENTERS SYSTEMS IN HIGH WATER STRESS REGIONS

Recently, the rapid expansion of data centers to generate AI and cloud computing coincides with increasing global water scarcity. In fact, around one-quarter of today's DCs buildings, and nearly one-third of those under construction, are in areas predictable to face a greater water scarcity by 2050. (MSCI, 2026). According to S&P Global Sustainable, Physical Risk and S&P 451 DCKB datasets Analysis, 43% of DCs globally are functioning in regions facing high-water stress in the current decade. All DCs in Middle Eastern countries, like some countries in Europe (Belgium, Greece and Spain) and certain Latin American countries (Chile, Peru and Mexico) are in areas with high water stress in the 2020s (CloudTech, 2025; Canada's national observer, 2026). Water stress issues is aggravated by climate change, reducing so that the industry's ability to reduce efficiently equipment temperature.

For Hyperscale DCs operators, hyper-localized stressors, reducing and managing consumption is not only an environmental priority but also a business requirement. If a hyperscale facility is installed in a water-stressed basin such as the American Southwest or parts of Spain and relies on evaporative cooling during peak summer months, it competes directly with municipal supplies (World Resources Institute (WRI), 2019; The Gardien, 2025).

DCs siting often prioritizes access to power, connectivity and land. Yet many of the world's largest clusters are in regions that could face worsening water scarcity. According to GeoSpatial analysis, about 14,000 data-center assets worldwide shows that about one in four existing facilities may experience more frequent scarcity days by 2050, especially in Chile, Brazil, Mexico, Turkey and Australia, where rapid digital expansion is intersecting with rising climate stress. About projects currently under construction, 30% are in regions where water scarcity is expected to intensify (S&P Global, 2026; MSCI GeoSpatial, 2026). This situation indicates that the next challenge of AI infrastructure could deepen resource risk if water considerations are not integrated into development decisions.

The optimal balance is highly location specific. In water-stressed regions, operators may favor air-based cooling or dry cooling systems that minimize water use, even at higher energy cost (Leila et al., 2022). In areas with limited power capacity or carbon-intensive grids, evaporative or liquid cooling may be more practical because they are more energy-efficient and require less electricity than air-based systems (Nuoa et al., 2022). As AI drives greater computational intensity, these trade-offs are becoming more material for operators and investors.

Figure 14. Number of Data Centers vs water Scarcity level (Source: Reboot Development: The Economics of a Livable Planet; Damania et al.,2025)

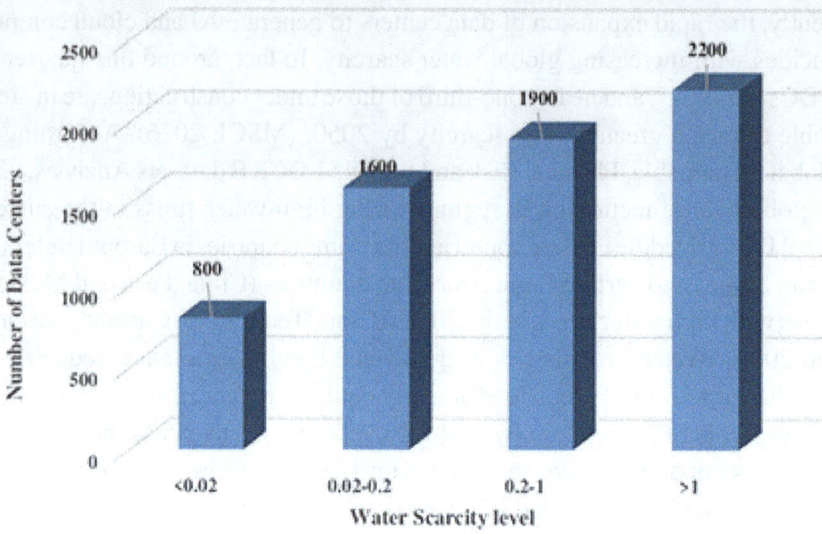

Figure 15. Water stress exposure in the top 12 countries by data center installed capacity

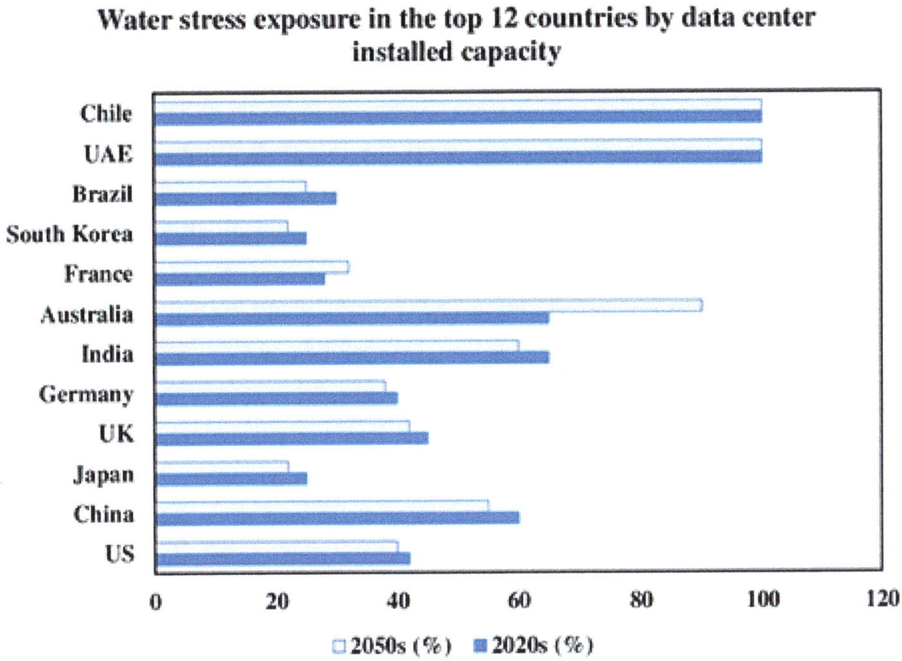

Source World Resources Institute (WRI), 2026).

6. COMPETITION FOR WATER RESOURCES: DATA CENTERS VERSUS TRADITIONAL SECTORS

Water demand, particularly in water-stressed regions, is dominated by traditional fields such as drinking, agriculture, textiles, chemicals, and energy production industry, and municipal use. Nevertheless, hyperscale DCs recently are emerging as an important competitor. In such arid and semi-arid regions, the installation of hyperscale data centers has raised concerns about prioritizing technological development over agriculture, and food production sustainability (Bashroush & Lawrence, 2020) (**Figure 16**). **Table 4** illustrated Daily Water use in various industries.

Interestedly, the implantation of hyperscale data centers building close to traditional industrial regions can increase water pressure shared water infrastructure producing higher cost (Vanham et al., 2018). This situation creates a public concern and regulatory scrutiny. For example, such communities that host major facilities

(Google and Microsoft…) have raised questions about the sustainability and equity of water allocation, particularly during drought conditions (Ren et al., 2021).

This is primarily attributed to the existence of extensive land in these locations, robust energy infrastructure, and strong network connectivity rather than for abundant water resources. Consequently, data centers may compete with local communities and natural ecosystems for limited water supplies, potentially leading to adverse environmental impacts such as reduced river flows and disruptions to aquatic biodiversity (Ristic et al., 2015).

Additionally, the temporal incompatibility between water availability and demand can aggravate conflicts.

Despite these efforts, the main issue of competition for water resources remains. As digital infrastructure continues to grow, good planning, regulation, and integrated water resource management will be an obligation to create a balance between data centers and traditional sectors. Policymakers must ensure that water allocation decisions are equitable, transparent, and aligned with long-term sustainability goals (Smart water magazine, 2026; Mytton, 2021).

Figure 16. Competition for Water Resources: Data Centers Vs. Traditional Sectors (Didactum Security, 2026).

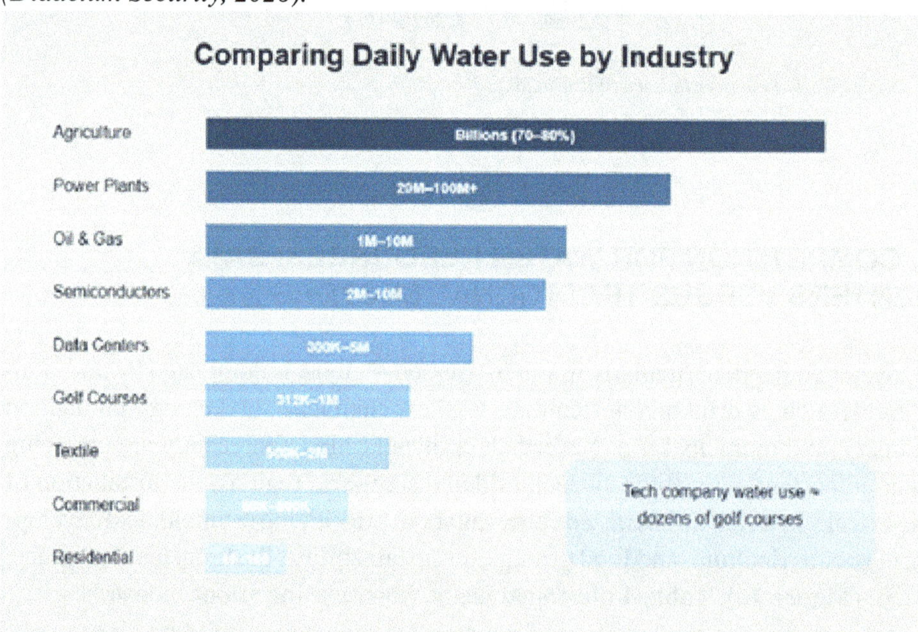

Table 4. Daily Water use in various industries (World Resources Institute (WRI), 2026)

Sector	Daily Water Use	Notes
Data Centers	300K – 5M gallons	Cooling-intensive, varies by size & location
Golf Courses	312K – 1M gallons	Irrigation-heavy, climate dependent
Agriculture	Billions of gallons	~70–80% of total U.S. freshwater use
Mining / Manufacturing	Millions of gallons	Processing, cooling, and cleaning
Power Plants	20M – 100M+ gallons	Thermoelectric cooling dominates
Oil & Gas Extraction	1M – 10M gallons	Hydraulic fracturing & processing
Food & Beverage Industry	1M – 5M gallons	Meat, dairy, and processing plants
Textile Industry	500K – 2M gallons	Dyeing and finishing are water-heavy
Semiconductor Fabrication	2M – 10M gallons	Ultra-pure water requirements
Commercial Buildings	50K – 500K gallons	HVAC, plumbing, landscaping
Residential Use	80–100 gallons per person/day	Household consumption

7. MITIGATING WATER SCARCITY IN CLOUD-INTENSIVE REGIONS

Dealing with water scarcity challenge, caused by DCs implantation, has become a critical issue for industry stakeholders and policymakers as well as consumers and traditional sectors. Numerous strategies have been emerged to reduce the water footprint of data centers while maintaining operational efficiency. Table 5 showed examples of mitigation strategies for reducing water consumption in data centers.

7.1 Adoption of Water-Efficient Cooling Technologies

One of the most active strategies to decrease water consumption in water-stressed areas is the deployment of water-efficient cooling technologies. Evaporative cooling systems, despite their efficiency in energy, are be highly water-intensive consumers. In this context, new alternative solutions such as liquid cooling (including direct-to-chip and immersion cooling), air-cooled systems, and other hybrid cooling techniques can be used to even reduce water use. Liquid cooling systems can provide higher thermal efficiency and are suited for high-density workloads such as AI and high-performance computing (HPC) (Zhang et al., 2021; Gao et al., 2022). These technologies can reduce reliance on evaporative processes and thus lower overall water consumption.

7.2 Implementation of Water Reuse and Recycling Systems

Water reuse and recycling represent alternative key strategy for improving water sustainability in data centers Hyperscale's. Cooling systems" Closed-loop" allow water recirculation with low losses, reducing significantly freshwater withdrawals. Treatment technologies, which include membrane filtration (Reverse osmosis…) and chemical conditioning, permit for the reuse of process water while keeping system performance and avoiding scaling or biofouling growth (Mytton, 2021).

7.3 Utilization of Alternative Water Sources

Data centers Hyperscale are progressively turning to alternative water sources, reducing their consumption of freshwater resources. These sources contain desalinated seawater, treated municipal wastewater (reclaimed water), industrial effluents. The use of treated municipal wastewater is particularly advantageous in urban environments, where it decreases struggle with potable water resources (Shehabi et al., 2016). Interestedly, desalination, can deliver a reliable water supply in coastal regions, especially when it is coupled with renewable energy sources such solar or wind energy to minimize environmental impacts (Jones et al., 2019). However, the economic and environmental trade-offs of these alternatives must be carefully evaluated (S&P Global, 2025).

7.4 Strategic Site Selection and Sustainable Siting Practices

Positioning decisions that reflect crisis sustainability, local water accessibility, and long-term weather plans can help minimize environmental influences (IEA, 2023). Data centers hyperscale geographical location is a critical element in determining their water footprint. In fact, the implantation of DC hyper scale's buildings in areas with available water resources or cooler climates can meaningly reduce the use of both direct and indirect water. Some workers in this filed are also traveling the co-location of data centers with renewable energy sources and non-conventional cooling environments, such as underwater or Arctic deployments, to enhance sustainability (Veolia, 2025).

An intelligent site selection for data centers based on potential forms of synergy with other nearby commercial and industrial developments will create opportunities to provide greater benefits for both the communities in which they are located and for the larger community as a whole. For example, in some locations there are parts of the existing foul water network that require replacement or upgrades, and the placement of a data center, if located appropriately, could potentially use treated effluent instead of drinking-quality water to be used in cooling its servers. There are many examples of high-quality water being produced and used at industrial sites where the reuse of this water as part of their operation may be restricted for products that are consumed by humans, but if treated properly, it may be possible for data centers to utilize this wastewater through some method of treatment.

7.5. Use AI for water consumption reduction

For new and existing facilities, AI-driven optimization and real-time analytics offer means to reduce water usage. Methods include systems that dynamically adjust cooling system parameters based on temperature, humidity and server load, thus minimizing unnecessary water and energy usage. Systems simulate operating conditions at various loads, enabling predictive maintenance and more efficient use of cooling resources. These technologies enhance responsiveness and reduce the risk of overheating or overcooling (S & P Global, 2025)

7.6. Integrated Policy and Governance Approaches

Water management in places with a lot of cloud cover is an important process to reduce water consumption. A good governance system is required. Controlling measures, such as water usage reporting, productivity standards, and incentives for sustainable practices, can drive industry-wide improvements. Additionally, collaboration between various actors in this sector such as governments, utilities, and private sector actors is essential to ensure equitable water allocation and long-term resource resilience (Hoekstra et al., 2011; Veolia, 2025; S&P, 2025).

Table 5a. Examples of mitigation strategies for reducing water consumption in data centres

Mitigation strategy	Description	Reported effectiveness	Best application conditions	Main limitations and trade-offs	References
Free cooling (air-side economization)	Uses cool outdoor air to dissipate heat from IT equipment, thereby reducing dependence on mechanical refrigeration systems.	Can lower cooling energy demand by approximately **20–60%**, particularly in favorable climatic conditions.	Most effective in **temperate or cold regions** with extended periods of low outdoor temperature and low humidity.	Performance declines in hot or humid climates; requires advanced air filtration and humidity regulation systems, with potential risks of airborne contamination.	Shehabi et al., 2016; IEA, 2025
Evaporative and water-based cooling systems	Relies on water evaporation to absorb and remove heat from the data center environment.	Can achieve **PUE values as low as 1.1–1.3**, substantially improving cooling efficiency.	Particularly suitable for **dry and semi-arid climates**, where evaporation processes are highly effective.	Requires substantial water consumption; less suitable for water-scarce regions and increasingly subject to environmental regulations.	Herrera et al., 2025; IEA, 2025
Direct liquid cooling and immersion cooling	Removes heat directly from processors using liquid-based thermal transfer systems, such as cold plates or full immersion techniques.	Provides **10–40% energy savings**, improves heat transfer efficiency, and supports high-density computing workloads such as AI training.	Highly effective for **high-performance computing (HPC)**, AI clusters, and hyperscale computing environments.	High initial investment, greater operational complexity, maintenance challenges, and still limited large-scale deployment.	Masanet et al., 2020; Pratikkumar et al., 2025

Table 5b. (Continued).

Workload shifting and edge computing	**Redistributes computational tasks geographically or closer to end users to improve efficiency and reduce centralized processing demand.**	**Improves operational efficiency and reduces network transmission energy, particularly for distributed digital services.**	**Well suited for 5G networks, IoT ecosystems, and latency-sensitive distributed applications.**	**Requires sophisticated orchestration, may increase infrastructure fragmentation, and introduces synchronization challenges.**	**Moderate reduction in energy use, limited reduction in water demand, and carbon benefits depending on regional electricity sources.**	**Chatlatanagulchai et al., 2024**
PUE optimization and facility design improvements	Includes architectural, mechanical, and operational design enhancements aimed at minimizing non-IT energy consumption.	Modern hyperscale facilities commonly achieve **PUE values of 1.1–1.2**, compared with approximately **2.0** in conventional legacy facilities.	Most applicable to **newly designed hyperscale data centers** with integrated efficiency-focused infrastructure.	Approaches thermodynamic efficiency limits, with diminishing returns as PUE values approach optimal thresholds.	Strong reduction in **energy use**, moderate reduction in **water consumption**, and significant reduction in **carbon emissions**.	Masanet et al., 2020; Herrera et al., 2025

7.7. Outlook

As the demand for digital services remains to raise, the water footprint of data centers is predictable to grow unless proactive measures are applied. A combination of technological innovation, resource-efficient project, and sustainable policy interferences will be essential to balance digital increase with environmental stewardship. By integrating water-conscious strategies and into the operated and planning of data centers, the industry can contribute to more sustainable and resilient cloud infrastructure systems (**Table 6** presented various Examples of selected policy measures governing data-center efficiency).

Table 6. Examples of selected policy measures governing data-center efficiency

Region/ Country	Regulation/Policy	Key requirements/Targets	Implementation timeline
European Union	Energy Efficiency Directive (revised)	Requires large data centers (installed power of 500kW or more) to disclose energy performance and water use	In effect since September 2024
Singapore	Green Data Centre Roadmap	Targets to achieve: PUE $\leq$ 1.3; WUE $\approx$ 2.0 m^3/MWh for new approvals	Launched May 2024
Germany	Energy Efficiency Act	PUE<1.2 for new centers from July 2026 ; renewable sourcing > 50%	Since 2024
China	Action Plan on the Green Developement of Data Centers (2024)	Natinal average PUE <1.5 by end-2025 ; increased renewable- energy utilization by 10% annulay	2024- 2030

Figure 17. Mitigating Water Scarcity in Cloud-Intensive Regions.

8. CONCLUSION

The rapid growth and development of cloud computing and hyperscale data centers becomes a defining feature of the modern digital economy, enabling unprecedented advances in big data analytics, artificial intelligence, Machine learning and global connectivity. Contrary to the descriptive assessments frequently provided in institutional reports, this chapter offer an integrated analytical perspective about the water–energy–digital nexus of hyperscale data centers. Exiting reports often address energy demand, water consumption, or regional resource stress as separate dimensions, the present chapter synthesizes these elements within a unified framework that connects global data center expansion, cooling technology evolution, environmental resource constraints, and policy responses.

Additionally, this chapter demonstrates that such technological development is often intertwined with significant environmental constraints, particularly energy consumption water scarcity. As data centers increasingly focus in regions already experiencing water- stress regions, their substantial and continuous water consumption for cooling emerges as a critical sustainability challenge.

The analysis indicates that, despite of the big improvements in energy efficiency sector, water demand has not declined proportionally. In contrast, the growing scale, density, and computational intensity of cloud infrastructures generate a continuous water demand growth. This situation is further accentuated and exacerbated by the widespread dependence on water-intensive cooling systems and the limited adoption of alternative water sources across the industry, which is based generally on fresh water resources. As result, data centers are becoming major localized stressors on freshwater systems, particularly in arid and semi-arid regions.

This chapter highlights the huge competition for fresh water resources use between traditional sectors such as agriculture, industry, and municipal and data centers. This competition generates an important issue related to environmental sustainability, social acceptance, and economic equity especially during periods of drought or water shortages. This challenge is accentuated by the mismatch between constant operational water demand and the seasonal variability of water availability, highlighting the need for more adaptive and context-specific resource management strategies.

The chapter also indicates that technological innovation cannot fully address the issue. Despite of solutions provided by advanced cooling systems, water recycling, and the use of alternative water sources to reduce water footprints, their effectiveness remains linked on local climatic, economic, and regulatory conditions. Interestedly, maintainable solutions have to be supported integrated and advanced approaches that combine engineering advancements, strategic site selection, transparent water governance, and cross-sectoral collaboration.

Looking ahead, the chapter highlights the importance of bring into line digital growth with responsible resource organization. Future progress in AI-driven optimization, climate-resilient infrastructure design, and predictive water modeling will play a crucial role in enhancing efficiency and reducing environmental impacts. Inversely, it is important to build stronger collaboration among various actors in this filed such as governments, industry stakeholders, researchers, and local communities in order to make sur that water resources are well managed equitably and sustainably.

In conclusion, treated the issue of water scarcity in regions hosting large cloud infrastructures needs a holistic and forward-looking plans. Making a Balance between the benefits of digital transformation with the protection of vital water resources is mandatory for accomplishing maintainable and hardy development. By mixing innovative technologies, policy intervention, and good fresh water resource governance, the growth of cloud infrastructure can be guided toward a future where digital progress does not come at the expense of environmental sustainability.

By combining technological, environmental, and policy considerations, this chapter aims to provide readers with a comprehensive framework for understanding the sustainability implications of cloud infrastructure growth and for guiding future decision-making toward resilient and water-conscious digital development.

REFERENCES

Apstech Advisors. (2026). Your Strategic Partner in the Water Industry, Data Center Water Consumption in the US: Challenges, Trends, and Market Opportunities for 2025–2030.

Al Kez, D., Foley, A. M., & Fadhli, W. B. M. (2025, October). Hasan Wong, Andrea Dolfi, and Geetha Srinivasan. 'AI-Driven Cooling Technologies for High-Performance Data Centres: State-of-the-Art Review and Future Directions'. *Sustainable Energy Technologies and Assessments*, *82*, 104511. DOI: 10.1016/j.seta.2025.104511

Alkrush, A. A., Salem, M. S., Abdelrehim, O., & Hegazi, A. A. (2024, April). O. Abdelrehim, and A. A. Hegazi. 'Data Centers Cooling: A Critical Review of Techniques, Challenges, and Energy Saving Solutions'. *International Journal of Refrigeration*, *160*, 246–262. DOI: 10.1016/j.ijrefrig.2024.02.007

Bolón-Canedo, V., Morán-Fernández, L., Cancela, B., & Alonso-Betanzos, A. (2024, September). 'A Review of Green Artificial Intelligence: Towards a More Sustainable Future'. *Neurocomputing*, *599*, 128096. DOI: 10.1016/j.neucom.2024.128096

Chatlatanagulchai, et al. (2024). Adaptive data-sharing and workload distribution in digital infrastructures. (ICT systems optimization studies).

Chen, H., Li, D., Wang, S., Chen, T., Zhong, M., Ding, Y., Li, Y., & Huo, X. (2023). 'Numerical Investigation of Thermal Performance with Adaptive Terminal Devices for Cold Aisle Containment in Data Centers'. *Buildings (Basel, Switzerland)*, *13*(2), 268. DOI: 10.3390/buildings13020268

Chen, X., Wang, X., Wang, L., Zheng, H., Ding, T., & Li, Z. (2025, January). 'Multistage Data Center Cooling System for Temperature Gradation and Matching'. *Applied Energy*, *377*, 124592. DOI: 10.1016/j.apenergy.2024.124592

Chini, C. M., Konar, M., & Stillwell, A. S. (2017). Direct and indirect urban water footprints of the U nited S tates. Water Resources Research, 53(1), 316-327.

Cisco. (n.d.). Cisco Annual Internet Report (2018-2023) White Paper.

Cowls, J., et al. (2023). Machine learning and energy optimization in data centers. Nature / AI sustainability literature (reviewed works).

Cowls, J., Tsamados, A., Taddeo, M., & Floridi, L. (2023). 'The AI Gambit: Leveraging Artificial Intelligence to Combat Climate Change—Opportunities, Challenges, and Recommendations'. *AI & Society*, *38*(1), 283–307. DOI: 10.1007/s00146-021-01294-x

Damania, R., Ebadi, E., Mayr, K., Russ, J., & Zaveri, E. (2025). *Reboot Development: The Economics of a Livable Planet*. World Bank., DOI: 10.1596/978-1-4648-2271-1

Dieter, C. A., ... (2018). Estimated use of water in the United States in 2015 U.S. Geol. Surv. Circ. 1441 65.

Fortune Business Insights. "Smart Strategies, Giving Speed to your Growth Trajectory," 2025. https://www.fortunebusinessinsights.com/data-center-market 109851 (accessed March 03, 2026).

Gao, W., Hu, Q., Ye, Z., ... 'Deep Learning Workload Scheduling in GPU Datacenters: Taxonomy, Challenges and Vision'. Version 3. Preprint, arXiv, 2022. DOI: 10.48550/ARXIV.2205.11913

Global, S. P. (2025). The World's Total Data Center Capacity is Shifting Rapidly to Hyperscale Operators RENO, NV, June 24, 2025).

Google. (2022–2023). Environmental Report. *Sustainability Report*, •••, 2021.

Hao, K. (2024). AI is taking water from the desert. The Atlantic. https://www.theatlantic.com/technology/archive/2024/03/ai-water-usage-data-centers-desert/677622/

Herrera, et al. (2025). Environmental impacts of hyperscale data centers. Data center sustainability research.

Herrera, M., Xie, X., Menapace, A., Zanfei, A., & Brentan, B. M. (2025, October). 'Sustainable AI Infrastructure: A Scenario-Based Forecast of Water Footprint under Uncertainty'. *Journal of Cleaner Production*, *526*, 146528. DOI: 10.1016/j.jclepro.2025.146528

. Hyperscale Data Centers Market Size. (n.d.). Share, Growth, and Industry Analysis, By Type (Colocation, cloud, enterprise), By Application (IT, telecom, eCommerce, AI and analytics), Regional Insights and Forecast to 2033.

Insider, B. (2025). The 'AI boom' pits neighbor against neighbor. https://www.businessinsider.com/data-centers-northern-virginia-noise-air-pollution-cost-2025-5

Insider, B. (2025). The true cost of data centers.

International Energy Agency (2023–2024). Data Centres and Data Transmission Networks (reports and tracking updates).

International Energy Agency (IEA). (2025). Data centres and data transmission networks. https://www.iea.org/reports/data-centres-and-data-transmission-networks

Jin, C., & Bai, X. (2019, August). 'The Study of Servers' Arrangement and Air Distribution Strategy under Partial Load in Data Centers'. *Sustainable Cities and Society, 49*, 101617. DOI: 10.1016/j.scs.2019.101617

Karimi, L., Shinh, R., Laisure-Pool, C., Green, M., Yacuel, L., Ashby, J., & Hickenbottom, K. L. (2025, October). 'Energy and Water Dynamics in Data Center Cooling: Insights from a Modeling Study in Hot-Arid Climates'. *Applied Thermal Engineering, 276*, 126802. DOI: 10.1016/j.applthermaleng.2025.126802

Khan, N. Inside Tech World (2025). https://insidetechworld.com/artificial-intelligence/chatgpt-water-consumption-and-environmental-impact/

Kim, J. H., Shin, D. U., & Kim, H. (2024). 'Data Center Energy Evaluation Tool Development and Analysis of Power Usage Effectiveness with Different Economizer Types in Various Climate Zones'. *Buildings (Basel, Switzerland), 14*(1), 299. DOI: 10.3390/buildings14010299

Lei, N., Cheng, Z., Cao, Z., & Masanet, E. (2021). 'An Integrated Decision-Making Framework for Sustainable Data Center Operation through Intelligent Load Scheduling'. *Journal of Physics: Conference Series, 2042*(1), 012091. DOI: 10.1088/1742-6596/2042/1/012091

Lei, N., Lu, J., Cheng, Z., Cao, Z., Shehabi, A., & Masanet, E. (2023). 'Geospatial Assessment of Water Footprints for Hyperscale Data Centers in the United States'. *Journal of Physics: Conference Series, 2600*(17), 172003. DOI: 10.1088/1742-6596/2600/17/172003

Lei, N., Lu, J., Shehabi, A., & Masanet, E. (2025, June). 'The Water Use of Data Center Workloads: A Review and Assessment of Key Determinants'. *Resources, Conservation and Recycling, 219*, 108310. DOI: 10.1016/j.resconrec.2025.108310

Masanet, E., Shehabi, A., Lei, N., Smith, S., & Koomey, J. (2020). Recalibrating global data center energy-use estimates. *Science, 367*(6481), 984–986. DOI: 10.1126/science.aba3758

Mavisclara, O. A., Oshobugie, I. I., Olufunmi, A. T., Bolaji, A. A., Olawale, A. O., Anulika, N. C., & Samuel, O. K. (2025). 'The AI-Driven Energy Surge: A Comprehensive Review of Sustainable Power Solutions for Data Centers'. *Global Journal of Engineering and Technology Advances, 24*(3), 328–344. DOI: 10.30574/gjeta.2025.24.3.0279

Mytton, D. (2021). 'Data Centre Water Consumption'. *NPJ Clean Water, 4*(1), 11. DOI: 10.1038/s41545-021-00101-w

Nahid, F. A., & Islam, M. M. Md. Kamruzzaman, and Ahmed Al Mansur. 'The Rising Energy Footprint of Data Centers: A Review of Global Trends, Challenges, and Opportunities'. 2025 IEEE 7th International Conference on Sustainable Technologies For Industry 5.0 (STI), 11 December 2025, 1–6. https://doi.org/DOI: 10.1109/STI69347.2025.11367578

Pomnar, A., Rajawat, A. S., Tatkar, N. S., & Bhaladhare, P. (2023). Sustainable Power Prediction and Demand for Hyperscale Datacenters in India. Engineering Proceedings, 59(1), 124.

PwC. (2025). A new era of load growth: How data centers are becoming part of our communities and energy system.

Ristic, B., Madani, K., & Makuch, Z. (2015). 'The Water Footprint of Data Centers'. *Sustainability (Basel)*, *7*(8), 11260–11284. DOI: 10.3390/su70811260

Rogoway, M. (2023). Google's water use is soaring in the Dalles, records show, with two more data centers to come. Oregon Live.

Saboori, B., et al. (2006). ICT sector emissions and carbon intensity dependency. Environmental economics literature.

Sachs, J. (2024). AI-driven optimization for energy systems and data centers. Sustainability and energy systems research.

Shehabi, A.. (2016). *Data center energy and cooling systems analysis. Lawrence Berkeley National Laboratory*. LBNL.

Siddik, M. A. B., Shehabi, A., & Marston, L. (2021). 'The Environmental Footprint of Data Centers in the United States'. *Environmental Research Letters*, *16*(6), 064017. DOI: 10.1088/1748-9326/abfba1

Silva-Llanca, L., Ponce, C., Bermúdez, E., Martínez, D., Díaz, A. J., & Aguirre, F. (2023, September). 'Improving Energy and Water Consumption of a Data Center via Air Free-Cooling Economization: The Effect Weather on Its Performance'. *Energy Conversion and Management*, *292*, 117344. DOI: 10.1016/j.enconman.2023.117344

Tang, Yu., Bai, X., Jin, C., Fu, Y., Mao, W., & Xu, X. (2022, March). 'Experimental and Numerical Study on the Thermal Environment of a Data Center'. *Building and Environment*, *212*, 108741. DOI: 10.1016/j.buildenv.2021.108741

Taylor Wiseman & Taylor. (2025). Your AI Assistant is Thirsty – Examining the Water Footprint of Data Centers.

Van Heddeghem, W., Lambert, S., Lannoo, B., Colle, D., Pickavet, M., & Demeester, P. (2014, September). 'Trends in Worldwide ICT Electricity Consumption from 2007 to 2012'. *Computer Communications*, *50*, 64–76. DOI: 10.1016/j.comcom.2014.02.008

Van Zetten, J. D., Cholette, M. E., & Bamdad, K. (2025, April). 'Data Centre Building Energy Optimization with Differential Temperature Control Utilizing I.T. Power Measurement'. *Journal of Building Engineering*, *100*, 111722. DOI: 10.1016/j.jobe.2024.111722

Vanham, D., Comero, S., Gawlik, B. M., & Bidoglio, G. (2018). 'The Water Footprint of Different Diets within European Sub-National Geographical Entities'. *Nature Sustainability*, *1*(9), 518–525. DOI: 10.1038/s41893-018-0133-x

Wang, Z., Zeng, C., Zhu, Z., Li, Y., Ma, X., & Zhao, X. (2025). 'Time-Series Machine Learning for Predictive Optimisation of a Highly Efficient Evaporative Cooling System'. *Building Services Engineering Research and Technology*, *46*(2), 189–202. DOI: 10.1177/01436244251315047

Zhang, Q., Ng, K. K. W., Kazer, C., Yan, S., Sedoc, J., & Liu, V. 'MimicNet: Fast Performance Estimates for Data Center Networks with Machine Learning'. *Proceedings of the 2021 ACM SIGCOMM 2021 Conference*, 9 August 2021, 287–304. https://doi.org/DOI: 10.1145/3452296.3472926

Zhao, R., Du, Y., Yang, X., Zhou, Z., Wang, W., & Yang, X. (2023, October). 'A Critical Review on the Thermal Management of Data Center for Local Hotspot Elimination'. *Energy and Building*, *297*, 113486. DOI: 10.1016/j.enbuild.2023.113486

Chapter 7
The Hidden Cost of Intelligence:
Water Sustainability in AI-Driven Data Ecosystems

Sonali Srivastava
http://orcid.org/0000-0003-0551-8442
Jaypee Institute of Information Technology, Noida, India

Manisha Singh
http://orcid.org/0000-0003-1741-0148
VIT Business School, VIT Bhopal University, Bhopal, India

ABSTRACT

The fast development of artificial intelligence (AI) has changed the world industries and has intensified healthcare, financial, manufacturing, education, and government services innovations. However, the hidden side of this mighty digital revolution is a very simple environmental fact: it takes a ton of water to cool AI data centers. This chapter is a critical analysis of the invisible water footprint of data ecosystems in the digital age powered by AI, and why the digital age is in dire need of sustainable water management strategies. The chapter suggests that to have sustainable AI the world needs a holistic approach that balances the growth of technologies and the ecological concern. To ensure that the water-smart data ecosystems will have minimal ecological footprint without reducing the pace of AI innovation, cooperation between technology firms, policymakers, environmental scientists, and the local communities is necessary.

DOI: 10.4018/979-8-3373-8434-4.ch007

1. INTRODUCTION

Artificial Intelligence (AI) is rapidly turning into a latent technology, which characterizes the contemporary economy, social and governance structures of the population. In healthcare, predictive analytics to algorithmic trading, in finance, and automation, in manufacturing, AI-based systems are becoming involved in critical decision making. The increase is facilitated by the fact that the volume of data being generated, hosted and processed has been increasing by huge amounts that are made possible by hyperscale data centers and cloud computing systems to a significant degree. Despite the generally accepted positive role of AI, the environmental footprint, particularly the energy use and carbon footprint are the aspects that have been widely debated by academics and policy makers, but its water footprint has received a relatively low level of research. The more recent literature has begun to highlight the environmental externalities of AI systems, and its large-scale model training and inference process is energy-consuming. However, the implications of AI infrastructure on freshwater resources, in particular, the cooling of the data center, and electricity generation is often not considered in these discussions. As the world is increasingly adopting AI, there has been an increasing necessity of effective thermal regulation systems which has led to an increased reliance on water-heavy cooling systems such as evaporative cooling systems and chilled water systems. This poses a complex, though concealed sustainability issue, especially where it is already a region where there is some water stress.

The environmental performance of AI infrastructure can thus be considered in the context of the bigger picture of the water-energy nexus, which is a condensation of the water reliance on energy generation and vice versa. Not only does the data center require electricity to make calculations, but they also require cooling down and even the production of this electricity, be it a thermal power plant, hydropower or renewable power, can be very water intensive. Therefore, there is direct (cooling) and indirect (energy supply chains) water consumption by AI systems. Regardless of this twofold influence, the current literature is still divided, technical literature has centered on the computational effectiveness, environmental literature has centered on resource use, and policy literature has centered on the gaps in governance, usually in isolation, and not as a system (Abbas Khan et al., 2024).

This fragmentation is addressed by this chapter through the elaboration of a single analytical prism that cuts across the technical, environmental, and governance aspects of the AI-enabled data ecosystems. It states that the carbon-based narratives cannot be applied to sustainability of AI, but water as a resource should be considered as influential and constraining. This is especially so in the face of growing water scarcity, global warming and competing needs of agriculture, industry and urbanization in this manner all over the world. The chapter can help us have a more comprehensive idea of digital sustainability in the light of every AI infrastructure, with the hydrological and ecological constraints.

The rapid expansion of the machine learning models, and in particular the deep learning models and the large language models is one of the most significant factors that drive the growing water demands in AI systems. The computation of these models is very computationally intensive, and may take thousands of high-performance GPUs 24 hours. This processing capability generates a great deal of heat and elaborate cooling systems are employed so that the performance will be effective and the hardware will not be damaged by the heat. Most situations have evaporative cooling systems due to their low cost and effectiveness in cooling systems, although in the majority of cases, they consume large quantities of freshwater. The quantity of water requested will increase significantly when the AIs workloads increase, particularly when the generative AI and applications of real-time inferences are adopted (Abdulameer et al., 2025).

In addition, the complexity is also contributed by geographical location of data centers. A significant agglomeration of AI infrastructure is often centered on the places with good economic prospects, energy resources and connectivity, and may not always be centered on places with a good water supply. Such a mismatch can contribute to water strains on the localities especially in arid and semi-arid areas. The recent experience demonstrates that the data centres can consume the scarce water resources in comparison to the local population and ecosystems which is threatening in the context of environmental justice, and even the fair distribution of resources. These dynamics highlight the importance of having location sensitive sustainability analysis that consider the regional hydrological conditions and governance structures.

Besides the operational consumption of water, the lifecycle water footprint of AI infrastructure encompasses upstream and downstream activities, including the production of hardware, supply chain logistics, and end-of-life disposal, in addition to the production of hardware. The production of semiconductors, e.g., is highly water-demanding, thus, ultra-pure water is required to produce them. Similarly, data centers often demand water usage, as electricity generation, which is used by these facilities, often needs water usage, depending on the energy mix. The overall assessment of environmental impacts of AI should therefore consider the direct and indirect water usage in the entire lifecycle of AI (Al Kez et al., 2022).

It is against this background that some of the contributions that this chapter tries to make are discussed below. First, it provides a conceptual map among AI systems, energy usage and water resources that provides a hierarchical view of evaluating the environmental impact of digital infrastructure. Second, it gives a critical review of interdisciplinary literature that discloses certain critical trends and methodological directions and research gaps. Third, it offers quantitative and lifecycle-based sustainability measures such as water usage efficiency and water footprint in the lifecycle, to prefer empirical analysis. Fourth, it is policy, governance and equity analysis in which there is the need of open reporting, regulatory regulation and the management of resources in an inclusive manner. Finally, it introduces research opportunities, i.e., scenario modelling, technological innovation and integrated sustainability assessment.

2. LITERATURE REVIEW

The literature on the Artificial Intelligence (AI) and sustainability has been accumulated, which has generated beneficial data on the environmental effects, which digital infrastructures may have. Nonetheless, the literature, though increased, remains cross-disciplinary and has little or no synthesis of the technical, environmental and policy-based views. The section presents a critical synthesis of literature that is available, in the form of categorizing it into three broad streams, including technical studies, environmental studies, and policy and social studies and identifies key gaps that should lead to a more holistic analytical process.

Figure 1. Energy water Framework

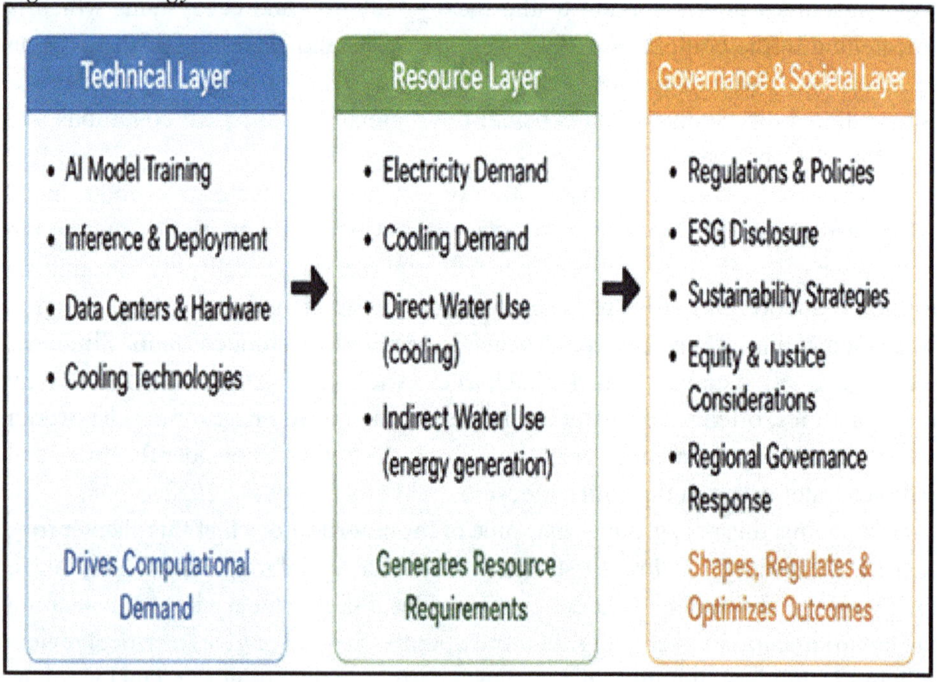

Source: Author owns developed

a. Technical Research: Cooling, optimization of infrastructure and efficiency

Technical AI sustainability has also been explored primarily to improve the computational performance, hardware optimization, and cooling performance. The energy efficiency of data centers, and in particular, how the architectural design and control of the workload can be used to reduce the amount of power used. Recent studies explores more advanced cooling systems, including liquid immersion cooling, free-air cooling, and thermal optimization systems based on AI, which will assist in saving energy and providing performance guarantees (Afoma et al., 2026).

Much of this literature also addresses algorithmic efficiency, including model compression, pruning and distributed computing schemes that seek to reduce the amount of computation. The techniques are particularly relevant because of the exponentially large machine learning models, which have a high processing power and generate significant thermal energy. However, a weakness of this stream is that it lacks the use of environmental indicators particularly the use of water. Even though energy efficiency is traditionally considered the main performance indicator which can be measured through such indicators as Power Usage Effectiveness (PUE) water-related factors are not often taken into account when developing system optimization models (Abulaiti & Liu, 2025). The cooling technologies are evaluated more on energy saving basis rather than on water intensity to provide a partial analysis of the trade-offs in sustainability. As such, technical research will tend to perceive environmental impacts as secondary to designing, rather than design specifications.

b. Environmental Research: Lifecycle Effects and Water Footprint

It has been increasingly popular to measure the water footprint of digital infrastructure, including AI-based data ecosystems, by environmental research. Direct water use (e.g., cooling in data centers) but also indirect water use that is incorporated in electricity production and the hardware manufactured. These studies are more likely to adopt the lifecycle assessment (LCA) approaches and variousiate blue, green and grey water footprints so as to encompass the whole picture of environmental impact.

In this literature, one can observe that water as a key element of AI infrastructure is often overlooked yet affects the infrastructure significantly, particularly in the regions, where evaporative cooling systems are commonly used. It also underlines the importance of geographical location, whereby the data center has a greater environmental impact based on local weather and water supply in addition to the sources of energy (Dada et al., 2024). To provide an example, data centers located in arid regions will be a source of water stress in the region, despite being able to be energy-efficient.

The environmental study can also be applied to the water-energy nexus knowledge and demonstrates the correlation of water usage with the process of energy production. This aspect of view is needed when we are referring to indirect effects since electricity production and especially thermal power plants can be water intensive. The combination of direct and indirect resource use provides a more in-depth view of the ecological footprint of AI via environmental studies.

c. Governance, Ethics and Sustainability in Policy and Social Studies

The third literature is related to the policy, governance and social facets of AI sustainability. Studies address such issues as regulatory frameworks, environmental responsibility, ethical considerations related to the use of the resources. This literature shows that there is need to have transparent reporting, standardized metrics and institutional control to ensure that the development of AI is geared towards the achievement of sustainability (Cazzetta et al., 2026).

Policy-based research also places a focus on Environmental, Social and Governance (ESG) frameworks, such as disclosure frameworks such as the Global Reporting Initiative (GRI) and new regulatory frameworks such as the Corporate Sustainability Reporting Directive (CSRD) in the European Union. These structures are aimed at increasing accountability whereby organizations are required to report on how they use their resources including how they use water. Additionally, the governance research is also concerned with the problem of cross-border regulation since digital infrastructure is international.

Equity and environmental justice issues are increasingly becoming incorporated in this literature with a special emphasis on disproportionality of environmental harms to regions and communities. The data centers are usually situated in areas where the economy is good and where water supply is varied which is a concern in terms of resource allocation, the effect on the community and the strength of the locals.

3. CONCEPTUAL FOUNDATIONS

In order to remedy the disjointedness of earlier research, this chapter develops a Conceptual AI Sustainability Framework that connects the technological processes, flow of resources and governance structures into one analytical framework. The existing studies have a tendency to investigate the efficiency of AI, its effects on the environment, or the policy implications independently, and little is known regarding the connections between these factors (Bhatia et al., 2025). This gap is addressed by the suggested framework that considers AI infrastructure as a multi-layered system, where the computational processes determine resource consumption, and institutional and societal forces determine and regulate this consumption. It can be seen through this holistic perspective that the water footprint of AI can be analyzed, particularly in relation to the water-energy nexus. The framework is organized into three layers, which are interrelated and that cover the main elements of AI-driven data ecosystems: Technical, Resource, and Governance and Societal.

a. Technical Layer: Computational Drivers Resource Demand

The fundamental layer of the framework is the technical layer and comprises the core computational processes to support the functionality of AI. This will involve model training, inference, data storage as well as data processing process which requires high computational power. Continued rapid growth of deep learning architectures and large-scale models has led to a rapid increase in processing intensity, often in units of floating-point operations (FLOPs) and GPU utilization (Bose et al., 2021).

Figure 2. Conceptual Framework of all integrated layers

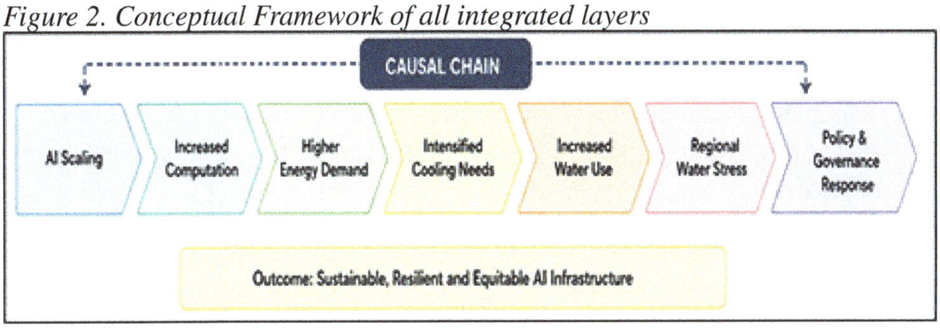

Source: Author owns developed

The above figure integrates all the layer which includes high-performance GPUs, TPUs and dedicated AI accelerators. When used, these elements generate a lot of heat and hence, elaborate cooling systems are needed to stabilize the systems and guarantee performance. The common air cooling as well as more sophisticated liquid immersion cooling technologies are critical aspects that can impact the trends in energy efficiency and consumption of water.

b. Resource Layer: Water Energy Interdependencies

Resource layer explains the physical resources that are required to sustain AI processes, and in particular, the energy consumption and water usage are highlighted. These are very much interconnected resources and they form the basis of water-energy nexus. It is not just the computations made in the data centers that require electricity, but also coolers and the electricity generation typically demands copious amounts of water, particularly in thermal power generation (Chu & Huang, 2023).

There are two primary ways of AI infrastructure water use:

Direct water use: Direct water use is largely associated with data center cooling systems, especially the evaporative cooling, which can consume large quantities of fresh water.

Indirect water use: It is part and parcel of the energy supply system that deals with water usage during the generation of electricity and mining of fuel.

Additionally, the lifecycle approach is also incorporated into the framework since the impact of water on water resources cannot be specific to the period of operation because it may be related to hardware production, the supply chain process, and disposal at the end of the life of the product as well (Ansorge & Stejskalová, 2025).

To enable quantitative analysis, quantitative measures of sustainability are added in this layer and they include:

- Water Usage Effectiveness (WUE) - water consumption versus load of IT.
- Power Usage Effectiveness (PUE) - indicating the data center energy efficiency.
- Lifecycle Water Footprint (LWF) - total amount of water use at each of the stages.

These measures allow comparing technologies, geographic locations, and strategies of operation and address one of the major flaws of the existing literature.

c. Government and Society Stratum: Economics, Regulation and Equity

The administrative and cultural strata describe the institutional and socio-economic context within which AI infrastructure is operating. Though the degree of environment impact is determined by the dynamics of the technical and resource, the structures of governance determine the nature and distribution of these impacts and also mitigation.

This layer includes:

policy frameworks, regulatory measures, e.g., water withdrawal restrictions, environmental compliance regulations, and e.g., compulsory disclosure requirements.

Economic tools, including the pricing mechanism, incentive to sustainable technologies and cost frameworks in the location of infrastructure.

Aspects of equity and justice, such as unfair distribution of environmental costs between regions and individuals.

According to one of the latest studies, the absence of transparency and non-standardized reporting and disclosure of water usage at AI and data center operations is a pressing concern. This presents a responsibility issue and limits the effectiveness of the interventions of governance. In addition, there is a common location of the data centres in regions with favourable economic conditions and varying water supply and raises the question of whether there are variations between the regions and environmental justice. The social component also entails the relations of the stakeholders, including the industrial water consumption and community demand, and the application of the popular opinion and activism in the construction of the infrastructure (Amanatidis et al., 2026).

d. Synthesized System Dynamics and Analytical Value

The benefit of proposed AI Energy Water Nexus framework is that it can tie together the technical, resources, and governance layers into a logical framework of cause-and-effect correlations, thereby, transforming a collection of perceived as a fragmented set of issues to an analytical model (He et al., 2019). Its core idea is that the technical layer, consisting of the design of AI models, computational intensity and cooling needs is the primary demand of resources. As the size of AI systems grows, particularly when dealing with large-scale machine learning models, the computational demands are very high, and hence have higher energy consumption. This growth is reflected in the resource layer where the demand of power is directly translated into electricity usage, and also water consumption, particularly through cooling systems and embedded water in the process of the generation of power. The process is mediated by the layer of governance and the society at that time and how these resource demands are met in the form of policies, regulations, economic incentives and sustainability frameworks.

This interconnection may be considered a causal cascade, in which the first step, which is scaling of AI models can cause the second step, which is computation, more energy requirements, more cooling requirements and lastly more water requirements. When these pressures are exerted under specific geographic and hydrological conditions, they contribute to the regional water stress and this in turn results in the policy and governance responses to control, optimize or redistribute the utilization of resources. This sequence can be directly mapped into key leverage points where interventions would likely be most effective using the framework. The framework also facilitates a systematic analysis of trade-offs e.g. water and energy efficiency trade-offs or economic cost and environmental sustainability trade-offs which are the subject of real life decision-making (Herrera et al., 2025).

The other significant advantage of such a framework is that it may be employed to support the evaluation of the regional vulnerabilities and systemic risks. It reiterates that even with the same technological systems, the sustainability outcomes can be radically different, as it links the infrastructure design to the local environmental conditions, and the capacity of governance (Huang et al., 2020. This is what renders the framework particularly helpful when considering the geographical distribution of the impacts and where the intervention is most in dire need.

4. DATA FLOWS AND SYSTEMS

Artificial intelligence (AI) ecosystems are built on multi-layered data flows and span across computation, storage, transmission, and processing across distributed digital infrastructures. Such ecosystems are not confined to discrete and solitary data centres, but instead they are defined to exist as networked systems of cloud platforms, edge devices and high-performance computing clusters, which can work together to help underpin large-scale data-intensive operations (Jayakumar et al., 2024). Even though much research has been conducted on the functional efficiency of these systems, their implication on resources, water and energy flows in particular require more concerted analysis.

Among the most significant contrasts, which can be drawn when it comes to the environmental footprint of AI ecosystems, it must be the direct and indirect water use. Direct water use will be largely associated with data center cooling systems that are of great importance in maintaining stability in operations. High-performance computing infrastructure also generates a significant amount of heat, and its cooling systems, particularly evaporative cooling systems, cool the heat by using fresh water. These systems can consume large volumes of water especially in hyperscale data centres and therefore a big part of the total water footprint of AI (Hassan et al., 2022).

Power consumption typically consumes a lot of water especially in the thermo-electric power production whereby water is used to produce steam and to cool. It is possible to connect the water needs even with the renewable energy sources, i.e., hydropower and some types of solar thermal energy. The second phase consists of data storage infrastructure that consists of servers, storage arrays and networking equipment. These parts are powered continuously, and cooled, and have a direct and indirect impact on the amount of water used in the operating cycle (Fonseca et al., 2016).

The final stage of the lifecycle is end-of-life disposal and recycling which is more problematic to the environment. The result of electronic gadgets (e-waste) generated as a result of the obsolescence of the hardware devices contains hazardous materials that must be used and treated with care. The recycling process itself might demand the water utilization and the use of energy and inappropriate disposal could pollute the water supplies. Even though this is a very important step, it is often poorly reflected in the sustainability analysis, which leads to the absence of the full picture of the lifecycle impacts.

An integrated analysis of data ecosystems and resource streams reveals that AI infrastructure is a closed loop transformation of resources and the inputs (data, energy, water) is continuously turned to outputs (computational results, heat, emissions). Not only is this system efficient on how it has been technologically developed but also in how it is run e.g. in the way it shares workload and in the way it is geographically located and resources optimised (Goyal et al., 2024). To provide an example, ambient temperature of a place could be used to place data centers in colder regions and therefore reduce cooling expenses, or the position of computational loads could be moved to places with an abundance of water or renewable power thereby reducing environmental impact.

5. WATER DATA CENTERS AND WATER CONSUMPTION

The AI infrastructure has data centers as its functional background, which consists of the computing infrastructure required to train models, make inferences, store and process data. As the number of people requesting AI-based services increases, the size and complexity of the data centers are increasing, becoming hyperscale hubs, or able to execute millions of simultaneous computations. Though the energy consumption and carbon emissions aspect has been a critical subject matter when talking about data centers, their dependence on freshwater resources, particularly in cooling, have become a major sustainability concern (Egbemhenghe et al., 2023).

The major use of water in data centers is in the form of a cooling system that is required to cool the heat generated by the high-performance computing equipment. One of the most popular methods of cooling is evaporative cooling because it involves the use of water to absorb heat and remove it. Despite the fact that the rate of energy consumption is lower in this approach compared to the traditional air conditioning, it can consume large volumes of water, especially in large plants during hot weather. To fulfill the requirements of the standardized assessment, certain significant sustainability measures have been developed:

Water Usage Efficiency (WUE)

A popular metric, WUE is calculated as the ratio of annual total water consumption to IT equipment energy consumption (liters per kWh) and is commonly used to gauge the efficiency of water consumption in data centers. It provides a direct indication of the quantity of water to sustain the activity of calculation. A low WUE implies that it consumes less water. However, WUE predominantly concentrates on water use at the operational level and it does not consider indirect and lifecycle effects of water (Mahadeva et al., 2023).

Power Usage (PUE) Effectiveness

PUE calculates the efficiency of a data center in energy consumption which is calculated as the total consumption of energy in a facility divided by the energy used by IT equipment. The PUE is not a direct measure of water use but has a direct relationship with water use, as cooling requirements are frequently influenced by the energy efficiency and hence the water use (Khalaj & Halgamuge, 2017). Data centers that have less optimized cooling systems also have lower values of PUE and consequently, less water.

Lifecycle Water Footprint (LWF)

LWF extends the operational measures to the total amount of water consumed in the data center infrastructure lifecycle. This includes the water used during the production of the hardware, energy production, cooling and disposal of the hardware. LWF employs a broader indicator of environmental impact as it also considers the direct and indirect use of water.

The combination of these measures enables to compare quantitatively in terms of regions, technologies, and operational strategies in order to overcome the critical limitation of the existing literature. To illustrate, a data center that is energy efficient but consuming a lot of water, low-PU and high-WU may be termed as data-efficient but water-intensive, hence the need to have the multi-dimensional performance evaluation. Correspondingly, lifecycle measurement depicts non-operational effects, which cannot be overcome by operational measures (Kaushik et al., 2026).

Table 1. Sustainability Metrics for AI Data Centers

Metric	Definition	Formula/Measurement	Significance
WUE	Water Usage Effectiveness	Liters/kWh	Measures water efficiency
PUE	Power Usage Effectiveness	Total facility energy / IT energy	Measures energy efficiency
LWF	Lifecycle Water Footprint	Total lifecycle water use	Captures indirect impacts
HSI	Hydrological Sensitivity Index	Regional stress indicator	Measures local vulnerability

The above table depicts the sustainability of data centers in dry or water starved regions may be compromised as the consumption of the water will be competing with the needs of the local communities and ecology. Conversely, the cooling of the facilities in cooler climates could be more reliant on the air-based cooling, which would reduce the use of water. All these disparities imply the necessity of taking care to make sure that assessment structures are context-sensitive and incorporate regional hydrological conditions.

Despite increased growth in measurement, availability and transparency of data have problems. Many companies do not disclose the exact data on water consumption, and no standard reporting models are associated with the AI loads. This limits the opportunities of scientists and policy-makers to assess the real magnitude of water consumption, and develop particular interventions. The adoption of WUE, PUE and LWF provides a well-grounded foundation of such analysis that enables making more informed decisions and making the transition to the digital infrastructure relying on water consumption (Masud et al., 2025).

6. NEW TECHNOLOGIES AND COOLING TECHNOLOGIES

AI-driven data centers strongly rely on cooling systems to define the energy use and the use of water. As computational intensity increases, particularly with the development of large-scale AI models, the significance of efficient thermal management has increased (Mahmoudian et al., 2026). Traditional methods of cooling are economically consuming and were proven to be effective that is why new technologies were created to attain a lower environmental impact.

Traditional Cooling Systems and limitations

The air-based or Computer Room Air Conditioning (CRAC) units or the evaporative or water-based cooling systems, which depend on the ability of water to absorb the heat, is usually the foundation of a traditional data center. Evaporative cooling is more efficient in terms of energy consumption; however it consumes a lot of fresh water and is therefore not sustainable in water scarce regions. Air-based systems, in turn, require less water, though, they are more likely to require more energy, and, therefore, water and energy efficiency is also a trade-off. The table below depicts the comparison between water, energy, carbon emission and cost across all cooling technologies.

Table 2. Water Energy Carbon Trade-Offs in Cooling Technologies

Cooling Technology	Water Consumption	Energy Consumption	Carbon Impact	Economic Cost
Traditional Water Cooling	High	Moderate	Moderate	Moderate
Air-Based Free Cooling	Low	Moderate–High	Moderate	Low
Liquid Immersion Cooling	Very Low	Low	Low	High
Hybrid Cooling Systems	Medium	Low–Moderate	Low	Moderate
AI-Driven Thermal Optimization	Low	Low	Low	Moderate–High

Liquid Immersion Cooling

Liquid immersion cooling is a process of immersing computing components in a thermally conductive liquid, which is also a good heat sink and heat transfer media. This method reduces significantly the need of air circulation, and can eliminate the use of water in the cooling process. It also maximizes energy efficiency, since it reduces overhead on cooling. However, implementation cost and compatibility to current infrastructure remains an issue (Mandal et al., 2025).

Air-Based Free Cooling

The cooling systems are free cooling systems that cool the air with the outer air of the environment, which is usually cooler and colder climates. The energy and water consumption can be significantly lowered through free cooling since it will be necessary to reduce the use of mechanical cooling systems (Mitchell-Jackson et al., 2003). The procedure is highly location based and is most applicable in regions where there are low ambient temperature and humidity.

Hybrid Cooling Systems

Hybrid systems combine air and liquid cooling systems though it dynamically switches between air and liquid cooling systems as per the environment and workload requirements. This flexibility will allow the data centres to optimise the use of resources, balancing between water and energy resources. Hybrid systems are starting to be used in modern facilities due to their flexibility and resistance.

AI-Driven Thermal Optimization

The AI itself is also being applied to optimize the cooling process by predictive analytics and real-time monitoring. Machine learning algorithms are able to know the patterns of temperature and work load distribution and the surrounding to change the cooling strategies dynamically. This improves efficiency, and lessens the waste of water and energy (Ojadi et al., 2025).

7. WATER STRESS, GEOGRAPHY, GLOBAL CONTEXT

The effects of AI-driven data ecosystems on the environment are not evenly distributed, but instead it is very geographically and hydrologically sensitive. Another factor that may predetermine the sustainability results is geographical location because of the climate conditions, the availability of water, and regulations, location is another aspect that contributes significantly to the quantity of water that data centers use (Nadjahi et al., 2018).

Climate has been known to be one of the greatest water use determinants. Data centers with warmer temperatures are also more energy intensive to cool to preferred operating temperatures with large volumes of evaporative cooling systems that use large amounts of water. Conversely, waterless cooling systems, such as air-based or free cooling systems in their plants, might also be applied in cooler climates, and would require much less water (Oró et al., 2015). This reliance on climate produces natural unequalities in resource use whereby the same amount of calculating activity could have a water footprint of monumental variation based on the geographic location.

Climate is directly related to the issue of water availability. Regions with abundant supplies of freshwater may be better placed to support data center operations that require a lot of water, but regions underwater pressure or underwater are at higher risk. The ratios of water withdrawal to supply available are the indices normally used to quantify water stress and it varies greatly across the world (Richards et al., 2023). To take just one example, water stress is already a reality in certain regions of the United States, India, and the Middle East, and the sustainability of the expanding AI infrastructure in these regions is an issue. In this case, the interests of agriculture, industry, and households can be at odds with the use of data centers, which further worsens the conflict over local resources.

Lifecycle Water Footprint and Indirect Impacts

To quantify the real environmental impact of AI-based data ecosystems, the lifecycle water footprint should be viewed as an integrated whole, and not merely the figures of operations. The AI infrastructure water consumption is in the form of a sequence of development processes, operations and activities that relate to indirect use of energy sources that lead to the total environmental impacts. The lifecycle stages covered in a single lifecycle approach will be more insightful and holistic in addressing water dependence of AI.

Figure 3. Lifecycle water footprint of AI Infrastructure

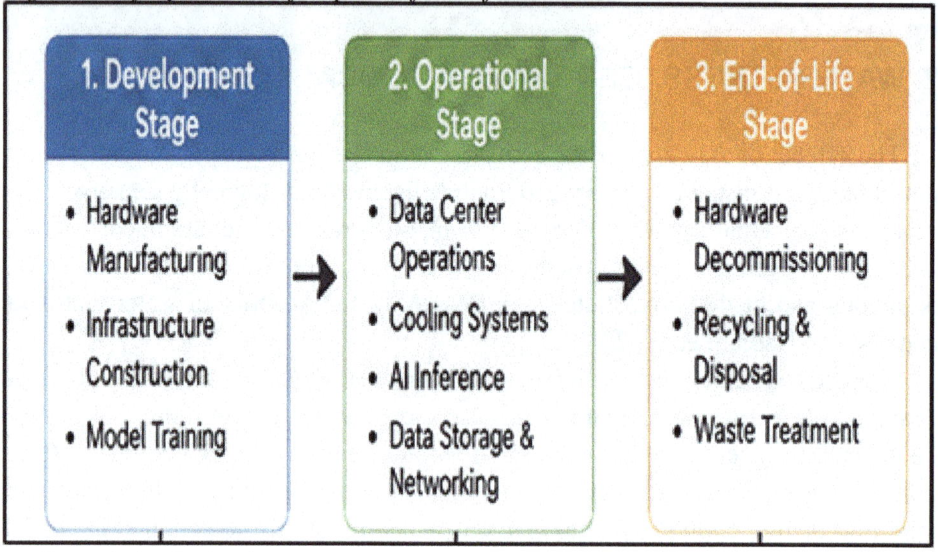

Source: Author developed

The above figure illustrates the chain of energy underlying AI functioning is incorporated within the indirect water usage. The operational stage is the most apparent stage of water usage as it involves data center cooling, real-time inference processes. As AI systems are implemented in applications, they require continuous computation and data processing and, therefore, continued energy use and heat dissipation. Evaporative based cooling systems and other types of cooling systems utilise water in order to maintain the best possible operating conditions (Raihan et al., 2024). Such consumption depends on aspects such as the severity of work, cooling technologies and geographical positioning. It is important to note that the operational water use level is not fixed and depends on the varying demand, complexity of the model and the growth of infrastructure.

Besides direct water use, there is an indirect step in the lifecycle perspective which includes the use of water in the form of electricity generation and energy supply chains. Information facilities usually use electricity produced in a way that requires water, especially in a thermoelectric power plant, whereby water is used to cool it down to create steam. Even though renewable energy sources may be less water-intensive, they may need the use of water depending on the technology (e.g., hydropower) (Pimenow et al., 2025).

Disposal and recycling of electronic equipment are also another critical part of the lifecycle, known as end-of-life management. As AI infrastructure advances, so does the volume of electronic waste (e-waste) to be recycled or discarded along with the obsolete hardware. It can result in other types of water consumption and the danger of water pollution in case of an inadequate conduction. Such a stage is not taken into account during the analysis but contributes to the total environmental impact of AI systems (Richards et al., 2023).

Lifecycle stages can be used to calculate Lifecycle Water Footprint (LWF), a comprehensive indicator that adds up the cumulative water usage of development, operation and indirect processes (Siddiqui et al., 2022). Advantageous is LWF where technologies can be compared, trade-offs can be assessed and opportunities to intervene can be identified. Using hardware as an example, when its efficiency is increased, the volume of water consumed during the operations may go down but may result in an increase in the quantity of water consumed in the production process hence illustrating the need to make balanced decisions. Water footprint of AI is impossible to measure in one step, but in a multi-step lifecycle approach that considers both direct and indirect impacts. The incorporation of development, operation and water use related to energy use makes this section fill the critical voids in the available literature and form the foundation of more comprehensive and sound sustainability analysis (Sharma et al., 2025).

8. POLICY, GOVERNANCE AND ECONOMIC ASPECTS

The innovation of technologies in isolation is also not enough to make the AI-based data ecosystems sustainable, and necessitates efficient policy frameworks, governance structures, and economic tools to regulate responsible consumption of resources. Some of the most apparent gaps that the existing literature has opened and that the reviewers have pointed out include the lack of operationalization of the governance strategies to drive the consumption of water in AI infrastructure. The section bridges that gap by cogitating on the regulatory standards, economic drivers and how water considerations can be integrated into the overall sustainability agendas (Ruiz Carrillo et al., 2026).

The implementation of the compulsory water disclosure policies of the data center operators is one of the most urgent demands. At present, voluntarily, certain organizations are also publishing sustainability reports, which include disclosure of water usage, however, there is no standardized and enforced reporting framework, which is specifically employed and applied to AI-related water usage. Such lack of transparency constrains accountability and it allows policymakers, researchers and stakeholders to evaluate the actual magnitude of environmental impact, which is hard to quantify. Compulsory reporting like the carbon reporting requirement would allow benchmarking, comparison and contrast and incentives to the organizations to be more water efficient (Singh et al., 2024).

Table 3. Governance and ESG Frameworks Relevant to AI Water Sustainability

Framework/Policy	Focus Area	Relevance to AI Data Centers
GRI 303	Water and effluents	Water disclosure and reporting
CSRD (EU)	Sustainability reporting	Mandatory environmental reporting
TCFD	Climate-related disclosure	Resource risk management
UN SDG 6	Clean water and sanitation	Sustainable water management
ESG Water Stewardship Programs	Corporate sustainability	Operational accountability

The table 3 integrate this disclosure policy with the existing climate and sustainability framework, including Environmental, Social, and Governance (ESG) standards. There are systems such as the Global Reporting Initiative (GRI 303: Water and Effluents) and the newly introduced tools of regulation such as the Corporate Sustainability Reporting Directive (CSRD) of the EU that provide systematized approaches to water accounting. However, these frameworks often lack AI-related indices, in particular, indices that would indicate the relationship between computational loads and water consumption (Sit et al., 2020). The digital infrastructure and sustainability governance would be more in line by incorporating AI-relevant metrics, such as water per training cycle, or water per inference workload.

Along with regulatory disclosure, the economic incentives play a critical role in the organizational behavior. The decisions available to the data center operators are typically cost, performance, and scalability. In this way, economic incentives should be coupled with sustainability goals in order to further encourage use of water saving technologies and practices. Policy instruments to encourage investment in solutions that are environmentally friendly include water pricing schemes, tax incentives on environmentally sustainable infrastructure, and subsidies on low-water cooling technologies (Tranfield et al., 2003). An example is water-stressed regions, where a system of varying water prices may be used to discourage large amounts of water use, and to promote saving.

9. SUSTAINABILITY MEASURES AND RISK EVALUATIONS

As AI-based data ecosystems spread, there is a growing need to have structured risk assessment frameworks that can systematically evaluate the environmental impacts- particularly risks in water. The literature available is more inclined towards specific problems such as cooling requirement or water scarcity in the area and does not offer a standardized list of indicators that can be contrasted across the infrastructures, geographies and model of operations (Vanitha et al., 2023). To address this gap, the given chapter suggests a set of essential sustainability metrics that would be able to capture the multidimensionality of water risks associated with AI.

The first indicator is hydrological sensitivity which shows how susceptible the activities of data centers to the conditions of water stresses in the locality are. This includes the presence of water in an area, seasonal variations and other alternative needs such as agricultural, industrial and domestic water. The second indicator is the water footprint intensity, or the amount of consumed water of one unit of computation output (e.g. one training cycle, one inference request, or one kWh of IT load). This indicator is anchored on the related metrics such as Water Usage Effectiveness (WUE) but is extended to functional benchmarking to enable a comparison of technologies and workloads. Strong water footprint implies poor use of the resources and an indicator of technological or operational optimization.

The third-important dimension is climate vulnerability due to the exposure of data centers to the risks of climate conditions such as droughts, heatwave, and fluctuations of precipitation. Most of them will increase the operating risks on water-dependent infrastructure due to climate change that will increase water shortages in them. In the absence of climate forecasting during the development of data centers, there is the possibility of disruption of water supply in the future, which affects the reliability and cost framework. A climate vulnerable risk assessment also ensures that there is a sustainability perspective in the future (Zhao et al., 2025).

The fourth measure is a governance effectiveness and this is the measure of the effectiveness of the regulatory frameworks, mechanism of enforcement and the capacity of the institutions in terms of water management. Regions that have established better systems of governance are likely to implement water-use limits, water efficiency regulations as well as transparency measures to minimize the environmental ri sks. Conversely, unsustainable practices can be sustained due to the existence of the poor governance conditions that strengthen the ecological and social vulnerabilities (Zimmermann et al., 2012).

10. MITIGATION STRATEGIES

The multi-layer mitigation plan to assist in reducing the water footprint of AI-driven data ecosystems must integrate technological innovation, operational optimization and policy intervention. The mitigation must be developed at the technical, operational, and governance level since the AI infrastructure is complicated, rather than by a single solution.

Technical Strategies

Among the most direct methods of water saving is on the technological level with improvements of cooling efficiency. The new cooling systems such as liquid immersion cooling, hybrid cooling systems and air based free cooling have the potential to reduce the quantity of water consumed significantly compared to the previous systems of evaporation cooling. These technologies do not only decrease the direct water consumption, but also enhance the energy efficiency, therefore, indirect t water consumption, which is linked with production of electricity (Ye et al., 2024).

The other significant technical step is the optimization of the computational processes by AI. Machine learning algorithms can be used to optimize the scheduling of workloads, thermal management, and energy consumption in real-time. To illustrate the point, predictive analytics may control the cooling level based on the workloads trends, minimizing unnecessary resources utilization. Besides, it is possible that the work on the algorithms in efficiency (model compression, sparsity, and hardware acceleration) leads to a drop in the load of the computations, thereby reducing energy and water footprint (Goyal et al., 2024).

Operational Strategies

The manner in which AI infrastructure is deployed and run is connected to operational strategies. One of the most promising methods through which the computational workload can be dynamically allocated to the data centers in the locations with lower water stress or more sustainable sources of energy is the geographic workload distribution (Bhatia et al., 2025). This plan will take advantage of the global nature of cloud infrastructure to leverage the resources across locations in order to reduce the local environment load.

Another important operation measure is Water recycling and reuse systems. Closed-loop cooling systems and wastewater treatment systems can also be of great help in reducing the amount of fresh water withdrawal since the water is used in the plant. Still other state-of-the-art data centers are going to be even water-positive, meaning that conservation and restoration initiatives will supply more water than they will use. In addition, site selection and infrastructure design is also relevant. Cooling can be reduced by locating the data centers in cooler climates, or where a natural source of renewable energy is high, indirect water consumption can be reduced as well. The combination of climate-friendly design principles will ensure that in the evolving environmental factors, the facilities can be operational (Dada et al., 2024.

Policy Strategies

The policy and governance interventions play critical roles in scaling mitigation activities. The governments can establish regulatory frameworks to have water-use efficiency goals, withdrawal limits, and to mandate environmental impact assessment of new data center projects. The water disclosure policies may be obligatory and may help to achieve a higher level of transparency and accountability that will enable benchmarking and make informed decisions (Huang et al., 2020).

In particular, transparency and reporting are needed that leads to alignment of industry practice towards sustainability. The joined system of reporting, including metrics of ESG, can be utilized to ensure that the organizations track water consumption and report it on the regular basis. This will not only help to comply with the regulations but also build corporate responsibility and stakeholder involvement. The issue of water impacts due to AI is multi-dimensional and requires a multi-faceted approach. The technical innovations, optimization of operations and interventions in policies should work together to reduce direct and indirect use of water (Kaushik et al., 2026). By organizing mitigation efforts on these three levels, the stakeholders can shift to more sustainable and resilient AI ecosystems.

11. RESEARCH GAPS AND FUTURE TREND

AI and water sustainability are a new area of research with the level of technological changes being extremely high and there is no empirical evidence. With the ever-growing AI architecture, some important trends and gaps in the literature start to emerge, which is why it is necessary to conduct additional research and development. Among the most significant ones, there is the application of AI-based optimization of sustainability, where AI technologies are applied to streamline the data centers process. This will entail monitoring of resource consumption in real-time, proactive maintenance and optimization of adaptive cooling and energy consumption. All these applications show that AI can be used to reduce its ecological footprint per se, in an efficiency multiplication loop.

Unlike a centralized data center, edge computing brings the computing services nearer to end-users. This can assist in decreasing the latency and network load, but new resources management and water consumption challenges come about since smaller, distributed facilities are not necessarily as efficient as large-scale data centers. The design of the green data center is in the developmental stage as well. These involve incorporation of renewable source of energy, new cooling technology and sustainable building. The future of AI systems has assumed a new form through innovations like modular data centers, passive cooling-designed and water-neutral infrastructure.

12. CONCLUSION

This chapter has demonstrated that on the one hand, Artificial Intelligence can change the world of economy and society, yet, its impact on the environment, in particular, its consumption of water is not analyzed and included in the sustainability discussion. The chapter provides more interdisciplinary and holistic perspective of the AI sustainability by examining AI-driven data ecosystems in the context of technical processes, resource flows, and governance structure and socio-economic impact. The key contribution of the work is the integration of both direct and indirect impact of water across lifecycle of AI infrastructure with a focus on the correlation of data processing, energy consumption, and hydrology. The discussion points out that the water use issue is not an issue of local operation problem, but a systemic problem that is determined by geography, climate, policy, and design of technology.

REFERENCES

Abbas Khan, M., Khan, H., Omer, M. F., Ullah, I., & Yasir, M. (2024). Impact of artificial intelligence on the global economy and technology advancements. In El Hajjami, S., Kaushik, K., & Khan, I. U. (Eds.), *Artificial general intelligence (AGI) security: Smart applications and sustainable technologies* (pp. 147–180). Springer Singapore., DOI: 10.1007/978-981-97-3222-7

Abdulameer, L., Al-Khafaji, M. S., Al-Awadi, A. T., Al-Mansoori, H., Hassan, A. K., & Mahmood, S. (2025). Artificial intelligence in climate-resilient water management: A systematic review of applications, challenges, and future directions. *Water Conservation Science and Engineering*, *10*(1), 44. Advance online publication. DOI: 10.1007/s41101-025-00371-2

Abulaiti, H., & Liu, Y. (2025). Ecological sensitivity assessment and driving force analysis of the Tarim river basin. *Scientific Reports*, *15*(1), 34630. Advance online publication. DOI: 10.1038/s41598-025-18270-w PMID: 41044162

Afoma, U. M., Singh, S., Mishra, A. K., Sharma, C. K., Gupta, K., Mishra, M. K., Roy, B., Verma, V. V., & Sharma, V. K. (2026). Integrating artificial intelligence in environmental monitoring: A paradigm shift in data-driven sustainability. *EcoHealth*, *23*(1), 38–57. DOI: 10.1007/s10393-025-01752-8 PMID: 40875110

Al Kez, D., Foley, A. M., Laverty, D., Furszyfer Del Rio, D., & Sovacool, B. (2022). Exploring the sustainability challenges facing digitalization and internet data centers. *Journal of Cleaner Production*, *371*, 133633. Advance online publication. DOI: 10.1016/j.jclepro.2022.133633

Amanatidis, P., Lyratzis, E., Angelopoulos, V., Kouloumpris, E., Skaperdas, E., Bassiliades, N., Vlahavas, I., Maris, F., Emmanouloudis, D., & Karampatzakis, D. (2026). Intelligent water management through edge-enabled IoT, AI, and big data technologies. *IoT*, *7*(1), 5. Advance online publication. DOI: 10.3390/iot7010005

Ansorge, L., & Stejskalová, L. (2025). Water footprint which is not the water footprint: Critical review of the article by Müller et al. (2024). *Journal of Environmental Management*, *374*, 124038. Advance online publication. DOI: 10.1016/j.jenvman.2025.124038 PMID: 39793505

Bhatia, A. B., Sharma, M., & Bhatia, B. (2025). AI and environmental data science. In Kulkarni, S. (Ed.), *Artificial intelligence-driven models for environmental management*. Wiley., DOI: 10.1002/9781394282555.ch3

Bose, R., Roy, S., Mondal, H., Ghosh, A., & Das, S. (2021). Energy-efficient approach to lower the carbon emissions of data centers. *Computing*, *103*(8), 1703–1721. DOI: 10.1007/s00607-020-00889-4

Cazzetta, A., Zangaro, F., Marcucci, F., Julius, O. T., Rainò, M., Shauer, M., Massaro, R., Semeraro, T., Basset, A., & Pinna, M. (2026). Applying artificial intelligence (AI) innovative tools for ecological research and monitoring of transitional water ecosystems: A systematic review. *Environments (Basel, Switzerland)*, *13*(4), 193. Advance online publication. DOI: 10.3390/environments13040193

Chu, J., & Huang, X. (2023). Research status and development trends of evaporative cooling air-conditioning technology in data centers. *Energy and Built Environment*, *4*(1), 86–110. DOI: 10.1016/j.enbenv.2021.08.004

Dada, M. A., Majemite, M. T., Obaigbena, A., Daraojimba, O. H., Oliha, J. S., & Nwokediegwu, Z. Q. S. (2024). Review of smart water management: IoT and AI in water and wastewater treatment. *World Journal of Advanced Research and Reviews*, *21*(1), 1373–1382. DOI: 10.30574/wjarr.2024.21.1.0171

Egbemhenghe, A. U., Ojeyemi, T., Iwuozor, K. O., Emenike, E. C., Ogunsanya, T. I., Anidiobi, S. U., & Adeniyi, A. G. (2023). Revolutionizing water treatment, conservation, and management: Harnessing the power of AI-driven ChatGPT solutions. *Environmental Challenges*, *13*, 100782. Advance online publication. DOI: 10.1016/j.envc.2023.100782

Fonseca, J. A., Nguyen, T.-A., Schlueter, A., & Marechal, F. (2016). City Energy Analyst (CEA): Integrated framework for analysis and optimization of building energy systems in neighborhoods and city districts. *Energy and Building*, *113*, 202–226. DOI: 10.1016/j.enbuild.2015.11.055

Goyal, M. K., Kumar, S., & Gupta, A. (2024). AI for water policy. In AI innovation for water policy and sustainability (pp. 59–84). SpringerBriefs in Water Science and Technology. Springer. https://doi.org/DOI: 10.1007/978-3-031-72014-7_4

Hassan, M., Malhotra, K., & Firdaus, M. (2022). Application of artificial intelligence in IoT security for crop yield prediction. ResearchBerg Review of Science and Technology, 2(1), 136–157. Retrieved from https://researchberg.com/index.php/rrst/article/view/150

He, G., Zhao, Y., Jiang, S., Zhu, Y., Li, H., & Wang, L. (2019). Impact of virtual water transfer among electric sub-grids on China's water sustainable developments in 2016, 2030, and 2050. *Journal of Cleaner Production*, *239*, 118056. Advance online publication. DOI: 10.1016/j.jclepro.2019.118056

Herrera, M., Xie, X., Menapace, A., Zanfei, A., & Brentan, B. M. (2025). Sustainable AI infrastructure: A scenario-based forecast of water footprint under uncertainty. *Journal of Cleaner Production*, *526*, 146528. Advance online publication. DOI: 10.1016/j.jclepro.2025.146528

Huang, P., Copertaro, B., Zhang, X., Shen, J., Löfgren, I., Rönnelid, M., Fahlen, J., Andersson, D., & Svanfeldt, M. (2020). A review of data centers as prosumers in district energy systems: Renewable energy integration and waste heat reuse for district heating. *Applied Energy*, *258*, 114109. Advance online publication. DOI: 10.1016/j.apenergy.2019.114109

Jayakumar, D., Bouhoula, A., & Al-Zubari, W. K. (2024). Unlocking the potential of artificial intelligence for sustainable water management focusing operational applications. *Water (Basel)*, *16*(22), 3328. Advance online publication. DOI: 10.3390/w16223328

Kaushik, A., Soundappan, T., Mujawar, M. A., & Bhansali, S. (2026). Standing of electrochemical sensors on the 250th anniversary of the USA. *Journal of the Electrochemical Society*, *173*(9), 097509. DOI: 10.1149/1945-7111/ae67af

Khalaj, A. H., & Halgamuge, S. K. (2017). A review on efficient thermal management of air- and liquid-cooled data centers: From chip to the cooling system. *Applied Energy*, *205*, 1165–1188. DOI: 10.1016/j.apenergy.2017.08.037

Mahadeva, R., Kumar, M., Gupta, V., Manik, G., Gupta, V., Alawatugoda, J., Manik, H., Patole, S. P., & Gupta, V. (2023). Water desalination using PSO-ANN techniques: A critical review. *Digital Chemical Engineering*, *9*, 100128. Advance online publication. DOI: 10.1016/j.dche.2023.100128

Mahmoudian, F., Khoramipoor, Z., Nabizadeh Chianeh, F., Valikhan Anaraki, M., & Farzin, S. (2026). Towards cleaner waters: A comprehensive review of pollutant, quality challenges, monitoring advances, and treatment strategies. *Physics and Chemistry of the Earth Parts A/B/C*, *144*(Part 1), 104441. Advance online publication. DOI: 10.1016/j.pce.2026.104441

Mandal, S., Yadav, A., Panwar, R., Kumar, A., Singh, V., Sharma, P., & Gupta, R. (2025). Smart water management for SDG 6: A review of AI and IoT-enabled solutions. *Water Conservation Science and Engineering*, *10*(2), 83. Advance online publication. DOI: 10.1007/s41101-025-00407-7

Masud, M. M., Shamem, A. S. M., Saif, A. N. M., Hossain, M. S., Rahman, M. M., & Islam, M. A. (2025). The role of artificial intelligence in sustainable water management in Asia: A systematic literature review with bibliographic network visualization. *International Journal of Energy and Water Resources*, *9*(1), 247–265. DOI: 10.1007/s42108-024-00319-7

Mitchell-Jackson, J., Koomey, J. G., Nordman, B., & Blazek, M. (2003). Data center power requirements: Measurements from Silicon Valley. *Energy*, *28*(8), 837–850. DOI: 10.1016/S0360-5442(03)00009-4

Nadjahi, C., Louahlia, H., & Lemasson, S. (2018). A review of thermal management and innovative cooling strategies for data center. *Sustainable Computing : Informatics and Systems*, *19*, 14–28. DOI: 10.1016/j.suscom.2018.05.002

Ojadi, J. O., Owulade, O. A., Odionu, C. S., & Onukwulu, E. C. (2025). AI-driven optimization of water usage and waste management in smart cities for environmental sustainability. *Engineering and Technology Journal*, *10*(3), 4284–4306. DOI: 10.47191/etj/v10i03.36

Oró, E., Depoorter, V., Garcia, A., & Salom, J. (2015). Energy efficiency and renewable energy integration in data centres: Strategies and modelling review. *Renewable & Sustainable Energy Reviews*, *42*, 429–445. DOI: 10.1016/j.rser.2014.10.035

Pimenow, S., Pimenowa, O., Prus, P., & Niklas, A. (2025). The impact of artificial intelligence on the sustainability of regional ecosystems: Current challenges and future prospects. *Sustainability (Basel)*, *17*(11), 4795. Advance online publication. DOI: 10.3390/su17114795

Raihan, A., Paul, A., Rahman, M. S., Islam, S., Paul, P., & Karmakar, S. (2024). Artificial intelligence (AI) for environmental sustainability: A concise review of technology innovations in energy, transportation, biodiversity, and water management. *Journal of Technology Innovations and Energy*, *3*(2), 64–73. DOI: 10.56556/jtie.v3i2.953

Richards, C. E., Tzachor, A., Avin, S., & Fenner, R. (2023). Rewards, risks and responsible deployment of artificial intelligence in water systems. *Nature Water*, *1*(5), 422–432. DOI: 10.1038/s44221-023-00069-6

Ruiz Carrillo, J. A., Huamaní Jordan, O., Mendoza Loor, E. G., & Espín Beltrán, C. X. (2026). Impact of artificial intelligence on the sustainable use of water resources. *Sustainability (Basel)*, *18*(8), 3864. Advance online publication. DOI: 10.3390/su18083864

Sharma, R., Whig, P., & Gupta, S. K. (2025). AI-driven environmental problem design for sustainable solutions. In Kulkarni, S. (Ed.), *Artificial intelligence-driven models for environmental management*. Wiley., DOI: 10.1002/9781394282555.ch8

Siddiqui, A., Siddiqui, M., & Kulkarni, N. (2022). Artificial intelligence in water conservation: A meta-analysis study. *Prabandhan: Indian Journal of Management, 15*(3), 24–41. DOI: 10.17010/pijom/2022/v15i3/160407

Singh, J., Goyal, S., Kumar Kaushal, R., Kumar, N., & Singh Sehra, S. (Eds.). (2024). *Applied data science and smart systems* (1st ed.). CRC Press., DOI: 10.1201/9781003471059

Sit, M., Demiray, B. Z., Xiang, Z., Ewing, G. J., Sermet, Y., & Demir, I. (2020). A comprehensive review of deep learning applications in hydrology and water resources. *Water Science and Technology, 82*(12), 2635–2670. DOI: 10.2166/wst.2020.369 PMID: 33341760

Tranfield, D., Denyer, D., & Smart, P. (2003). Towards a methodology for developing evidence-informed management knowledge by means of systematic review. *British Journal of Management, 14*(3), 207–222. DOI: 10.1111/1467-8551.00375

Vanitha, S., Radhika, K., & Boopathi, S. (2023). Artificial intelligence techniques in water purification and utilization. In Vasant, P., Litvinchev, I., & Rodríguez-Aguilar, R. (Eds.), *Human agro-energy optimization for business and industry* (pp. 202–218). IGI Global., DOI: 10.4018/978-1-6684-4118-3.ch010

Ye, Z., Yin, S., Cao, Y., Zhang, X., Wang, H., Li, J., & Chen, Q. (2024). AI-driven optimization of agricultural water management for enhanced sustainability. *Scientific Reports, 14*(1), 25721. Advance online publication. DOI: 10.1038/s41598-024-76915-8 PMID: 39468244

Zhao, T., Song, C., Yu, J., Xing, L., Xu, F., Li, W., & Wang, Z. (2025). Leveraging immersive digital twins and AI-driven decision support systems for sustainable water reserves management: A conceptual framework. *Sustainability (Basel), 17*(8), 3754. Advance online publication. DOI: 10.3390/su17083754

Zimmermann, S., Meijer, I., Tiwari, M. K., Paredes, S., Michel, B., & Poulikakos, D. (2012). Aquasar: A hot water cooled data center with direct energy reuse. *Energy, 43*(1), 237–245. DOI: 10.1016/j.energy.2012.04.037

KEY TERMS AND DEFINITIONS

AI-Driven Data Ecosystems: Integrated digital environments where AI processes and analyzes data to generate insights and support decision-making.

Artificial Intelligence (AI): Technology that enables machines to perform tasks requiring human-like intelligence: learning, and decision-making.

Data Centre Cooling: The processes and technologies used to remove heat and maintain optimal operating temperatures in data centers.

Environmental Governance: The framework of policies: regulations, and institutions that guide environmental management and sustainability.

Lifecycle Assessment (LCA): A method for evaluating the environmental impacts of a product, process, or system throughout its entire life cycle.

Resource Efficiency: The effective use of resources to achieve maximum output while minimizing waste and environmental impact.

Sustainable AI: The design and use of AI systems that minimize environmental impacts while maximizing social and economic benefits.

Water Footprint: The total amount of freshwater consumed directly and indirectly by an activity: product, or organization.

Water Sustainability: The efficient management of water resources to meet current and future environmental: social, and economic needs.

Water-Energy Nexus: The interconnected relationship between water and energy systems: where each depends on the other for production and operation.

Chapter 8

Water as a Structural Requirement for AI Data Centers:

Thermodynamics, Workload Intensity, and the Limits of Air Cooling

Krishna Chaitanya Sunkara
http://orcid.org/0009-0009-6159-4280
Oracle, USA

ABSTRACT

AI data centers do not use water because they are inefficient; they use it because physics demands it. GPU-heavy workloads turn most electrical power into heat at the silicon layer, and that heat has to be removed continuously. Once rack densities cross roughly 40–50 kW, air cooling struggles to keep systems stable, so liquid cooling becomes the practical path. A large share of facility energy already goes into cooling, and it keeps rising as AI shifts to long training runs and steady inference. Water here is not a consumable, it acts as a heat transport medium, carrying energy away from the rack. That energy is not lost and can often be reused. This chapter looks at water through thermodynamics and real rack power, using measures like liters per inference and temperature rise to show that water demand follows workload intensity and time, not just facility size. Unlike prior work that treats cooling as a secondary optimization problem, this analysis frames it as a first order constraint coupled directly to workload duration and compute intensity

DOI: 10.4018/979-8-3373-8434-4.ch008

INTRODUCTION

The sustained growth of artificial intelligence has forced the data center industry to confront a thermodynamic reality that earlier generations of computing were able to defer: every watt of electrical energy consumed by a computing system eventually exits as heat, and at the power densities produced by modern GPU-dense AI racks, the rate of heat generation exceeds the physical removal capacity of air-based cooling systems by a large margin. This chapter addresses that reality quantitatively, building from chip-level thermal specifications through rack-level heat generation, facility-scale cooling infrastructure, and regional water resource implications.

For most of the data center industry's history, cooling was treated as a supporting function. Server racks drew 3 to 8 kW, workloads were intermittent, and air-cooled systems with modest airflow rates maintained acceptable thermal conditions. Artificial intelligence workloads, specifically GPU-accelerated training and inference jobs, altered every assumption in that picture. A rack populated with 48 H100 GPUs draws 47 kW continuously during a training run, produces approximately 160,500 BTU/hr of heat, and does so without the utilization troughs that earlier workload types provided for thermal recovery (Sunkara & Narukulla, 2025). This combination of sustained load and high density removes the margin that air cooling previously relied on. The mass-flow heat transfer equations demonstrate it unambiguously, and Section 3 of this chapter evaluates the result explicitly.

Liquid cooling, specifically water-based direct liquid cooling (DLC) delivered through cold plates in contact with GPU die surfaces, resolves the thermal challenge through the far superior volumetric heat capacity of water: approximately 3,400 times that of air under standard conditions (Cengel & Cimbala, 2018). At 32 gallons per minute per rack, a water-cooled loop removes the same heat load that would require physically impossible airflow rates and fan power vastly exceeding compute power to approximate through air. The selection is driven by physical constraints rather than design preferences.

This chapter has the following objectives. First, it derives the heat generation rate of a standard AI GPU rack from component-level thermal specifications. Second, it demonstrates through mass-flow analysis that air cooling fails at these densities. Third, it establishes liquid cooling requirements and the fluid dynamics governing their practical implementation. Fourth, it introduces the TTIR-based workload-coupled water model and applies it to training runs, annual per-rack demand, and hyperscale campus totals at 300 MW and 1 GW. Fifth, it examines water governance and heat recovery to complete the resource picture. All results trace back to measurable physical constants and published operational parameters.

Unlike prior work that treats cooling as a secondary optimization problem, this analysis frames it as a first order constraint coupled directly to workload duration and compute intensity.

BACKGROUND

The thermodynamic relationship between computing and heat dates to the earliest digital computers. Landauer's principle establishes that any logically irreversible computation must dissipate a minimum amount of energy as heat (Masanet et al., 2020). In practice, real computing systems dissipate vastly more than this theoretical minimum through resistive losses in transistor switching, interconnect resistance, voltage regulation, and power supply conversion. The fraction of electrical energy that converts to useful information-theoretic work is negligible; nearly all of it becomes heat.

Data center cooling has evolved through several architectural generations in response to rising rack densities. Through the 1990s and early 2000s, raised-floor plenum systems supplied cool air beneath the floor and extracted warm exhaust above server rows, managing densities below 5 kW per rack without difficulty. The adoption of hot-aisle cold-aisle containment in the mid-2000s, documented in ASHRAE's TC 9.9 thermal guidelines (American Society of Heating, Refrigerating and Air-Conditioning Engineers, 2021), pushed practical air-cooled density limits to 10–15 kW per rack by reducing mixing of hot and cold airstreams. High-velocity in-row cooling units and rear-door heat exchangers extended air-based approaches further still, reaching 20–25 kW in well-managed deployments.

GPU-accelerated AI workloads disrupted this incremental progression. The NVIDIA H100 SXM5, representing the current standard for production AI training infrastructure, carries a thermal design power of 700 W, a figure more than triple that of a high-end server CPU and substantially higher than the 400 W TDP of its predecessor H100 PCIe variant (NVIDIA Corporation, 2024). At 48 GPUs per rack, the GPU subsystem alone exceeds what the entire previous generation of air-cooled rack infrastructure was designed to manage. The International Energy Agency reported that data center energy consumption reached approximately 200–250 TWh globally in 2022, with AI workloads representing a rapidly growing fraction (International Energy Agency, 2023). Cooling accounts for approximately 30–40% of that total, and the proportion grows as rack densities rise.

Earlier published work (Sunkara & Narukulla, 2025) established the quantitative baseline for AI rack power consumption and heat dissipation that the present chapter extends to cooling system design, workload-coupled water demand modeling, and hyperscale infrastructure analysis.

The present chapter builds directly on those power and heat figures to derive the thermodynamic case for liquid cooling and its water requirements.

Water consumption in AI infrastructure has only recently been treated as a research problem, distinct from the energy analyses that dominated earlier data center literature. This gap was highlighted in Li et al. (2023) which measured both direct and indirect water footprints over the training and inference lifecycle of large language models. Their figures are striking: training GPT-3 at Microsoft's U.S. data centers evaporated roughly 700,000 liters of freshwater directly, and they projected global AI water withdrawal climbing to between 4.2 and 6.6 billion cubic meters annually by 2027. What distinguishes that work methodologically is its grounding in operational facility measurements rather than modeled approximations, and its identification of what the authors termed spatial-temporal water efficiency diversity, the observation that identical inference workloads draw different volumes of water depending on where and when they run, because local cooling plant design and ambient temperature both mediate the relationship between compute and consumption. The present chapter takes a different path to a complementary destination: rather than working backward from facility-level water logs, it derives water demand from the governing thermodynamic equations and ties the result to job duration through the TTIR framework, producing a per-rack water quantity that generalizes across workload types without requiring inference request data any further. The engineering literature has meanwhile been building an empirical picture of how cooling architecture choices translate into water and energy outcomes at the component and facility scale. Borkowski and Piłat (2022) tested customized liquid cooling loops under conditions of substantial outdoor temperature variation and found that stable rack-level thermal performance was maintained while cooling plant power draw fell a result that puts measured numbers behind the thermodynamic efficiency argument this chapter constructs from first principles. The water side of that trade-off received direct quantification in Brown and Patel (2023), who reported in IEEE Transactions on Industry Applications that high-density deployments using liquid cooling consumed up to 80 percent less water than equivalent air-side evaporative systems, because eliminating the cooling tower removes the single largest source of consumptive loss. That finding maps precisely onto the distinction this chapter draws between the gross circulation volumes Equation (18) derives and the far smaller consumptive fraction captured by the WUE metric in Equation (43). Lu et al. (2024) extended the analysis downstream, modelling waste heat recovery configurations for liquid-cooled facilities in Applied Energy and demonstrating viable technical pathways for redirecting warm secondary loop water into district heating and absorption cooling networks, the same recovery pathways this chapter scales to the 300 MW and 1 GW campus cases.

Figure 1. 3-D system architecture overview: AI workloads drive GPU rack heat generation (160,500 BTU/hr), cold plates extract heat through a 32 GPM water loop, warm water returns through the distribution manifold to the facility heat exchanger and chiller plant, which supplies cooled water back to the rack. Heat recovery options shown on right

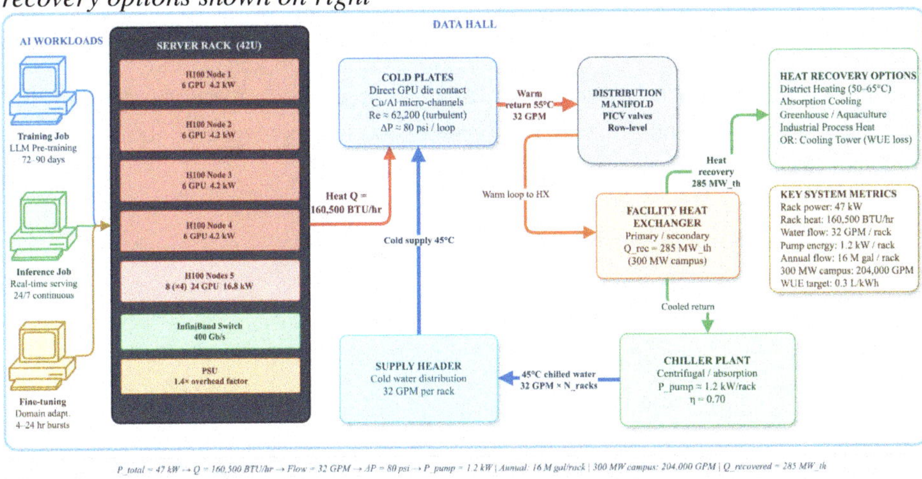

THERMODYNAMIC ANALYSIS OF AI RACK HEAT GENERATION

Rack-Level Power and Heat Derivation

The thermal load of an AI data center originates at the chip level. The H100 SXM5 GPU carries a thermal design power of 700 W, representing the sustained power dissipation at which the chip operates during production training workloads. A standard high-density AI rack configuration accommodates 8 compute nodes in a dual-socket arrangement, each node carrying 6 H100 GPUs, yielding 48 GPUs per rack. Rack-level GPU power is:

$$P_{\text{rack}} = N_{\text{GPU}} \times P_{\text{GPU}} \tag{1}$$

where P_rack is the GPU-only power draw (W), N_GPU is the number of GPU accelerators per rack, and P_GPU is the thermal design power per GPU.

$$P_{\text{rack}} = 48 \times 700 \text{ W} = 33,600 \text{ W} = 33.6 \text{ kW} \tag{2}$$

Total rack power adds CPU processors (~350 W each, two per node), NVLink switches, system memory, NVMe storage, 400 Gb/s network interface cards, and power supply conversion losses. An overhead factor of 1.4 is consistent with operational measurement data from deployed AI clusters:

$$P_{total} = 1.4 \times P_{rack} \tag{3}$$

The overhead factor of 1.4 accounts for CPU processors, memory, network interface cards, storage, and power supply conversion losses measured in operational AI cluster deployments.

$$P_{total} = 1.4 \times 33{,}600 = 47{,}040 \text{ W} \approx 47 \text{ kW} \tag{4}$$

Energy consumed over any operating interval t:

$$E = P_{total} \times t \tag{5}$$

where E is energy consumed (Wh) and t is the operating time interval (hr).

At steady state, essentially all electrical energy converts to heat with no useful mechanical work output and no significant energy storage in a running server:

$$Q = E = P_{total} \times t \tag{6}$$

At steady state, no useful mechanical work exits the server; the first law of thermodynamics requires that all electrical energy converts to heat.

Applying the standard engineering thermal conversion:

$$1 \text{ W} = 3.412 \text{ BTU/hr} \tag{7}$$

$$Q_{rack} = 47{,}040 \times 3.412 = 160{,}504 \text{ BTU/hr} \approx 160{,}500 \text{ BTU/hr} \tag{8}$$

This 160,500 BTU/hr represents the continuous, steady-state heat generation rate the cooling system must match without interruption for the full duration of a training or inference workload. Unlike enterprise server workloads that exhibit utilization variability, AI training jobs operate at near-constant maximum GPU utilization from job start to completion, eliminating the thermal recovery periods that earlier cooling system designs relied upon.

Figure 2. Thermo-computational flow: AI workload power consumption heat generation (160,500 BTU/hr) liquid cooling loop (32 GPM) heat recovery or rejection. Air cooling branch shown with failure flag (ΔT = 297°F). Key equations annotated at base.

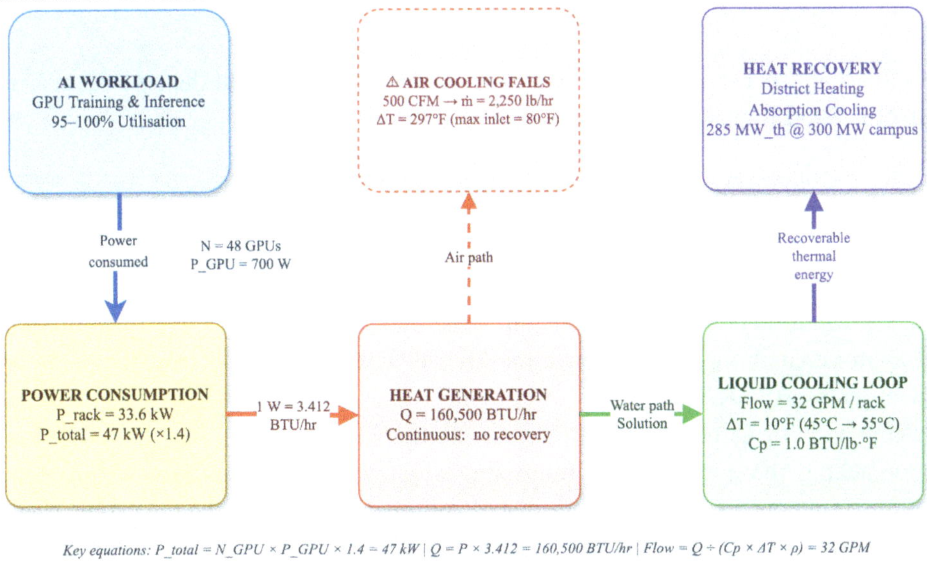

Why Air Cooling Fails at AI Rack Densities

Convective heat removal by any moving fluid follows:

$$Q = \dot{m}_{air} \times C_{p,air} \times \Delta T \tag{9}$$

where Q is heat removed (BTU/hr), _air is mass flow rate (lb/hr), C_p,air = 0.24 BTU/lb·°F is specific heat of air at constant pressure, and ΔT is the air temperature rise across the rack (°F). Mass flow rate relates to volumetric flow V (CFM) and air density ρ:

$$\dot{m}_{air} = \rho \times V \times 60 \tag{10}$$

At standard data center inlet conditions (65°F, sea level), ρ = 0.075 lb/ft^3. A high-capacity in-row cooling unit delivering 500 CFM, a value representative of aggressive air cooling practice:

$$\dot{m}_{air} = 0.075 \times 500 \times 60 = 2,250 \text{ lb/hr} \tag{11}$$

The temperature rise when this airflow confronts the full 160,500 BTU/hr rack load:

$$\Delta T = \frac{Q}{\dot{m}_{air} \times C_{p,air}} \tag{12}$$

Rearranging Equation (9) to solve for the temperature rise of the air stream as it absorbs the full rack thermal load.

$$\Delta T = \frac{160,500}{2,250 \times 0.24} \approx 297°F \tag{13}$$

This exceeds allowable operating conditions by several multiples, which means the constraint is not marginal, it is but structural. Air entering at 65°F would exit at approximately 362°F, far above the 80.6°F maximum server inlet temperature specified by ASHRAE A1 class equipment (American Society of Heating, Refrigerating and Air-Conditioning Engineers, 2021). The equipment would fail before approaching these conditions. Increasing airflow does not resolve the problem because fan power scales with the cube of volumetric flow rate, the fan affinity law:

$$P_{fan} \propto V^3 \tag{14}$$

The fan affinity law states that fan shaft power scales with the cube of volumetric flow rate; doubling airflow requires eight times the fan power.

Reducing ΔT to an acceptable 15°F requires approximately 9,800 CFM per rack, a factor of ~20 increase. Fan power would scale by $20^3 = 8,000\times$, consuming far more energy than the compute load being cooled. Physical chassis constraints also limit airflow to roughly 1,200 CFM before acoustic, pressure, and turbulence limits are exceeded. Air cooling at 47 kW per rack is not a design challenge to be engineered around; but it becomes infeasible under practical airflow and energy constraints.

Liquid Cooling: Thermodynamic Efficiency and Flow Requirements

Water's specific heat at constant pressure is 1.0 BTU/lb·°F and its density at 60°F is 62.4 lb/ft^3, yielding a volumetric heat capacity of 62.4 BTU/ft^3·°F versus 0.018 BTU/ft^3·°F for air, a ratio of approximately 3,467 (Cengel & Cimbala, 2018). Applying the heat transfer equation to the water-cooled rack with a design temperature differential of 10°F, typical of direct liquid cooling systems with 45°C supply water:

$$Q = \dot{m}_{water} \times C_{p,water} \times \Delta T \tag{15}$$

The identical governing heat transfer equation applied to water, where C_p,water = 1.0 BTU/lb·°F replaces C_p,air = 0.24 BTU/lb·°F, yielding a specific heat advantage of 4.17× before accounting for density.

$$C_{p,water} = 1.0 \text{ BTU/lb} \cdot {}^{\circ}\text{F} \tag{16}$$

$$\dot{m}_{water} = \frac{Q}{C_{p,water} \times \Delta T} = \frac{160,500}{1.0 \times 10} = 16,050 \text{ lb/hr} \tag{17}$$

Converting to gallons per minute (water density = 8.34 lb/gal):

$$Flow_{rack} = \frac{16,050}{8.34 \times 60} = 32.1 \text{ GPM} \approx 32 \text{ GPM per rack} \tag{18}$$

where 8.34 lb/gallon is the density of water at 60°F and the factor of 60 converts hours to minutes; 32 GPM flows comfortably through a standard 1.5-inch Schedule 40 distribution pipe.

Thirty-two GPM flows through a standard 1.5-inch Schedule 40 pipe with no exotic infrastructure. Direct liquid cooling (DLC) cold plates machined for direct GPU die contact extract heat before it enters the rack air space; rear-door heat exchangers (RDHx) intercept warm exhaust air at the rack exit; and full immersion in dielectric fluid maximizes coverage at the cost of increased fluid management complexity. For high-density AI racks, GPU cold plates with supplemental air for ancillary components represent the prevalent deployment architecture. The flow rate from derived as in. Eq (18) shows that flow rate sits comfortably within standard piping limits, shifting the problem from feasibility to implementation

Table 1. Comparative cooling medium properties against the thermal load of a single 47 kW AI rack (ΔT = 10°F for liquid options; 15°F target for air).

Cooling Medium	Specific Heat (BTU/lb·°F)	Density (lb/ft³)	Required Flow, 160,500 BTU/hr
Air	0.24	0.075	~9,800 CFM (physically impractical)
Water, DLC	1.00	62.4	32 GPM
Dielectric Fluid	~0.35	~100	~45 GPM

Fluid Dynamics, Pressure Drop, and Pump Energy

The thermal analysis establishes that 32 GPM is required; the fluid dynamic analysis establishes whether that flow is mechanically achievable and at what energy cost. The Darcy-Weisbach equation governs frictional pressure loss in pipe flow:

$$\Delta P = f \times \frac{L}{D} \times \frac{\rho v^2}{2} \tag{19}$$

The Darcy-Weisbach equation, where f is the dimensionless Darcy friction factor, L is the equivalent pipe length (ft) including fittings, D is the internal pipe diameter (ft), ρ is water density (lb/ft^3), and v is mean flow velocity (ft/s).

where f is the dimensionless Darcy friction factor, L is equivalent pipe length in feet, D is

internal pipe diameter in feet, ρ is fluid density (lb/ft^3), and v is mean flow velocity (ft/s).

Figure 3. Side-by-side 3-D comparison: air cooling path (left) yields ΔT = 297°F equipment failure; liquid cooling path (right) yields ΔT = 10°F at 32 GPM full heat removal. Volumetric heat capacity advantage (×3,467) annotated.

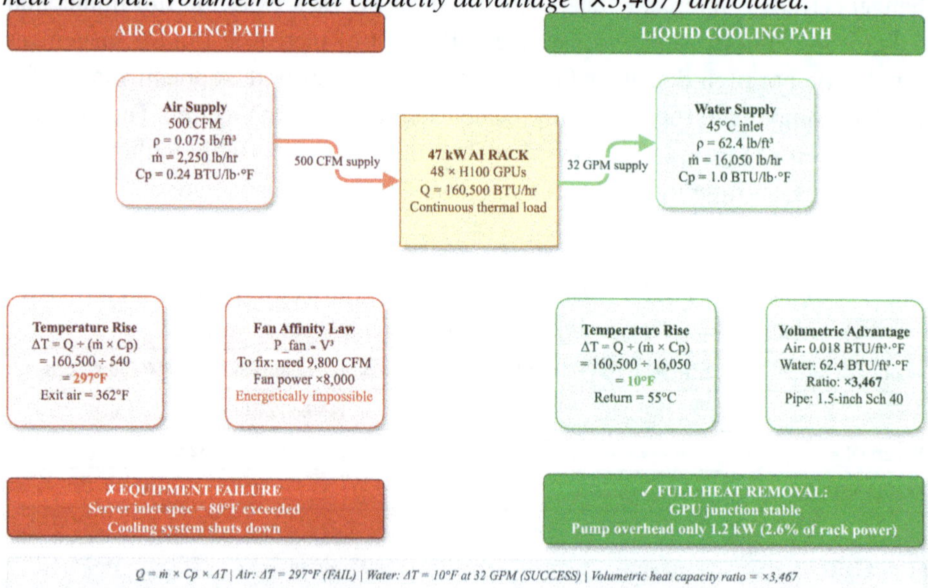

The Reynolds number characterizes flow regime:

$$Re = \frac{\rho \times v \times D}{\mu} \tag{20}$$

The Reynolds number characterizes the flow regime, where μ is the dynamic viscosity of water (6.72×10^{-4} lb/ft·s at 60°F); values above approximately 4,000 indicate turbulent flow.

where μ is dynamic viscosity (6.72×10^{-4} lb/ft·s for water at 60°F). For 32 GPM through 1.5-inch Schedule 40 pipe (D = 0.1342 ft), mean velocity is:

$$v = \frac{Q}{A} = \frac{0.0712\,\text{ft}^3/\text{s}}{\frac{\pi}{4} \times 0.1342^2} \approx 5.0\text{ft/s} \tag{21}$$

where Q is the volumetric flow rate (ft^3/s) converted from 32 GPM, and A is the cross-sectional area of 1.5-inch Schedule 40 pipe with internal diameter D = 0.1342 ft.

$$Re = \frac{62.4 \times 5.0 \times 0.1342}{6.72 \times 10^{-4}} \approx 62,200 \text{ [turbulent]} \tag{22}$$

Turbulent flow (Re > 4,000) increases the convective heat transfer coefficient at cold plate surfaces, a benefit. For a 50-foot equivalent rack-level loop (including cold plate, fittings, and valves), with Moody-chart friction factor f = 0.020:

$$\Delta P_{\text{loop}} = 0.020 \times \frac{50}{0.1342} \times \frac{62.4 \times 5.0^2}{2} \approx 80 \text{ psi} \tag{23}$$

Applying f = 0.020 from the Moody chart for smooth commercial pipe at Re = 62,200 over a 50-foot equivalent loop length representative of a single rack cooling circuit including cold plate, manifold connections, and control valving.

Pump power at centrifugal pump efficiency η = 0.70:

$$P_{\text{pump}} = \frac{\Delta P \times Q}{\eta_{\text{pump}}} \approx 1.2 \text{ kW per rack} \tag{24}$$

where η_pump = 0.70 is centrifugal pump efficiency at design point; the factor 144 converts psi to lb/ft^2, and Q is in ft^3/s.

The 1.2 kW pump overhead is 2.6% of the 47 kW rack load, a modest contribution that keeps overall facility Power Usage Effectiveness (PUE) in the range of 1.10 – 1.15 for well-designed hyperscale deployments (Uptime Institute, 2023). Pressure-independent control valves (PICVs) at each rack branch maintain design flow regardless of position in the distribution loop.

WORKLOAD-COUPLED WATER DEMAND: THE TTIR FRAMEWORK

Time-to-Inference-Ready Model

Flow rate analysis establishes a steady-state cooling requirement; workload analysis connects that rate to volumetric water demand over time. Two racks with identical power draws but different workload durations consume fundamentally different volumes of water per operational cycle. Time-to-inference-ready (TTIR) captures this temporal dimension, defined as the elapsed time from training job submission to deployment of the resulting model for inference:

$$TTIR = t_{\text{ready}} - t_{\text{start}} \tag{25}$$

Throughout this interval, the GPU cluster runs at sustained utilization. Total rack energy consumption:

$$E_{\text{TTIR}} = P_{\text{total}} \times TTIR \tag{26}$$

Since all energy becomes heat removed by the water loop, the total water volume circulated per rack over the TTIR interval is:

$$W_{\text{TTIR}} = \frac{P_{\text{total}} \times TTIR}{C_{\text{p,water}} \times \Delta T \times \rho_{\text{water}}} \tag{27}$$

where ρ_water = 8.34 lb/gallon is the density of water at 60°F, converting mass flow to volumetric gallons over the full TTIR duration.

Substituting Q_rack = 160,500 BTU/hr, C_p = 1.0 BTU/lb·°F, ΔT = 10°F, ρ_water = 8.34 lb/gal:

$$W_{\text{TTIR}} = \frac{160,500 \times TTIR}{1.0 \times 10 \times 8.34} = 1,925 \times TTIR \text{ [gallons per rack]} \tag{28}$$

The linear dependence on time makes workload duration the dominant driver of total water movement

For a 4-hour fine-tuning job:

$$W_{\text{TTIR}}(4 \text{ hr}) = 1,925 \times 4 = 7,700 \text{ gallons per rack} \tag{29}$$

For a 24-hour incremental pre-training checkpoint:

$$W_{TTIR}(24\text{ hr}) = 1{,}925 \times 24 = 46{,}200 \text{ gallons per rack} \tag{30}$$

The TTIR formulation highlights a system level relationship. Water demand does not scale with footprint, it scales with how long and how often workloads run, basically job frequency to be more precise. An organization that doubles its AI development cadence by running twice as many training jobs per month doubles its water demand without adding a single new rack. Water consumption is coupled to the velocity of AI development cycles, a dynamic that conventional capacity planning tools do not capture (Patterson et al., 2021).

The TTIR framework is more general than per-inference water metrics such as those reported by Li et al. (2023), and that generality follows directly from how the governing equations are structured. Equation (28) derives water demand from two quantities only: rack-level power draw and elapsed operating time. Neither quantity is workload-type-specific. A rack running a 72-hour pre-training job, a rack serving continuous real-time inference, and a rack executing a 4-hour fine-tuning pass all enter Equation (28) identically, producing water volumes of 138,600 gallons, 16 million gallons per year, and 7,700 gallons respectively, as Table 2 and Equation (36) show.

Per-inference metrics, by contrast, require knowledge of per-inference energy consumption and inference request volume, both of which are difficult to isolate at facility scale and vary with model size, batch configuration, and hardware generation. TTIR requires only power draw and operating duration, which are directly readable from facility power meters and job scheduler logs. The framework therefore provides a single unified instrument for water accounting across the full lifecycle of an AI system, from initial training through continuous production inference, without requiring workload decomposition.

Training Workload Water Accounting

Pre-training a frontier language model runs continuously across thousands of GPU racks for weeks to months. For a single rack over a 72-hour training window:

$$\begin{aligned} E_{train} = P_{total} \times T_{train} &= 47{,}040 \times 72 = 3{,}386{,}880 \text{ Wh} \\ &= 3{,}387 \text{ kWh} \end{aligned} \tag{31}$$

where T_train = 72 hours is the representative training window; this is the minimum energy commitment for a standard production training checkpoint.

Figure 4. TTIR workload-water coupling: timeline from t_start to t_ready with continuous rack operation, volumetric water scaling by duration (W = 1,925 × TTIR gallons), six example scenarios from fine-tuning to annual sustained inference, and development-cadence multiplier effect.

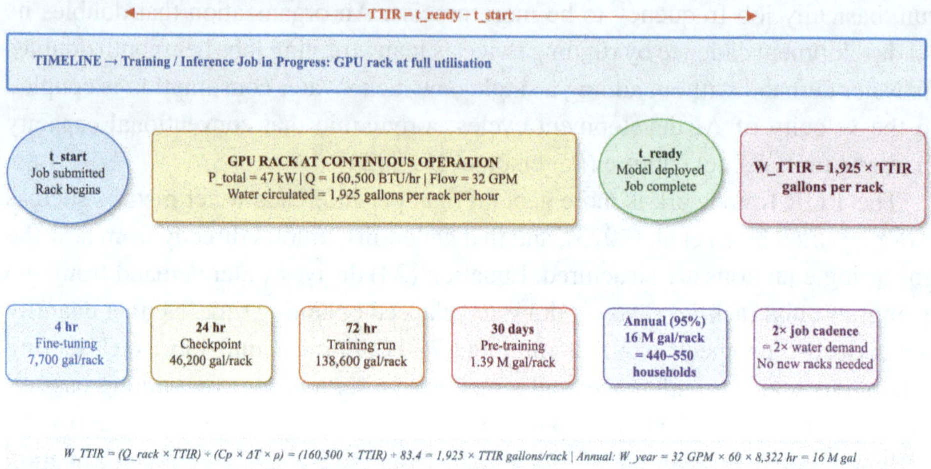

$$Q_{\text{train}} = 3{,}387 \text{ kWh} \times 3{,}412 \text{ BTU/kWh} = 11{,}557{,}644 \text{ BTU} \tag{32}$$

Converting from kWh to BTU using the exact thermal equivalence from Equation (7) to maintain unit consistency with the cooling flow analysis.

$$W_{\text{train}}(72 \text{ hr}) = 1{,}925 \times 72 = 138{,}600 \text{ gallons per rack} \tag{33}$$

The consumptive fraction, water genuinely lost to atmosphere through evaporative cooling towers rather than returned to the distribution system, is typically 1–3% of circulated volume. For the 72-hour training run, actual consumptive loss per rack is approximately 1,400–4,200 gallons, a figure two orders of magnitude less than the gross circulation total that sometimes appears in environmental reporting. Table 2 summarizes water demand across representative training durations.

Table 2. Water circulation and consumptive loss per AI rack across training durations (P_total = 47 kW, ΔT = 10°F, 2% evaporative loss).

Training Duration	Energy per Rack (kWh)	Water Circulated (gallons)	Consumptive Loss at 2% (gallons)
4 hr, fine-tuning	188	7,700	154
24 hr, checkpoint	1,129	46,200	924
72 hr, standard run	3,387	138,600	2,772
30 days, pre-training	33,869	1,385,600	27,712
90 days, frontier model	101,606	4,156,800	83,136

Annual Water Flow Per Rack

For facility planning and water rights assessment, annual water flow per rack at sustained operation is the relevant unit. An AI inference rack serving production traffic operates at approximately 95% annual utilization:

$$H_{\text{year}} = 8,760 \times 0.95 = 8,322 \text{ hours per year} \tag{34}$$

Assuming 95% annual utilization to account for planned maintenance windows and hardware replacement cycles; inference racks typically achieve this utilization level in production AI deployments.

$$W_{\text{year}} = Flow_{\text{rack}} \times 60 \times H_{\text{year}} \tag{35}$$

where Flow_rack = 32 GPM (from Equation 18) and the factor of 60 converts GPM to gallons per hour; H_year is annual operating hours from Equation (34).

$$W_{\text{year}} = 32 \times 60 \times 8,322 = 15,978,240 \text{ gallons/yr} \approx 16 \text{ million gallons per rack per year} \tag{36}$$

Sixteen million gallons per rack per year in cooling water circulation, equivalent to the annual domestic water use of approximately 440 – 550 American households, based on U.S. Geological Survey figures for average household consumption of roughly 80 – 100 gallons per day (Leaney et al., 2022). This is working fluid executing thermodynamic transport, not water being consumed, but the infrastructure required to handle these flow rates is real, physical, and subject to engineering, permitting, and maintenance requirements that must be planned for.

HYPERSCALE CAMPUS WATER INFRASTRUCTURE

300 MW and 1 GW Campus Flow Calculations

Hyperscale AI campus deployments, including the Stargate project and comparable initiatives by major cloud operators, specify facilities in the hundreds-of-megawatts to gigawatt range (International Energy Agency, 2023). Per-rack water analysis scales linearly to campus totals. Number of AI racks at facility power capacity P_facility:

$$N_{racks} = \frac{P_{facility}}{P_{total}} \tag{37}$$

where P_facility is total installed power capacity of the campus (W) and P_total = 47,040 W is per-rack power from Equation (4).

For a 300 MW campus:

$$N_{racks}(300\text{ MW}) = \frac{300,000,000}{47,040} \approx 6,378\text{ racks} \tag{38}$$

$$Flow_{total}(300\text{ MW}) = 6,378 \times 32 = 204,096\text{ GPM} \approx 204,000\text{ GPM} \tag{39}$$

For a 1 GW hyperscale campus:

$$N_{racks}(1\text{ GW}) = \frac{1,000,000,000}{47,040} \approx 21,260\text{ racks} \tag{40}$$

$$Flow_{total}(1\text{ GW}) = 21,260 \times 32 = 680,320\text{ GPM} \approx 680,000\text{ GPM} \tag{41}$$

The Water Usage Effectiveness (WUE) metric, defined as liters of water consumed per kWh of IT load, captures actual consumptive loss (Leaney et al., 2022). Leading hyperscale operators achieve WUE values of 0.2–0.5 L/kWh. For a 300 MW campus at WUE = 0.3:

$$W_{consumed} = WUE \times P_{IT} \times H_{year} \times k_{conv} \tag{42}$$

where WUE is the Water Usage Effectiveness in L/kWh, P_IT is IT equipment power (kW), H_year is annual operating hours, and k_conv = 0.000264 gal/L converts liters to gallons.

$$W_{consumed}$$

$$= 0.3 \times 300,000 \times 8,322 \times 0.000264 \approx 197\text{ million gallons/yr} \tag{43}$$

Table 3. Water circulation and consumptive loss scaling from single rack to 1 GW hyperscale campus (32 GPM/rack, 95% utilization, WUE = 0.3 L/kWh for campus rows).

Scale	Racks	Total Cooling Flow (GPM)	Annual Circulation (billion gal)	Annual Consumption, WUE 0.3 (million gal)
Single rack	1	32	0.000016	-
Row, 20 racks	20	640	0.00032	-
Pod, 100 racks	100	3,200	0.0016	-
Campus, 300 MW	6,378	204,000	1.01	197
Hyperscale, 1 GW	21,260	680,000	3.38	656

Figure 5. Hyperscale 3-D scaling cascade: rack (32 GPM) row pod 300 MW campus (204,000 GPM, 197 M gal/yr consumed) 1 GW hyperscale (680,000 GPM, 656 M gal/yr). WUE, regional governance, and heat recovery panels annotated below cascade.

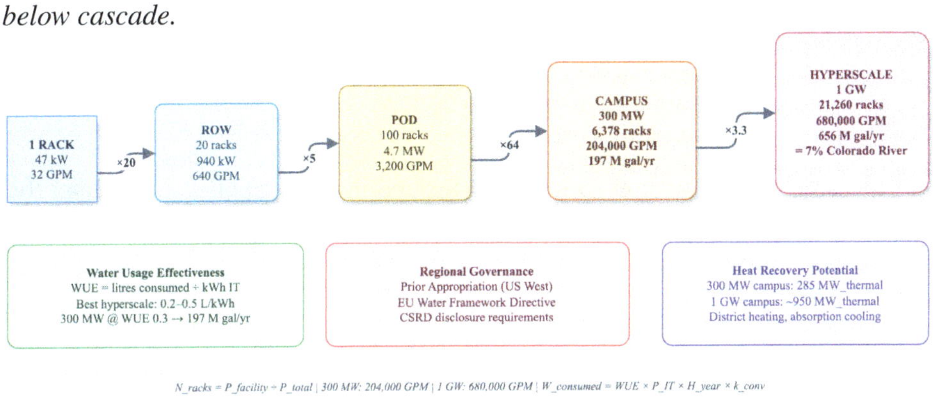

Water Infrastructure and Regional Governance

At hyperscale cooling flow rates, data center water demand intersects directly with regional hydrology, agricultural water rights, municipal supply systems, and environmental regulation. In the western United States, the prior appropriation doctrine assigns water rights by historical use priority, meaning new large-scale industrial users must acquire rights before drawing on surface or groundwater (Shehabi et al., 2018). European deployments face the EU Water Framework Directive and national permitting requirements that assess withdrawal impacts on aquifer health.

Engineering responses to water scarcity pressure include zero-liquid-discharge (ZLD) systems that recover make-up water through mechanical vapor recompression, reclaimed municipal effluent as the primary cooling tower make-up source, and adiabatic dry coolers that eliminate evaporative water loss in cooler climates. The European Green Deal and the EU Corporate Sustainability Reporting Directive now require large data center operators to disclose water consumption metrics alongside energy, a trajectory pointing toward mandatory reduction targets (European Commission, 2020). Water is transitioning from a utility input to a governed resource allocation decision.

HEAT RECOVERY AND CIRCULAR THERMAL ENERGY SYSTEMS

The thermal energy carried in the cooling water loop is not waste; it is hot water at 40–60°C. The recoverable thermal power from the secondary cooling loop of a 300 MW campus:

$$Q_{\text{recovered}} = \dot{m}_{\text{water}} \times C_{\text{p,water}} \times \Delta T \tag{44}$$

The same heat transfer equation used for cooling now applied in reverse: the thermal power available for recovery equals the product of water mass flow rate, specific heat, and temperature differential across the heat exchanger secondary loop.

$$Q_{\text{recovered}}(300\text{ MW}) = 300\text{ MW} \times 0.95 = 285\text{ MW}_{\text{thermal}} \tag{45}$$

The factor 0.95 reflects that approximately 5% of facility electrical input is consumed by lighting, building management systems, and other non-IT overhead that does not enter the cooling water loop.

District heating networks offer the most mature recovery pathway. Stockholm's Fortum Varme network receives waste heat from server installations at 50–65°C, distributing it to residential and commercial buildings and displacing natural gas combustion. Analogous schemes operate in Helsinki, Amsterdam, and Dublin (Uptime Institute, 2023). Absorption chillers driven by waste heat at 80–90°C can reduce mechanical chiller electrical consumption by 20–35%. Agricultural co-location, greenhouses and aquaculture operations using waste heat for temperature maintenance, represents smaller-scale but genuine circular use.

The economic case for heat recovery improves as energy prices rise and as carbon pricing mechanisms assign a cost to atmospheric heat rejection. Treating the cooling water loop as a thermal utility rather than a waste stream changes the infrastructure investment calculus: heat exchangers, district heating connections, and absorption cooling equipment become assets with positive returns rather than optional green gestures.

It is important to situate direct liquid cooling within a broader cooling technology landscape that continues to evolve. Single-phase and two-phase full immersion cooling, in which server hardware is submerged in dielectric fluid rather than cooled through contact with water-bearing cold plates, eliminates the cooling tower make-up water requirement entirely because the dielectric fluid loop is hermetically closed and does not undergo evaporative loss. The thermodynamic basis for this is straightforward: evaporative water consumption in conventional data centers arises at the cooling tower, where water is deliberately vaporized to reject heat to atmosphere. Immersion systems reject heat through a secondary closed water loop to a dry cooler or heat exchanger without exposing process fluid to the atmosphere, reducing the WUE consumptive component toward zero under favorable ambient conditions. Haghshenas et al. (2023) quantified the energy efficiency advantage of immersion cooling through direct experimental comparison, reporting PUE values of 1.02 to 1.04 for immersion systems representing 10 to 50 percent less total energy consumption than equivalent air-cooled deployments, an advantage that also reduces the indirect water footprint associated with electricity generation. Rear-door heat exchangers represent a transitional architecture that can be retrofitted to existing air-cooled racks and recover a meaningful fraction of thermal output without full cold-plate replacement. Advanced high-density airflow optimization using computational fluid dynamics-guided containment can push effective air-cooled rack density to 25 to 30 kW in new build deployments, though this remains substantially below the 47 kW baseline this chapter analyses and does not eliminate the thermodynamic constraints equations (9) through (14) demonstrate. The selection among these architectures involves capital cost, fluid management complexity, hardware compatibility, and site water availability, all of which vary by deployment context. The thermodynamic analysis this chapter presents establishes the physical basis for that selection decision rather than prescribing a single outcome.

Closed-Loop Versus Open-Loop Secondary Cooling: The Consumption Distinction

The water analysis in preceding sections draws a distinction between water circulated and water consumed, and that distinction is determined entirely by whether the secondary cooling loop at the facility level is open or closed. At the rack level, direct liquid cooling is always closed: the same water circulates continuously through the cold plates and returns to the distribution manifold without contacting the atmosphere. Consumptive loss arises only in the secondary loop, specifically at the cooling tower, where water is deliberately evaporated to reject heat to atmosphere. A facility that replaces its evaporative cooling tower with a dry cooler or adiabatic pre-cooler operates a closed secondary loop, and its WUE approaches zero. Water still circulates at 32 GPM per rack from Equation (18), but none is consumed. The 197 million gallons per year from Equation (43) is therefore not a fixed cost of liquid cooling but a consequence of the secondary loop architecture choice. In water-stressed regions, closing the secondary loop is the single engineering decision that decouples AI compute scaling from freshwater consumption entirely, while preserving all the thermal benefits of direct liquid cooling that Equations (15) through (24) establish.

Economic Derivation: Water Cost Per Training Run and Per Inference Request

The volumetric water demand from the TTIR framework translates directly into economic cost through a regional water tariff. Industrial water pricing in the United States averages approximately \$0.004 per gallon for large-volume users, ranging from \$0.001 per gallon in water-abundant regions to \$0.012 per gallon or higher in water-stressed markets where scarcity premiums apply (Leaney et al., 2022). The water cost of a single training run on one rack is:

$$C_{\text{train}} = W_{\text{TTIR}} \times c_{\text{w}} \tag{46}$$

where C_train is the water cost of the training run (USD), W_TTIR is the circulated volume from Equation (28) in gallons, and c_w is the local industrial water tariff (USD/gallon). For a representative 72-hour training run at the US average tariff:

$$C_{\text{train}}(72\text{ hr}) = 138{,}600\text{ gal} \times \$0.004/\text{gal} = \$554\text{ per rack} \tag{47}$$

At a typical frontier training cluster of 1,000 racks running the same job simultaneously, the water cost of a single training run reaches \$554,000, a figure currently invisible in most AI infrastructure cost models, which account for electricity, hardware amortization, and networking but rarely for water as a direct input cost. The annual water cost per rack at sustained inference utilization follows from Equation (36):

$$C_{\text{rack,year}} = W_{\text{year}} \times c_{\text{w}} = 16{,}000{,}000\text{ gal} \times c_{\text{w}} \tag{48}$$

At the US average tariff, C_rack_year = \$64,000 per rack per year. A 300 MW campus of 6,378 racks carries an annual water cost of approximately \$408 million at average US industrial tariffs, rising above \$1.2 billion in high-stress water markets. These figures exclude water rights acquisition costs, regulatory compliance costs, and the scarcity premiums that markets in the American Southwest are already beginning to impose. A further useful unit connects training water cost to the productive output of the trained model: the Water Cost per Inference Request (WCIR):

$$WCIR = \frac{C_{\text{train}}}{N_{\text{inf}}} \tag{49}$$

where N_inf is the total number of inferences requests the trained model serves over its deployed lifetime before the next retraining cycle. For a model serving 10 million requests, trained on a 1,000-rack cluster over 72 hours, WCIR = \$554,000 / 10,000,000 = \$0.055 per 1,000 inference requests. As development cadence accelerates and models are retrained more frequently, WCIR rises proportionally, creating a direct economic incentive to extend model deployment lifetimes and reduce unnecessary retraining cycles, a dimension of AI operational efficiency that is distinct from compute cost optimization but coupled to it through the TTIR framework.

Environmental Impact Metric: Water Stress-Weighted Consumption

Raw water consumption volume is an inadequate measure of environmental impact because the same volume withdrawn in a water-abundant region has fundamentally different ecological consequences from the same withdrawal in a water-stressed basin. The withdrawal-to-availability ratio (WTA), defined as annual water withdrawal divided by total renewable freshwater availability in a given watershed, provides a regionally calibrated multiplier that converts raw consumption into an environmental impact score (Shehabi et al., 2018). A WTA below 0.1 indicates low stress; 0.1 to 0.4 moderate; 0.4 to 0.8 high; above 0.8 severe. The water-stress-weighted environmental impact of a data center deployment is:

$$EI = W_{\text{consumed}} \times WTA_{\text{region}} \tag{50}$$

where EI is the environmental impact score in stress-adjusted gallon-equivalents, W_consumed is annual consumptive water loss from Equation (43), and WTA_region is the dimensionless withdrawal-to-availability ratio for the deployment watershed. For the 300 MW campus consuming 197 million gallons per year:

$$EI\left(\text{Arizona},\ WTA_{\text{region}} = 0.80\right) = 197 \times 10^{6} \times 0.80$$
$$= 157.6 \times 10^{6} \quad (51)$$

$$EI\left(\text{Norway},\ WTA_{\text{region}} = 0.05\right) = 197 \times 10^{6} \times 0.05 = 9.9 \times 10^{6} \quad (52)$$

Table 4 applies Equation (50) to ten of the world's most significant AI data center locations, using WTA values derived from the WRI Aqueduct 4.0 Water Risk Atlas (Kuzma et al., 2023), which provides peer-reviewed, sub-catchment-level baseline water stress scores on a 0 to 5 scale. The EI score in the final column is calculated as W_consumed × WTA for a 300 MW campus consuming 197 million gallons per year, expressed in millions of stress-adjusted gallon-equivalents.

Table 4. Water Stress Context in major AI Data Centre locations: WRI Aqueduct 4.0 Baseline Water Stress Score, WTA Ratio and Environmental Impact Score for a 300 MW Campus. WTA = withdrawal-to-availability ratio; EI = W_consumed × WTA_region (Equation 50). Sources: WRI Aqueduct 4.0 (Kuzma et al., 2023); FAO AQUASTAT SDG Indicator 6.4.2 (FAO, 2024); UN-Water Progress Report on Water Stress (UN-Water, 2024); USGS National Water Information System (U.S. Geological Survey, 2023); Central Water Commission, India (Central Water Commission, 2023); Public Utilities Board Singapore (PUB, 2023); European Environment Agency Water Stress Report (EEA, 2023). All WTA values are basin-level approximations; sub-catchment stress may differ from country averages. EEA, 2023). All WTA values are basin-level approximations; sub-catchment stress may differ from country averages.

Location	Primary DC Hub	WTA (approx.)	Aqueduct Score (0–5)	Stress Category	EI Score (M gal-equiv/yr)*
Arizona, USA	Phoenix Metro	~0.80	4.5 / 5	Extremely High	157.6
Texas, USA	San Antonio / DFW	~0.45	3.0 / 5	High	88.7
Northern Virginia, USA	Ashburn / Loudoun	~0.12	1.5 / 5	Low–Medium	23.6
Pacific NW, USA	Portland / Seattle	~0.05	1.0 / 5	Low	9.9
Mumbai, India	Navi Mumbai	~0.72	4.0 / 5	High–Extreme	141.8
Singapore	Jurong / Tuas	~0.85	4.5 / 5	Extremely High	167.5

continued on following page

Table 4. Continued

Location	Primary DC Hub	WTA (approx.)	Aqueduct Score (0–5)	Stress Category	EI Score (M gal-equiv/yr)*
Netherlands	Amsterdam AMS-IX	~0.35	2.5 / 5	Medium–High	69.0
Ireland	Dublin	~0.08	1.0 / 5	Low	15.8
Sweden	Stockholm	~0.03	0.5 / 5	Very Low	5.9
Norway	Oslo	~0.05	0.5 / 5	Very Low	9.9

EI score calculated as W_consumed × WTA_region for a 300 MW campus (W_consumed = 197 million gallons/yr from Equation 43). WTA values are basin-level approximations from WRI Aqueduct 4.0 (Kuzma et al., 2023). Singapore manages extremely high baseline stress through desalination and water recycling infrastructure, which reduces operational risk but does not change the physical stress indicator. Northern Virginia (Loudoun County) WTA reflects the Potomac basin; sub-county aquifer drawdown risk is higher than the basin average suggests.

The identical 300 MW campus carries a water environmental impact score 16 times higher when sited in Arizona than in Norway, despite identical power draw, cooling architecture, and absolute water consumption volume. Equation (50) provides a quantitative instrument for site selection that incorporates regional hydrological carrying capacity alongside power availability. It also gives regulators a location-adjusted threshold tool: facilities in high-WTA watersheds would face tighter WUE requirements or mandatory closed secondary loops to bring EI below a regional limit, while facilities in low-WTA regions qualify for more permissive treatment. Multiplying EI by the local water tariff yields a Water Environmental Cost that can be integrated into AI lifecycle assessments and carbon-equivalent reporting frameworks.

Heat Recovery Economics: Breakeven Analysis and District Heating Quantification

The thermal energy in the cooling water loop of a 300 MW campus, 285 MW_thermal from Equation (45), carries direct market value in any district heating network. European district heating networks price delivered thermal energy in the range of €20 to €40 per MWh_thermal depending on network density and local gas price benchmarks. The Stockholm Fortum Värme network, Helsinki Energy, and analogous schemes in Amsterdam and Dublin operate within this range, with prices shifting toward the upper end following the energy price volatility of 2022 to 2024. The annual thermal energy revenue from connecting a 300 MW campus to a district heating offtake agreement is:

$$R_{\text{heat}} = Q_{\text{recovered}} \times H_{\text{year}} \times p_{\text{heat}} \tag{53}$$

where R_heat is annual heat revenue (USD), Q_recovered = 285 MW_thermal from Equation (45), H_year = 8,322 operating hours per year from Equation (34), and p_heat is the district heating offtake price per MWh_thermal. At a conservative p_heat = \$25/MWh_th, consistent with long-term northern European offtake agreements:

$$R_{\text{heat}} = 285\text{ MW} \times 8,322\text{ hr} \times \$25/\text{MWh} = \$59.3\text{ million per year} \tag{54}$$

The capital cost of heat exchanger infrastructure connecting the facility secondary loop to a district heating network typically ranges from \$5 million to \$15 million for a 300 MW campus, depending on distance to the nearest network injection point and required transfer capacity. The payback period on this investment is:

$$T_{\text{payback}} = \frac{CAPEX_{\text{HX}}}{R_{\text{heat}}} \tag{55}$$

At CAPEX_HX = \$10 million, T_payback = \$10M / \$59.3M = 0.17 years, approximately two months. At a conservative CAPEX of \$50 million for a new distribution main to a remote network, payback remains under one year. At 1 GW scale, Q_recovered reaches approximately 950 MW_thermal and R_heat rises to approximately \$198 million per year, sufficient to fund a dedicated urban district heating network serving 200,000 to 400,000 residents. The water-borne thermal energy loop therefore closes a complete economic circuit. Water enters the facility as a thermal transport medium, absorbs compute heat at the cold plates, delivers that heat to urban infrastructure that would otherwise combust natural gas to produce it, and returns as cooled water to restart the cycle. In a closed secondary loop configuration with active heat recovery, the net consumptive water loss of the system approaches zero while generating measurable revenue and displacing carbon emissions from the heating sector. This is the outcome that reframes water from a liability in AI infrastructure planning to an asset: not water consumed by computation, but water working for computation, for communities, and for the climate simultaneously.

FUTURE RESEARCH DIRECTIONS

Several areas warrant further investigation as AI rack densities continue increasing beyond the 47 kW baseline analyzed here. GPU accelerator roadmaps from major manufacturers indicate TDP values approaching 1,000–1,200 W per chip within the next two to three product generations, which would push rack power toward 70–90 kW and require revisiting both cold plate thermal interface design and distribution loop hydraulics.

Figure 6. Circular heat recovery loop: GPU rack heat source cold plate heat exchanger district heating / absorption cooling / industrial reuse (recovery path) or cooling tower (rejection path) chiller plant return to rack.

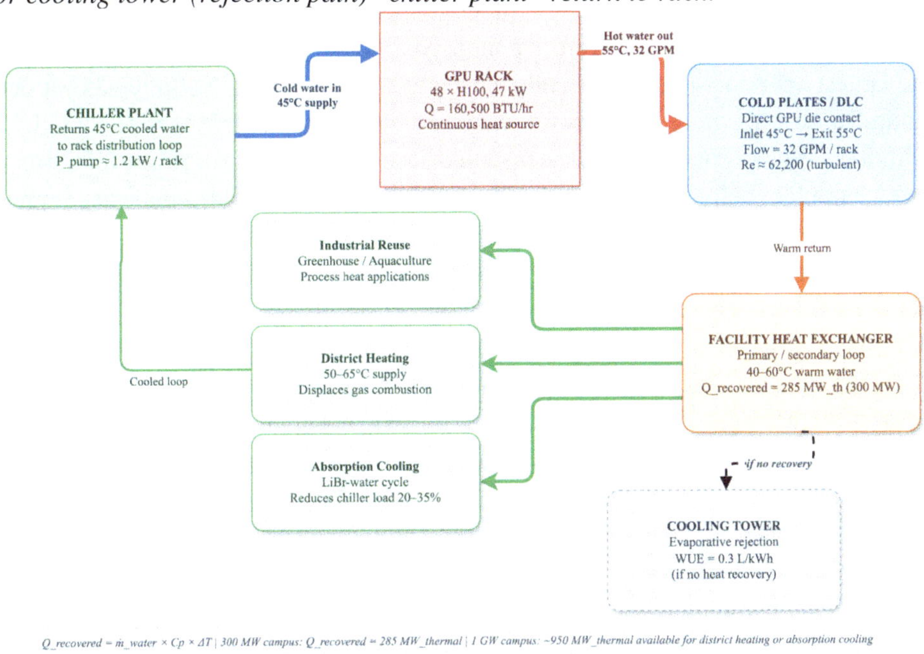

The TTIR framework introduced here treats training duration as a fixed input. Future work could model TTIR as a probabilistic function of model architecture, dataset size, and cluster configuration, yielding a stochastic water demand model that supports more rigorous infrastructure sizing under uncertainty. Integration of this model with WUE forecasting under regional climate scenarios, accounting for the projected drying of the American Southwest and similar water-stressed regions, would provide site selection tools that are currently absent from the literature.

Heat recovery efficiency from DLC warm water streams below 60°C presents an open engineering challenge. The temperature gap between available waste heat and the 80 – 90°C absorption chiller input requirement needs bridging through heat cascade or thermal storage designs that have not been optimized for the specific operating profile of AI training clusters. Research into thermally matched co-location facility designs, pairing AI campuses with industrial users whose process heat requirements align with available waste temperatures, could demonstrate economically viable circular energy systems at scale (Patterson et al., 2021).

How long the industry will remain dependent on conventional direct liquid cooling deserves serious attention as a research question. Swapping out the cooling plant of an operating hyperscale campus is not a scheduling exercise. Capital must be committed years in advance, hardware must be validated for compatibility, supply chains for immersion tanks and dielectric fluids are nowhere near the maturity of chilled-water distribution, and the campus cannot go dark while the work happens. Fleet-wide replacement is therefore a multi-year undertaking at minimum, and for the largest operators it is realistically a decade-scale program. Nobody has yet put a number on what that lag means for water resources: how many years of continued withdrawal at current rates does the installed base of direct liquid cooling infrastructure guarantee, before immersion or dry-cooled alternatives hold majority share? Answering that question would give regulators a time-bounded impact window rather than an open-ended liability, and it would give water rights administrators in data-center-heavy jurisdictions something they currently lack, a defensible planning horizon.

A separate and arguably more promising line of attack goes further upstream. Every watt that enters a GPU rack and does not exit as a photon, a signal, or mechanical work exits as heat. Silicon substrates conduct that heat poorly, and present GPU die designs push heat flux densities to levels where the substrate itself becomes the bottleneck. Materials researchers have been chipping away at this for years. Wang et al. (2025) replaced conventional silicon interposers in 3D-stacked GPU–HBM packages with hexagonal boron nitride, reporting in MRS Communications a 20°C drop in hotspot temperature and a 22 percent reduction in power leakage, with no changes to device logic, the substrate did the work. Kang et al. (2021) went further down the stack, integrating boron arsenide directly beneath gallium nitride high-electron-mobility transistors and demonstrating in Nature Electronics that boron arsenide outperforms both diamond and silicon carbide as a heat spreader, which had been the benchmark materials for thermal management in high-power devices. The implication is not subtle: substrate material choice is a design variable with first-order consequences for how much heat a chip produces per unit of useful work. Deeper still, 2D materials, graphene, transition metal dichalcogenides, hexagonal boron nitride, are being built into neuromorphic device architectures whose operating energy budgets sit orders of magnitude below anything achievable with CMOS (Wu et al., 2023). Should neuromorphic hardware reach production scale for AI inference, the equation W_TTIR = 1,925 × TTIR holds its form, but P_total in Equation (3) shrinks by orders of magnitude, and the cooling water demand and environmental impact scores of Equations (50) through (52) collapse with it. Tracking the TTIR and WUE profiles of these emerging architectures from laboratory demonstration through to production deployment would give planners a water-demand trajectory for the handover period between today's GPU-dense installations and whatever succeeds them.

On the software side, the TTIR framework introduced here turns every algorithmic efficiency gain into a water number. Cut TTIR and W_TTIR falls by the same proportion through Equation (28). Cut retraining frequency and C_rack_year from Equation (48) falls with it. Knowledge distillation compresses a frontier model into a student that runs a fraction of the parameters while holding its ground on downstream accuracy benchmarks (Springer, 2025). Less compute for training means shorter TTIR; less compute for inference means more requests served per unit of cooling water. Continual and incremental learning methods attack the problem from a different angle: instead of retraining from random weights each time the data distribution shifts, they update only the parameters the new distribution touches, while the rest of the model's knowledge stays put (Wang et al., 2024). Plugged into the TTIR model, a 4-hour incremental update uses 7,700 gallons per rack against the 138,600 a 72-hour full retraining run demands, 94 percent less water, delivered through a compiler and a training script rather than a cooling tower replacement. Architecture is beginning to matter here too. Models that allocate reasoning compute to task difficulty rather than burning tokens uniformly show up to 40 percent compute reductions while holding benchmark scores, and measured energy-per-token varies substantially across LLM architectures tackling identical tasks (EuroMLSys, 2025). Expressing these gains in TTIR-coupled water terms, building a Water Cost per Inference Request as a function of architecture family, training regime, and refresh cadence, would hand development teams a concrete dial for trading model performance against environmental footprint before a cluster is ever provisioned, not after it has been running for two years.

CONCLUSION

This chapter, from the derived values, has demonstrated through first-principles thermodynamic analysis that water-based liquid cooling is a structural requirement of modern AI data centers, not an optional enhancement. The 47kW continuous thermal load produced by a standard 48 GPU H100 rack, 160,500 BTU/hr, exceeds the physical capacity of air cooling by margins that no amount of engineering optimization can close. The mass-flow heat transfer equations give an air temperature rise of 297°F at practical airflow rates; the fan affinity law makes adequate airflow energetically catastrophic. Water, at 32 GPM per rack, solves the problem that physics imposes.

The TTIR-coupled water model connects this per-rack flow rate to actual volumetric demand: approximately 1,925 gallons per rack per hour of operation, rising to 138,600 gallons per rack for a 72-hour training run and 16 million gallons per rack annually at sustained operation. At 300 MW campus scale, cooling water flows at 204,000 GPM with 197 million gallons per year of genuine consumptive loss. At 1 GW, those figures reach 680,000 GPM and 656 million gallons per year. These numbers establish water as a co-equal infrastructure resource alongside electrical power in AI campus planning.

The governing relationship this analysis establishes is: AI Scale = f(Power, Time, Water). Any site selection, permitting, or infrastructure investment process that addresses power without addressing water is working from an incomplete model. The thermal energy captured in cooling water loops, 285 MW_thermal available at a 300 MW campus, is recoverable, and treating it as such transforms the water cooling system from a cost center into a thermal utility that can serve district heating, absorption cooling, and industrial heat applications. Managing water intelligently, governing its use responsibly, and recovering the thermal value it carries are the engineering and policy tasks that the AI infrastructure buildout now requires.

REFERENCES

American Society of Heating, Refrigerating and Air-Conditioning Engineers. (2021). *Thermal guidelines for data processing environments* (5th ed.). ASHRAE.

Borkowski, M., & Piłat, A. K. (2022). Customized data center cooling system operating at significant outdoor temperature fluctuations. *Applied Energy, 306*, 117975. DOI: 10.1016/j.apenergy.2021.117975

Brown, R., & Patel, C. (2023). Liquid cooling for sustainable data centers. *IEEE Transactions on Industry Applications, 59*(3), 2154–2163. DOI: 10.1109/TIA.2022.3209874

Cengel, Y. A., & Cimbala, J. M. (2018). *Fluid mechanics: Fundamentals and applications* (4th ed.). McGraw-Hill Education.

Central Water Commission. (2023). *National water availability assessment report 2023*. Ministry of Jal Shakti, Government of India. https://cwc.gov.in/national-water-availability-report

European Commission. (2020). *A European strategy for data*(COM/2020/66 final). European Commission. https://eur-lex.europa.eu/legal-content/EN/TXT/?uri=CELEX:52020DC0066

European Environment Agency. (2023). *Water resources across Europe: Confronting water stress*. EEA Report No. 2/2023. European Environment Agency. https://doi.org/DOI: 10.2800/000124

Food and Agriculture Organization of the United Nations. (2024). *Progress on level of water stress: 2024 update, SDG indicator 6.4.2*. FAO on behalf of UN-Water. DOI: 10.4060/cc8962en

Haghshenas, K., Setz, B., Blosch, Y., & Aiello, M. (2023). Enough hot air: The role of immersion cooling. *Energy Informatics, 6*(1), 14. DOI: 10.1186/s42162-023-00269-0

International Energy Agency. (2023). *Data centres and data transmission networks*. IEA. https://www.iea.org/energy-system/buildings/data-centres-and-data-transmission-networks

Kang, J. S., Li, M., Wu, H., Nguyen, H., Aoki, T., & Hu, Y. (2021). Integration of boron arsenide cooling substrates into gallium nitride devices. *Nature Electronics, 4*(6), 416–423. DOI: 10.1038/s41928-021-00595-9

Kuzma, S., Bierkens, M. F. P., Lakshman, S., Luo, T., Saccoccia, L., Sutanudjaja, E. H., & Van Beek, R. (2023). *Aqueduct 4.0: Updated decision-relevant global water risk indicators*. World Resources Institute., DOI: 10.46830/writn.23.00061

Leaney, J., Bilton, M., & Sheridan, C. (2022). *Water usage effectiveness in data centers: Benchmarking and best practice*(White Paper 69). The Green Grid. https://www.thegreengrid.org

Li,P., Yang,J., Islam,M. A., & Ren, S. (2023). Making AI less “thirsty” Uncovering and addressing the secret water footprint of AI models. arXiv preprint arXiv:2304.03271. https://arxiv.org/abs/2304.03271

Lu, T., Lü, X., Välisuo, P., Zhang, Q., & Clements-Croome, D. (2024). Innovative approaches for deep decarbonization of data centers and building space heating networks: Modelling and comparison of novel waste heat recovery systems for liquid cooling systems. *Applied Energy*, *357*, 122499. DOI: 10.1016/j.apenergy.2023.122473

Masanet, E., Shehabi, A., Lei, N., Smith, S., & Koomey, J. (2020). Recalibrating global data center energy-use estimates. *Science*, *367*(6481), 984–986. DOI: 10.1126/science.aba3758 PMID: 32108103

NVIDIA Corporation. (2024). *NVIDIA H100 Tensor Core GPU architecture: Technical overview*. NVIDIA Developer Documentation. https://developer.nvidia.com/blog/nvidia-hopper-architecture-in-depth/

Patterson, D., Gonzalez, J., Le, Q., Liang, C., Munguia, L.-M., Rothchild, D., So, D., Texier, M., & Dean, J. (2021). Carbon considerations for large language model training. *arXiv preprint arXiv:2104.10350*. https://arxiv.org/abs/2104.10350

Public Utilities Board. (2023). *Singapore's water story: Sustaining our water supply*. PUB, Singapore's National Water Agency. https://www.pub.gov.sg/watersupply/singaporeswaterstory

Shehabi, A., Smith, S. J., Masanet, E., & Koomey, J. (2018). Data center growth in the United States: Decoupling the demand for services from electricity use. *Environmental Research Letters*, *13*(12), 124030. DOI: 10.1088/1748-9326/aaec9c

Springer, A., & Hammerla, N. (2025). Knowledge distillation and dataset distillation of large language models: Emerging trends, challenges, and future directions. *Artificial Intelligence Review*, *58*(4), 1–42. DOI: 10.1007/s10462-025-11423-3

Sunkara, K. C., & Narukulla, K. (2025). Power consumption and heat dissipation in AI data centers: A quantitative thermodynamic framework. *International Journal of Innovative Research in Science, Engineering and Technology*, *14*(3), 1–12. DOI: 10.15680/IJIRSET.2025.1403001

UN-Water. (2024). *Progress on level of water stress: Mid-term status of SDG indicator 6.4.2 and acceleration needs*. United Nations Water. https://www.unwater.org/publications/progress-level-water-stress-2024-update

Uptime Institute. (2023). *Global data center survey results 2023*. Uptime Institute LLC. https://uptimeinstitute.com/resources/research-and-reports

U.S. Energy Information Administration. (2024). *Annual energy outlook 2024 with projections to 2050*. U.S. Department of Energy. https://www.eia.gov/outlooks/aeo/

U.S. Geological Survey. (2023). *National water information system: Data from web interface*. U.S. Department of the Interior. DOI: 10.5066/F7P55KJN

Wang, E. H., Yan, W., & Huang, R. (2025). Thermal analysis of 3D GPU-memory architectures with boron nitride interposer. *MRS Communications*, *15*, 1–8. DOI: 10.1557/s43579-025-00913-8

Wang, Z., Zhang, Z., Lee, C.-Y., Zhang, H., Sun, R., Ren, X., Su, G., Perot, V., Dy, J., & Karypis, G. (2024). A comprehensive survey of continual learning: Theory, method, and application. *IEEE Transactions on Pattern Analysis and Machine Intelligence*, *46*(8), 5362–5383. DOI: 10.1109/TPAMI.2024.3367329 PMID: 38407999

Wu, X., Kim, J. G., Hou, A., & Wang, S. (2023). Advanced AI computing enabled by 2D material-based neuromorphic devices. *npj Unconventional Computing*, 1, 1–9. DOI: 10.1038/s44335-025-00023-7

ADDITIONAL READING

Barroso, L. A., Clidaras, J., & Hölzle, U. (2013). *The datacenter as a computer: An introduction to the design of warehouse-scale machines* (2nd ed.). Morgan & Claypool. DOI: 10.1007/978-3-031-01741-4

Dayarathna, M., Wen, Y., & Fan, R. (2016). Data center energy consumption modeling: A survey. *IEEE Communications Surveys and Tutorials*, *18*(1), 732–794. DOI: 10.1109/COMST.2015.2481183

Garimella, S. V., Persoons, T., Weibel, J. A., & Gektin, V. (2017). Electronics thermal management in information and communications technologies: Challenges and future directions. *IEEE Transactions on Components, Packaging, and Manufacturing Technology*, *7*(8), 1191–1205. DOI: 10.1109/TCPMT.2016.2603600

Jones, N. (2018). How to stop data centres from gobbling up the world's electricity. *Nature*, *561*(7722), 163–166. DOI: 10.1038/d41586-018-06610-y PMID: 30209383

Van Heddeghem, W., Lambert, S., Lannoo, B., Colle, D., Pickavet, M., & Demeester, P. (2014). Trends in worldwide ICT electricity consumption from 2007 to 2012. *Computer Communications*, *50*, 64–76. DOI: 10.1016/j.comcom.2014.02.008

KEY TERMS AND DEFINITIONS

Boron Nitride Interposer: A substrate layer made from hexagonal boron nitride (h-BN) that sits between stacked dies in 3D GPU–HBM packages. Silicon conducts heat poorly; h-BN does not. Wang et al. (2025) swapped the silicon interposer for h-BN in a 3D-stacked AI chip architecture and recorded a 20°C drop in hotspot temperature alongside a 22 percent cut in power leakage, with no modifications to device logic. The result reframes interposer material not as a packaging detail but as a first-order thermal design decision, one with direct consequences for the P_total value from which this chapter's cooling water equations derive.

Closed-Loop Cooling: Any cooling arrangement in which the heat-carrying fluid moves through a sealed circuit and never contacts the atmosphere. At the rack level, direct liquid cooling is closed by construction, the same water goes round and round through the cold plates and back to the manifold. Whether the facility as a whole is closed depends on what happens at the secondary loop. Replace the evaporative cooling tower with a dry cooler or adiabatic pre-cooler and the secondary loop closes too, driving the consumptive fraction of the WUE metric toward zero without changing the 32 GPM per rack circulation rate from Equation (18).

Continual Learning: Training without starting over. When new data arrives, continual learning methods update the parameters that need to change and leave the rest alone, sidestepping the catastrophic forgetting that causes naive incremental fine-tuning to degrade earlier knowledge (Wang et al., 2024). The TTIR numbers make the water consequence concrete: a 4-hour incremental update costs 7,700 gallons per rack; the 72-hour full retraining run it replaces costs 138,600. The 94 percent reduction comes from the algorithm, not the plumbing.

Darcy-Weisbach Equation: The governing equation for frictional pressure loss in pipe flow ($\Delta P = f \times L/D \times \rho v^2/2$), used to determine pump sizing and energy overhead for liquid cooling distribution loops.

Direct Liquid Cooling (DLC): A thermal management approach in which water-cooled cold plates make direct contact with heat-generating components such as GPU dies, extracting heat before it enters the rack air volume.

District Heating: A city-scale thermal distribution system that pipes heat from a central generation point to homes and commercial buildings for space heating and hot water. For an AI data center with a working heat recovery loop, district heating converts what would otherwise be cooling tower waste into a revenue-producing product. At a 300 MW campus, the 285 MW_thermal recovered through Equation (45) fed into a district heating offtake agreement at conservative European tariff rates yields approximately $59 million per year, as Equation (54) shows, with heat exchanger capital typically recovered in under twelve months.

Environmental Impact Score (EI): Defined in Equation (50) as consumptive water loss multiplied by the regional WTA. The purpose is to make visible what raw volumetric figures hide: two campuses withdrawing identical amounts of water can sit at opposite ends of the ecological damage spectrum depending on where their watersheds fall on the stress scale. The EI score gives site-selection teams and regulators a single number that carries both the consumption volume and the regional scarcity context, enabling thresholds to be set and compared across jurisdictions.

Fan Affinity Law: The engineering relationship stating that fan power scales with the cube of volumetric airflow rate ($P \propto V^3$), which makes large increases in airflow energetically prohibitive.

Heat Recovery: The capture and reuse of thermal energy extracted from server equipment by cooling water, redirecting it to district heating networks, absorption cooling systems, or industrial processes rather than rejecting it to atmosphere.

Hyperscale Data Center: A facility designed to support very large-scale computing infrastructure, typically defined by power capacity exceeding 100 MW and housing tens of thousands of servers managed by a single operator.

Knowledge Distillation: A training strategy in which a smaller student network learns not from raw labelled data but from the output distributions of a larger, already-trained teacher. The student ends up with a fraction of the parameters and compute cost while matching the teacher closely enough on inference benchmarks to be production-viable (Springer, 2025). In TTIR terms the arithmetic is straight forward: a distilled model trains faster, so TTIR shrinks, and W_TTIR and C_train fall in direct proportion through Equations (28) and (46).

Neuromorphic Computing: Hardware that computes the way biology does, through the firing and suppression of spiking signals across synthetic synapses and neurons, rather than through the clocked multiply-accumulate operations of conventional CMOS circuits. The devices are built from 2D materials: graphene, transition metal dichalcogenides, hexagonal boron nitride. Their energy budgets per operation sit orders of magnitude below anything silicon can achieve today (Wu et al., 2023). If neuromorphic inference reaches production scale, W_TTIR = 1,925 × TTIR holds its form but P_total in Equation (3) shrinks by orders of magnitude, and the cooling water demand and environmental impact scores of Equations (50) through (52) collapse proportionally.

Open-Loop Cooling: The counterpart to closed-loop: a facility arrangement in which water reaches a cooling tower and is deliberately evaporated into the atmosphere to shed heat. That evaporation is not a side effect, it is the mechanism. It is also the sole origin of the consumptive water loss that Equation (43) calculates and that the WUE metric tracks. Close the tower and the consumption disappears; keep it open and the 197 million gallons per year figure for a 300 MW campus is the consequence.

Power Usage Effectiveness (PUE): A data center energy efficiency metric defined as total facility power divided by IT equipment power; a PUE of 1.0 represents perfect efficiency with no overhead.

Reynolds Number (Re): A dimensionless parameter characterizing flow regime in a fluid system ($Re = \rho vD/\mu$); values above approximately 4,000 indicate turbulent flow, which is the normal operating condition for data center cooling loops.

Thermal Design Power (TDP): The maximum sustained power dissipation, in watts, that a processor or GPU is designed to produce during normal operation, establishing the baseline for cooling system sizing.

Time-to-Inference-Ready (TTIR): The elapsed wall-clock time from the initiation of a model training job to the deployment of the resulting model for live inference requests; used in this chapter as the primary temporal unit for workload-coupled water demand modeling.

Water Cost per Inference Request (WCIR): An economic unit built directly on the TTIR framework. Take the total water cost of a training run, divide it by the number of inferences requests the deployed model handles before the next retraining cycle, and the result is WCIR (Equation 49). The figure is not static, push models through faster refresh cycles and WCIR climbs, because each deployed lifetime gets shorter while the training water cost stays the same.

Water Usage Effectiveness (WUE): A data center sustainability metric defined as the volume of water consumed (liters) per unit of IT energy delivered (kWh); captures evaporative and blowdown losses in cooling tower systems.

Withdrawal-to-Availability Ratio (WTA): The ratio of total annual freshwater withdrawn in a watershed to the total renewable freshwater that watershed receives. The scale runs from below 0.1, which signals low stress, through 0.1 – 0.4 for moderate and 0.4 – 0.8 for high, to above 0.8 where conditions are severe. This chapter uses the WTA as a regional weight in Equation (50), multiplying raw consumption against local scarcity to produce a stress-adjusted environmental impact score rather than treating a liter withdrawn in Norway as equivalent to one withdrawn in Arizona.

Conclusion

TOWARDS SUSTAINABLE AI INFRASTRUCTURE AND RESPONSIBLE WATER STEWARDSHIP

The unprecedented growth of artificial intelligence has transformed the digital landscape, creating new opportunities for innovation, economic development, scientific discovery, and societal advancement. At the heart of this transformation lies an expanding network of AI-driven data centers that provide the computational power necessary to support increasingly sophisticated applications. While these developments have generated remarkable technological progress, they have also revealed significant environmental challenges that demand immediate attention. Among these challenges, the growing impact of AI infrastructure on global water resources has emerged as one of the most critical yet underexamined issues of the digital age.

This edited volume, *The Growth of AI Data Centers and the Impact on Global Water Resources*, has sought to explore this complex relationship through a multidisciplinary examination of technological, environmental, economic, legal, and policy dimensions. The contributions presented throughout the book collectively demonstrate that the sustainability of artificial intelligence cannot be assessed solely through computational efficiency or energy consumption metrics. Rather, a comprehensive understanding of AI sustainability must also account for water usage, cooling requirements, ecosystem impacts, governance structures, and resource allocation challenges.

Several important insights emerge from the chapters included in this volume. First, the rapid expansion of AI infrastructure is intensifying demand for water resources in many regions across the world. As data centers become larger and more computationally intensive, the water required for cooling systems continues to increase. In water-stressed areas, this trend has the potential to exacerbate existing environmental and social pressures, creating competition among industrial, agricultural, ecological, and domestic water users.

Second, the volume highlights the importance of recognizing the interconnected nature of water and energy systems. AI data centers operate within a broader water-energy nexus where resource consumption in one domain directly influences sustainability outcomes in another. Addressing these interdependencies requires integrated planning approaches that consider energy efficiency, water conservation, carbon reduction, and environmental resilience as interconnected objectives rather than isolated concerns.

Third, the chapters demonstrate that technological innovation can play a significant role in mitigating environmental impacts. Advances in cooling technologies, predictive analytics, intelligent resource management systems, and sustainable infrastructure design offer promising pathways toward reducing water consumption while maintaining operational performance. The application of AI itself to water resource management further illustrates how emerging technologies can contribute to more efficient and sustainable resource utilization.

The volume also underscores the growing importance of governance, transparency, and public accountability. As AI infrastructure expands globally, policymakers, regulatory agencies, technology companies, and local communities must work collaboratively to ensure that data center development aligns with broader sustainability goals. Transparent reporting of water consumption, environmental impact assessments, responsible siting decisions, and equitable resource allocation mechanisms will become increasingly important components of sustainable digital infrastructure planning.

A recurring theme throughout this book is the need for interdisciplinary collaboration. The challenges associated with AI-driven water consumption cannot be addressed by any single discipline alone. Engineers, computer scientists, environmental researchers, economists, legal scholars, policymakers, and industry leaders must collectively contribute to the development of solutions that balance technological advancement with environmental stewardship. Such collaboration is essential for creating frameworks capable of supporting long-term sustainability while fostering continued innovation.

The contributions in this volume also reveal several important avenues for future research. Additional studies are needed to quantify the full water footprint of AI systems across their life cycles, evaluate the effectiveness of emerging cooling technologies, assess regional variations in water-related risks, and develop standardized metrics for measuring and reporting water consumption in digital infrastructure. Future research should also explore the social and ethical dimensions of water use in AI development, particularly in regions experiencing increasing resource scarcity.

Looking ahead, the continued growth of artificial intelligence is inevitable. The question is not whether AI will expand, but how that expansion can be managed responsibly. The future success of AI will depend not only on breakthroughs in algorithms and computing power but also on the ability of governments, industries, and communities to ensure that technological progress remains compatible with environmental sustainability and resource conservation.

Ultimately, water is more than a resource that supports technological infrastructure; it is a fundamental prerequisite for life, economic development, and ecological stability. As society embraces the transformative potential of artificial intelligence, safeguarding water resources must become an integral component of digital sustainability strategies. Achieving this balance will require innovation, informed policymaking, responsible corporate practices, and sustained global cooperation.

The chapters presented in this volume represent an important step toward understanding the hidden environmental dimensions of the AI revolution. By illuminating the relationship between AI data centers and global water resources, this book contributes to an emerging body of knowledge that will shape future discussions on sustainable technology development.

It is our hope that the insights presented herein will encourage further research, foster meaningful dialogue among stakeholders, and inspire practical actions that support both technological advancement and responsible environmental stewardship. Only through such efforts can society ensure that the future of artificial intelligence is not only powerful and transformative but also sustainable, equitable, and resilient for generations to come.

The Editors

Compilation of References

Abbas Khan, M., Khan, H., Omer, M. F., Ullah, I., & Yasir, M. (2024). Impact of artificial intelligence on the global economy and technology advancements. In El Hajjami, S., Kaushik, K., & Khan, I. U. (Eds.), *Artificial general intelligence (AGI) security: Smart applications and sustainable technologies* (pp. 147–180). Springer Singapore., DOI: 10.1007/978-981-97-3222-7

Abdulameer, L., Al-Khafaji, M. S., Al-Awadi, A. T., Al-Mansoori, H., Hassan, A. K., & Mahmood, S. (2025). Artificial intelligence in climate-resilient water management: A systematic review of applications, challenges, and future directions. *Water Conservation Science and Engineering*, *10*(1), 44. Advance online publication. DOI: 10.1007/s41101-025-00371-2

Abulaiti, H., & Liu, Y. (2025). Ecological sensitivity assessment and driving force analysis of the Tarim river basin. *Scientific Reports*, *15*(1), 34630. Advance online publication. DOI: 10.1038/s41598-025-18270-w PMID: 41044162

Accenture. (2024). *Digital transformation through intelligent infrastructure and analytics*. https://www.accenture.com

Accenture. (2024). *Technology vision 2024: Human by design*. https://www.accenture.com

Afoma, U. M., Singh, S., Mishra, A. K., Sharma, C. K., Gupta, K., Mishra, M. K., Roy, B., Verma, V. V., & Sharma, V. K. (2026). Integrating artificial intelligence in environmental monitoring: A paradigm shift in data-driven sustainability. *EcoHealth*, *23*(1), 38–57. DOI: 10.1007/s10393-025-01752-8 PMID: 40875110

Agarwal, A., & Beharry, E. (2026). A Transparency-Based Approach to Regulating the Resource Footprint of US Data Centers. *GRACE. Global Review of AI Community Ethics*, *4*(1).

Aggarwal, C. C. (2017). *Outlier analysis* (2nd ed.). Springer., DOI: 10.1007/978-3-319-47578-3

Ahrabi, R. R., Mousavi, A., Mohammadi, E., Wu, R., & Chen, A. K. (2025, April). AI-driven data center energy profile, power quality, sustainable sitting, and energy management: A comprehensive survey. In *2025 IEEE Conference on Technologies for Sustainability (SusTech)* (pp. 1-8). IEEE. DOI: 10.1109/SusTech63138.2025.11025802

Aili, A., Long, W., Cao, Z., & Wen, Y. (2024). Radiative free cooling for energy and water saving in data centers. *Applied Energy*, *359*, 122672. DOI: 10.1016/j.apenergy.2024.122672

Al Kez, D., Foley, A. M., Wong, F. W. B. H., Dolfi, A., & Srinivasan, G. (2025). AI-driven cooling technologies for high-performance data centres: state-of-the-art review and future directions. Sustainable Energy Technologies and Assessments, 82, 104511.

Al Kez, D., Foley, A. M., & Fadhli, W. B. M. (2025, October). Hasan Wong, Andrea Dolfi, and Geetha Srinivasan. 'AI-Driven Cooling Technologies for High-Performance Data Centres: State-of-the-Art Review and Future Directions'. *Sustainable Energy Technologies and Assessments*, *82*, 104511. DOI: 10.1016/j.seta.2025.104511

Al Kez, D., Foley, A. M., Laverty, D., Del Rio, D. F., & Sovacool, B. (2022). Exploring the sustainability challenges facing digitalization and internet data centers. *Journal of Cleaner Production*, *371*, 133633. DOI: 10.1016/j.jclepro.2022.133633

Alkrush, A. A., Salem, M. S., Abdelrehim, O., & Hegazi, A. A. (2024). Data centers cooling: A critical review of techniques, challenges, and energy saving solutions. *International Journal of Refrigeration*, *160*, 246–262. DOI: 10.1016/j.ijrefrig.2024.02.007

Amanatidis, P., Lyratzis, E., Angelopoulos, V., Kouloumpris, E., Skaperdas, E., Bassiliades, N., Vlahavas, I., Maris, F., Emmanouloudis, D., & Karampatzakis, D. (2026). Intelligent water management through edge-enabled IoT, AI, and big data technologies. *IoT*, *7*(1), 5. Advance online publication. DOI: 10.3390/iot7010005

Ansorge, L., & Stejskalová, L. (2025). Water footprint which is not the water footprint: Critical review of the article by Müller et al. (2024). *Journal of Environmental Management*, *374*, 124038. Advance online publication. DOI: 10.1016/j.jenvman.2025.124038 PMID: 39793505

Apstech Advisors. (2026). Your Strategic Partner in the Water Industry, Data Center Water Consumption in the US: Challenges, Trends, and Market Opportunities for 2025–2030.

ASHRAE. (2019). *ASHRAE great energy predictor III dataset*. American Society of Heating, Refrigerating and Air-Conditioning Engineers.

ASHRAE. (2021). *Thermal guidelines for data processing environments* (5th ed.).

Belady, C. (2007). In the data center, power and cooling costs more than the IT equipment it supports. *Electronics Cooling*, *13*(1), 24–27.

Bertino, E., & Jahanshahi, M. R. (2018, March). Adaptive and Cost-Effective Collection of High-Quality Data for Critical Infrastructure and Emergency Management in Smart Cities—Framework and Challenges. *ACM Journal of Data and Information Quality*, *10*(1), 1–6. DOI: 10.1145/3190579

Bhatia, A. B., Sharma, M., & Bhatia, B. (2025). AI and environmental data science. In Kulkarni, S. (Ed.), *Artificial intelligence-driven models for environmental management*. Wiley., DOI: 10.1002/9781394282555.ch3

Bolón-Canedo, V., Morán-Fernández, L., Cancela, B., & Alonso-Betanzos, A. (2024, September). 'A Review of Green Artificial Intelligence: Towards a More Sustainable Future'. *Neurocomputing*, *599*, 128096. DOI: 10.1016/j.neucom.2024.128096

Borkowski, M., & Piłat, A. K. (2022). Customized data center cooling system operating at significant outdoor temperature fluctuations. *Applied Energy*, *306*, 117975. DOI: 10.1016/j.apenergy.2021.117975

Bose, R., Roy, S., Mondal, H., Ghosh, A., & Das, S. (2021). Energy-efficient approach to lower the carbon emissions of data centers. *Computing*, *103*(8), 1703–1721. DOI: 10.1007/s00607-020-00889-4

Box, G. E. P., Jenkins, G. M., Reinsel, G. C., & Ljung, G. M. (2015). *Time series analysis: Forecasting and control* (5th ed.). Wiley.

Boyle, D. (2022). *SCADA systems and telemetry in industrial automation*. Elsevier.

Brownlee, J. (2018). *Deep learning for time series forecasting*. Machine Learning Mastery.

Brown, R., & Patel, C. (2023). Liquid cooling for sustainable data centers. *IEEE Transactions on Industry Applications*, *59*(3), 2154–2163. DOI: 10.1109/TIA.2022.3209874

Calvin, K., Dasgupta, D., Krinner, G., ... (2023) IPCC, 2023: Climate Change 2023: Synthesis Report. Contribution of Working Groups I, II and III to the Sixth Assessment Report of the Intergovernmental Panel on Climate Change (Core Writing Team, H. Lee and J. Romero (eds.)). IPCC, Geneva, Switzerland. Intergovernmental Panel on Climate Change (IPCC)

Candanedo, L. M., Feldheim, V., & Deramaix, D. (2017). Data-driven prediction models of energy use of appliances in a low-energy house. *Energy and Building*, *140*, 81–97. DOI: 10.1016/j.enbuild.2017.01.083

Cazzetta, A., Zangaro, F., Marcucci, F., Julius, O. T., Rainò, M., Shauer, M., Massaro, R., Semeraro, T., Basset, A., & Pinna, M. (2026). Applying artificial intelligence (AI) innovative tools for ecological research and monitoring of transitional water ecosystems: A systematic review. *Environments (Basel, Switzerland)*, *13*(4), 193. Advance online publication. DOI: 10.3390/environments13040193

CDP. (2026). *Turning Transparency to Action.* https://www.cdp.net/en. Accessed 22 Feb. 2026.

Cengel, Y. A., & Cimbala, J. M. (2018). *Fluid mechanics: Fundamentals and applications* (4th ed.). McGraw-Hill Education.

Central Water Commission. (2023). *National water availability assessment report 2023.* Ministry of Jal Shakti, Government of India. https://cwc.gov.in/national-water-availability-report

Chandola, V., Banerjee, A., & Kumar, V. (2009). Anomaly detection: A survey. *ACM Computing Surveys*, *41*(3), 1–58. DOI: 10.1145/1541880.1541882

Chang, Q., Huang, Y., Liu, K., Xu, X., Zhao, Y., & Pan, S. (2024). Optimization control strategies and evaluation metrics of cooling systems in data centers: A review. *Sustainability (Basel)*, *16*(16), 7222. DOI: 10.3390/su16167222

Chatlatanagulchai, et al. (2024). Adaptive data-sharing and workload distribution in digital infrastructures. (ICT systems optimization studies).

Chen, H., Li, D., Wang, S., Chen, T., Zhong, M., Ding, Y., Li, Y., & Huo, X. (2023). 'Numerical Investigation of Thermal Performance with Adaptive Terminal Devices for Cold Aisle Containment in Data Centers'. *Buildings (Basel, Switzerland)*, *13*(2), 268. DOI: 10.3390/buildings13020268

Chen, T., & Guestrin, C. (2016). XGBoost: A scalable tree boosting system. In *Proceedings of the 22nd ACM SIGKDD International Conference on Knowledge Discovery and Data Mining* (pp. 785–794). https://doi.org/DOI: 10.1145/2939672.2939785

Chen, X., Wang, X., Wang, L., Zheng, H., Ding, T., & Li, Z. (2025, January). 'Multistage Data Center Cooling System for Temperature Gradation and Matching'. *Applied Energy*, *377*, 124592. DOI: 10.1016/j.apenergy.2024.124592

Chini, C. M., Konar, M., & Stillwell, A. S. (2017). Direct and indirect urban water footprints of the U nited S tates. Water Resources Research, 53(1), 316-327.

Chu, J., & Huang, X. (2023). Research status and development trends of evaporative cooling air-conditioning technology in data centers. *Energy and Built Environment*, *4*(1), 86–110. DOI: 10.1016/j.enbenv.2021.08.004

Cisco. (2023). *Global networking trends report*. https://www.cisco.com

Cisco. (2023). *Internet of Things (IoT) and data center modernization report*. https://www.cisco.com

Cisco. (n.d.). Cisco Annual Internet Report (2018-2023) White Paper.

Cowls, J., et al. (2023). Machine learning and energy optimization in data centers. Nature / AI sustainability literature (reviewed works).

Cowls, J., Tsamados, A., Taddeo, M., & Floridi, L. (2023). 'The AI Gambit: Leveraging Artificial Intelligence to Combat Climate Change—Opportunities, Challenges, and Recommendations'. *AI & Society*, *38*(1), 283–307. DOI: 10.1007/s00146-021-01294-x

Dada, M. A., Majemite, M. T., Obaigbena, A., Daraojimba, O. H., Oliha, J. S., & Nwokediegwu, Z. Q. S. (2024). Review of smart water management: IoT and AI in water and wastewater treatment. *World Journal of Advanced Research and Reviews*, *21*(1), 1373–1382. DOI: 10.30574/wjarr.2024.21.1.0171

Dai, Y., Huang, Z., Khan, N., & Labbo, M. S. (2025). Smart water management: Governance innovation, technological integration, and policy pathways toward economic and ecological sustainability. *Water (Basel)*, *17*(13), 1932. DOI: 10.3390/w17131932

Damania, R., Ebadi, E., Mayr, K., Russ, J., & Zaveri, E. (2025). *Reboot Development: The Economics of a Livable Planet*. World Bank., DOI: 10.1596/978-1-4648-2271-1

de Vries-Gao, A. (2026). The carbon and water footprints of data centers and what this could mean for artificial intelligence. *Patterns (New York, N.Y.)*, *7*(1), 101430. DOI: 10.1016/j.patter.2025.101430 PMID: 41583976

Deloitte. (2024). *AI-powered predictive analytics for infrastructure optimization*. Deloitte Insights. https://www2.deloitte.com

Dieter, C. A., ... (2018). Estimated use of water in the United States in 2015 U.S. Geol. Surv. Circ. 1441 65.

Egbemhenghe, A. U., Ojeyemi, T., Iwuozor, K. O., Emenike, E. C., Ogunsanya, T. I., Anidiobi, S. U., & Adeniyi, A. G. (2023). Revolutionizing water treatment, conservation, and management: Harnessing the power of AI-driven ChatGPT solutions. *Environmental Challenges*, *13*, 100782. Advance online publication. DOI: 10.1016/j.envc.2023.100782

European Commission. (2020). *A European strategy for data*(COM/2020/66 final). European Commission. https://eur-lex.europa.eu/legal-content/EN/TXT/?uri=CELEX:52020DC0066

European Environment Agency. (2023). *Water resources across Europe: Confronting water stress*. EEA Report No. 2/2023. European Environment Agency. https://doi.org/DOI: 10.2800/000124

Fonseca, J. A., Nguyen, T.-A., Schlueter, A., & Marechal, F. (2016). City Energy Analyst (CEA): Integrated framework for analysis and optimization of building energy systems in neighborhoods and city districts. *Energy and Building*, *113*, 202–226. DOI: 10.1016/j.enbuild.2015.11.055

Food and Agriculture Organization of the United Nations. (2024). *Progress on level of water stress: 2024 update, SDG indicator 6.4.2*. FAO on behalf of UN-Water. DOI: 10.4060/cc8962en

Food and Agriculture Organization. (2023). *The state of the world's land and water resources for food and agriculture*. https://www.fao.org

Fortune Business Insights. "Smart Strategies, Giving Speed to your Growth Trajectory," 2025. https://www.fortunebusinessinsights.com/data-center-market 109851 (accessed March 03, 2026).

Gao, W., Hu, Q., Ye, Z., ... 'Deep Learning Workload Scheduling in GPU Datacenters: Taxonomy, Challenges and Vision'. Version 3. Preprint, arXiv, 2022. DOI: 10.48550/ARXIV.2205.11913

Gartner. (2024). *Top strategic technology trends for 2024*. https://www.gartner.com

Global, S. P. (2025). The World's Total Data Center Capacity is Shifting Rapidly to Hyperscale Operators RENO, NV, June 24, 2025).

Gnibga, W. E., Chien, A. A., Blavette, A., & Orgerie, A. C. (2024, June). Flex-CoolDC: datacenter cooling flexibility for harmonizing water, energy, carbon, and cost trade-offs. In *Proceedings of the 15th ACM International Conference on Future and Sustainable Energy Systems* (pp. 108-122). DOI: 10.1145/3632775.3661936

Goodfellow, I., Bengio, Y., & Courville, A. (2016). *Deep learning*. MIT Press. https://www.deeplearningbook.org

Google. (2022–2023). Environmental Report. *Sustainability Report*, •••, 2021.

Goyal, M. K., Kumar, S., & Gupta, A. (2024). AI for water policy. In AI innovation for water policy and sustainability (pp. 59–84). SpringerBriefs in Water Science and Technology. Springer. https://doi.org/DOI: 10.1007/978-3-031-72014-7_4

Green Grid. (2022). *Water usage effectiveness (WUE): A comprehensive metric for data centers*. https://www.thegreengrid.org

Gubbi, J., Buyya, R., Marusic, S., & Palaniswami, M. (2013). Internet of Things (IoT): A vision, architectural elements, and future directions. *Future Generation Computer Systems*, *29*(7), 1645–1660. DOI: 10.1016/j.future.2013.01.010

Haghshenas, K., Setz, B., Blosch, Y., & Aiello, M. (2023). Enough hot air: The role of immersion cooling. *Energy Informatics*, *6*(1), 14. DOI: 10.1186/s42162-023-00269-0

Hao, K. (2024). AI is taking water from the desert. The Atlantic. https://www.theatlantic.com/technology/archive/2024/03/ai-water-usage-data-centers-desert/677622/

Hassan, M., Malhotra, K., & Firdaus, M. (2022). Application of artificial intelligence in IoT security for crop yield prediction. ResearchBerg Review of Science and Technology, 2(1), 136–157. Retrieved from https://researchberg.com/index.php/rrst/article/view/150

He, G., Zhao, Y., Jiang, S., Zhu, Y., Li, H., & Wang, L. (2019). Impact of virtual water transfer among electric sub-grids on China's water sustainable developments in 2016, 2030, and 2050. *Journal of Cleaner Production*, *239*, 118056. Advance online publication. DOI: 10.1016/j.jclepro.2019.118056

Hennessy, J. L., & Patterson, D. A. (2019). *Computer architecture: A quantitative approach* (6th ed.). Morgan Kaufmann.

Herrera, et al. (2025). Environmental impacts of hyperscale data centers. Data center sustainability research.

Herrera, M., Xie, X., Menapace, A., Zanfei, A., & Brentan, B. M. (2025). Sustainable AI infrastructure: A scenario-based forecast of water footprint under uncertainty. *Journal of Cleaner Production*, *526*, 146528. DOI: 10.1016/j.jclepro.2025.146528

Hiremath, R. B. (2024, November). AI-Embedded data centres: promoting sustainability and reducing water footprint. In *2024 first international conference on data, computation and communication (ICDCC)* (pp. 40-44). IEEE. DOI: 10.1109/ICDCC62744.2024.10961316

Hlabisa, S. (2025). The ecology of artificial intelligence: Energy, water, materials, and land limits of digital systems. *Carbon Neutral Systems*, *1*(1), 19. DOI: 10.1007/s44438-025-00018-8

Hnayno, M., Chehade, A., Klaba, H., Bauduin, H., Polidori, G., & Maalouf, C. (2022). Performance analysis of new liquid cooling topology and its impact on data centres. *Applied Thermal Engineering*, *213*, 118733. DOI: 10.1016/j.applthermaleng.2022.118733

Hochreiter, S., & Schmidhuber, J. (1997). Long short-term memory. *Neural Computation*, *9*(8), 1735–1780. DOI: 10.1162/neco.1997.9.8.1735 PMID: 9377276

Hoerl, A. E., & Kennard, R. W. (1970). Ridge regression: Biased estimation for nonorthogonal problems. *Technometrics*, *12*(1), 55–67. DOI: 10.1080/00401706.1970.10488634

Huang, P., Copertaro, B., Zhang, X., Shen, J., Löfgren, I., Rönnelid, M., Fahlen, J., Andersson, D., & Svanfeldt, M. (2020). A review of data centers as prosumers in district energy systems: Renewable energy integration and waste heat reuse for district heating. *Applied Energy*, *258*, 114109. Advance online publication. DOI: 10.1016/j.apenergy.2019.114109

IBM Institute for Business Value. (2024). *The CEO's guide to generative AI.* https://www.ibm.com

IBM. (2024). *What Is Predictive Analytics?* https://www.ibm.com/think/topics/predictive-analytics

IDC. (2024). *Worldwide artificial intelligence infrastructure forecast, 2024–2028.* https://www.idc.com

Insider, B. (2025). The 'AI boom' pits neighbor against neighbor. https://www.businessinsider.com/data-centers-northern-virginia-noise-air-pollution-cost-2025-5

Insider, B. (2025). The true cost of data centers.

International Electrotechnical Commission. (2023). *Industrial communication networks and system integration (SCADA standards).* https://www.iec.ch

International Energy Agency (2023–2024). Data Centres and Data Transmission Networks (reports and tracking updates).

International Energy Agency (IEA). (2025). Data centres and data transmission networks. https://www.iea.org/reports/data-centres-and-data-transmission-networks

International Energy Agency (IEA). *Data Centres and Data Transmission Networks.* https://www.iea.org/reports/data-centres-and-data-transmission-networks

International Energy Agency. (2023). *Data centres and data transmission networks.* https://www.iea.org

International Energy Agency. (2023). *Data centres and data transmission networks*. IEA. https://www.iea.org/energy-system/buildings/data-centres-and-data-transmission-networks

International Monetary Fund (IMF) (2025). *Climate Change and the Role of the IMF*. https://www.imf.org/en/Topics/climate-change

International Monetary Fund (IMF) (2025). *Climate Change Indicators Dashboard*. https://www.imf.org/en/Topics/climate-change/ccid

International Monetary Fund (IMF) (2025). *Government Finance Statistics*. https://www.imf.org/en/Data/Statistics/GFS

International Monetary Fund (IMF) (2025). *World Economic Outlook Database*. https://www.imf.org/en/Publications/WEO/weo-database

International Telecommunication Union. (2023). *Digital infrastructure and smart systems for sustainability*. https://www.itu.int

Jadhav, S. A., & Liu, Z. (2026). Digital Twin-Based Cooling System Optimization for Data Center. *arXiv preprint arXiv:2603.01198.*

Jadhav, S., & Liu, Z. (2026). *Machine Learning guided cooling system optimization for data center. arXiv*. Cornell University.,

James, G., Witten, D., Hastie, T., & Tibshirani, R. (2021). *An introduction to statistical learning* (2nd ed.). Springer., DOI: 10.1007/978-1-0716-1418-1

Jareño, J., Aragón-Jurado, J. M., De La Torre, J. C., Ruiz, P., & Dorronsoro, B. (2026). Energy-efficient large language models, Future Generation Computer Systems, Volume 182, 108483, ISSN 0167-739X, DOI: 10.1016/j.future.2026.108483

Jayakumar, D., Bouhoula, A., & Al-Zubari, W. K. (2024). Unlocking the potential of artificial intelligence for sustainable water management focusing operational applications. *Water (Basel)*, *16*(22), 3328. Advance online publication. DOI: 10.3390/w16223328

Jiang, F., Duan, C., & Chen, B. (2025). Facility-level energy-driven water footprint and scarcity implications of Chinese data centers: A bottom-up analysis and scenario-based projection. *Applied Energy*, *399*, 126522. DOI: 10.1016/j.apenergy.2025.126522

Jin, C., & Bai, X. (2019, August). 'The Study of Servers' Arrangement and Air Distribution Strategy under Partial Load in Data Centers'. *Sustainable Cities and Society*, *49*, 101617. DOI: 10.1016/j.scs.2019.101617

Jing, Y., Xie, L., Li, F., Zhan, Z., Wang, Z., Yang, F., Fan, J., Zhu, Z., Zhang, H., Zhao, C., Shi, Z., Jiang, H., Xie, X., & Jiang, Y. (2024). Field test of cooling systems in two air-cooled data centers: Various regions, air distributions and evaporative cooling technologies. *Applied Thermal Engineering*, *248*, 123189. DOI: 10.1016/j.applthermaleng.2024.123189

Jones, N. (2018). How to stop data centres from gobbling up the world's electricity. *Nature*, *561*(7722), 163–166. DOI: 10.1038/d41586-018-06610-y PMID: 30209383

Jouppi, N. P.. (2017). In-datacenter performance analysis of a tensor processing unit. In *Proceedings of the 44th Annual International Symposium on Computer Architecture* (pp. 1–12). https://doi.org/DOI: 10.1145/3079856.3080246

Kang, J. S., Li, M., Wu, H., Nguyen, H., Aoki, T., & Hu, Y. (2021). Integration of boron arsenide cooling substrates into gallium nitride devices. *Nature Electronics*, *4*(6), 416–423. DOI: 10.1038/s41928-021-00595-9

Karimi, L., Shinh, R., Laisure-Pool, C., Green, M., Yacuel, L., Ashby, J., & Hickenbottom, K. L. (2025). Energy and water dynamics in data center cooling: Insights from a modeling study in hot-arid climates. *Applied Thermal Engineering*, *276*, 126802. DOI: 10.1016/j.applthermaleng.2025.126802

Karimi, L., Yacuel, L., Degraft-Johnson, J., Ashby, J., Green, M., Renner, M., & Hickenbottom, K. L. (2022). Water-energy tradeoffs in data centers: A case study in hot-arid climates. *Resources, Conservation and Recycling*, *181*, 106194. DOI: 10.1016/j.resconrec.2022.106194

Katal, A., Wazid, M., & Goudar, R. H. (2023). Big data analytics and IoT in smart infrastructure systems. *Journal of Network and Computer Applications*, *216*, 103556. DOI: 10.1016/j.jnca.2023.103556

Kaushik, A., Soundappan, T., Mujawar, M. A., & Bhansali, S. (2026). Standing of electrochemical sensors on the 250th anniversary of the USA. *Journal of the Electrochemical Society*, *173*(9), 097509. DOI: 10.1149/1945-7111/ae67af

Kelleher, J. D., Tierney, B., & D'Arcy, A. (2020). *Data science* (2nd ed.). MIT Press.

Khalaj, A. H., & Halgamuge, S. K. (2017). A review on efficient thermal management of air- and liquid-cooled data centers: From chip to the cooling system. *Applied Energy*, *205*, 1165–1188. DOI: 10.1016/j.apenergy.2017.08.037

Khan, N. Inside Tech World (2025). https://insidetechworld.com/artificial-intelligence/chatgpt-water-consumption-and-environmental-impact/

Kijak, R., & Woźnicki, J. (2026). Solutions for the Data Centers' Sustainability. In *The Palgrave Encyclopedia of Sustainable Resources and Ecosystem Resilience* (pp. 1–10). Springer Nature Switzerland. DOI: 10.1007/978-3-030-67776-3_81-1

Kim, J. H., Shin, D. U., & Kim, H. (2024). 'Data Center Energy Evaluation Tool Development and Analysis of Power Usage Effectiveness with Different Economizer Types in Various Climate Zones'. *Buildings (Basel, Switzerland)*, *14*(1), 299. DOI: 10.3390/buildings14010299

Kirk, D. B., & Hwu, W.-M. W. (2016). *Programming massively parallel processors* (3rd ed.). Morgan Kaufmann.

Koomey, J. (2011). Growth in data center electricity use 2005 to 2010. *A report by Analytical Press, completed at the request of The New York Times, 9*(2011), 161.

Kuzma, S., Bierkens, M. F. P., Lakshman, S., Luo, T., Saccoccia, L., Sutanudjaja, E. H., & Van Beek, R. (2023). *Aqueduct 4.0: Updated decision-relevant global water risk indicators*. World Resources Institute., DOI: 10.46830/writn.23.00061

Latif, I., Shafique, M. A., Ullah, H., Newkirk, A. C., Yu, X., & Munir, A. (2025). Cooling Matters: Benchmarking Large Language Models and Vision-Language Models on Liquid-Cooled Versus Air-Cooled H100 GPU Systems. *arXiv preprint arXiv:2507.16781*.

Lawrence Berkeley National Laboratory. (2025). *Energy and Water Efficiency in Data Centers*. https://eta.lbl.gov/publications/energy-and-water-efficiency

Lawrence Berkeley National Laboratory. (2025). *Energy and Water Modeling for Data Centers*. https://eta.lbl.gov/publications

Leaney, J., Bilton, M., & Sheridan, C. (2022). *Water usage effectiveness in data centers: Benchmarking and best practice*(White Paper 69). The Green Grid. https://www.thegreengrid.org

Lei, N., & Masanet, E. (2021). Climate- and Technology-Specific PUE and WUE Predictions for U.S. Data Centers using a Physics-Based Approach, PREPRINT (Version 1) available at Research Square. DOI: 10.21203/rs.3.rs-769999/v1

Lei, N., Cheng, Z., Cao, Z., & Masanet, E. (2021). 'An Integrated Decision-Making Framework for Sustainable Data Center Operation through Intelligent Load Scheduling'. *Journal of Physics: Conference Series, 2042*(1), 012091. DOI: 10.1088/1742-6596/2042/1/012091

Lei, N., Lu, J., Cheng, Z., Cao, Z., Shehabi, A., & Masanet, E. (2023). 'Geospatial Assessment of Water Footprints for Hyperscale Data Centers in the United States'. *Journal of Physics: Conference Series, 2600*(17), 172003. DOI: 10.1088/1742-6596/2600/17/172003

Lei, N., Lu, J., Shehabi, A., & Masanet, E. (2025). The water use of data center workloads: A review and assessment of key determinants. *Resources, Conservation and Recycling, 219*, 108310. DOI: 10.1016/j.resconrec.2025.108310

Lei, N., & Masanet, E. (2022). Climate-and technology-specific PUE and WUE estimations for US data centers using a hybrid statistical and thermodynamics-based approach. *Resources, Conservation and Recycling, 182*, 106323. DOI: 10.1016/j.resconrec.2022.106323

Li, P., Yang, J., Islam, M. A., & Ren, S. (2023). Making AI less“thirsty”: Uncovering and addressing the secret water footprint of AI models. arXiv preprint arXiv:2304.03271. https://arxiv.org/abs/2304.03271

Li, P., Yang, J., Islam, M. A., & Ren, S. (2025). Making ai less' thirsty'. *Communications of the ACM, 68*(7), 54–61. DOI: 10.1145/3724499

Liu, C., Hao, N., Zhang, T., Wang, D., Li, Z., & Bian, W. (2024). Optimization of data-center immersion cooling using liquid air energy storage. *Journal of Energy Storage, 90*, 111806. DOI: 10.1016/j.est.2024.111806

Liu, C., & Yu, H. (2021). Evaluation and optimization of a two-phase liquid-immersion cooling system for data centers. *Energies, 14*(5), 1395. DOI: 10.3390/en14051395

Lu, R., & Wang, D. (2025). A Thermal-aware Workload Scheduler for High-performance LLM Inference in Cooling-regulated Datacenters. *ACM SIGENERGY Energy Informatics Review, 5*(2), 98–104. DOI: 10.1145/3757892.3757906

Lu, T., Lü, X., Välisuo, P., Zhang, Q., & Clements-Croome, D. (2024). Innovative approaches for deep decarbonization of data centers and building space heating networks: Modelling and comparison of novel waste heat recovery systems for liquid cooling systems. *Applied Energy, 357*, 122499. DOI: 10.1016/j.apenergy.2023.122473

Machine Learning Repository, U. C. I. (2023). *Appliance energy prediction dataset.* https://archive.ics.uci.edu

Macknick, J., Newmark, R., Heath, G., & Hallett, K. C. (2011). A review of operational water consumption and withdrawal factors for electricity generating technologies.

Mahadeva, R., Kumar, M., Gupta, V., Manik, G., Gupta, V., Alawatugoda, J., Manik, H., Patole, S. P., & Gupta, V. (2023). Water desalination using PSO-ANN techniques: A critical review. *Digital Chemical Engineering*, *9*, 100128. Advance online publication. DOI: 10.1016/j.dche.2023.100128

Mahmoudian, F., Khoramipoor, Z., Nabizadeh Chianeh, F., Valikhan Anaraki, M., & Farzin, S. (2026). Towards cleaner waters: A comprehensive review of pollutant, quality challenges, monitoring advances, and treatment strategies. *Physics and Chemistry of the Earth Parts A/B/C*, *144*(Part 1), 104441. Advance online publication. DOI: 10.1016/j.pce.2026.104441

Mandal, S., Yadav, A., Panwar, R., Kumar, A., Singh, V., Sharma, P., & Gupta, R. (2025). Smart water management for SDG 6: A review of AI and IoT-enabled solutions. *Water Conservation Science and Engineering*, *10*(2), 83. Advance online publication. DOI: 10.1007/s41101-025-00407-7

Ma, S., Liu, Y., Liu, Y., Wang, J., Fang, Q., & Huang, Y. (2025). Artificial intelligence-enabled predictive energy saving planning of liquid cooling system for data centers. *Advanced Engineering Informatics*, *65*, 103283. DOI: 10.1016/j.aei.2025.103283

Masanet, E., Shehabi, A., Lei, N., Smith, S., & Koomey, J. (2020). Recalibrating global data center energy-use estimates. *Science*, *367*(6481), 984–986. DOI: 10.1126/science.aba3758 PMID: 32108103

Masud, M. M., Shamem, A. S. M., Saif, A. N. M., Hossain, M. S., Rahman, M. M., & Islam, M. A. (2025). The role of artificial intelligence in sustainable water management in Asia: A systematic literature review with bibliographic network visualization. *International Journal of Energy and Water Resources*, *9*(1), 247–265. DOI: 10.1007/s42108-024-00319-7

Mavisclara, O. A., Oshobugie, I. I., Olufunmi, A. T., Bolaji, A. A., Olawale, A. O., Anulika, N. C., & Samuel, O. K. (2025). 'The AI-Driven Energy Surge: A Comprehensive Review of Sustainable Power Solutions for Data Centers'. *Global Journal of Engineering and Technology Advances*, *24*(3), 328–344. DOI: 10.30574/gjeta.2025.24.3.0279

McKinsey & Company. (2024) *Advanced Analytics in Infrastructure Operations*. https://www.mckinsey.com/capabilities/operations/our-insights

McKinsey & Company. (2024). *The economic potential of generative AI and infrastructure demand*. https://www.mckinsey.com

Mekonnen, M. M., & Hoekstra, A. Y. (2016). Four billion people facing severe water scarcity. *Science Advances*, *2*(2), e1500323. DOI: 10.1126/sciadv.1500323 PMID: 26933676

Miciuła, I., Wojtaszek, H., Mastalerz, M., Wysocki, W., Plecka, P., Czaplewski, M., Doskocz, J., & Raba-Schulze, A. (2026). Optimization of Electricity Consumption by Information Technology Devices in Accordance with Environmental, Social, Governance and Sustainable Development Principles. *Energies*, *19*(5), 1139. DOI: 10.3390/en19051139

Microsoft. (2023). *AI for sustainable data centers: Water and energy optimization*. https://www.microsoft.com

MirhoseiniNejad, S. M., Badawy, G., & Down, D. G.MirhoseiniNejad. (2021). Holistic thermal-aware workload management and infrastructure control for heterogeneous data centers using machine learning. *Future Generation Computer Systems*, *118*, 208–218. DOI: 10.1016/j.future.2021.01.007

Mitchell-Jackson, J., Koomey, J. G., Nordman, B., & Blazek, M. (2003). Data center power requirements: Measurements from Silicon Valley. *Energy*, *28*(8), 837–850. DOI: 10.1016/S0360-5442(03)00009-4

Moore, J. D., Chase, J. S., Ranganathan, P., & Sharma, R. K. (2005, April). Making Scheduling" Cool": Temperature-Aware Workload Placement in Data Centers. In *USENIX ATC, General Track* (pp. 61-75).

Murino, T., Monaco, R., Nielsen, P. S., Liu, X., Esposito, G., & Scognamiglio, C. (2023). Sustainable energy data centres: A holistic conceptual framework for design and operations. *Energies*, *16*(15), 5764. DOI: 10.3390/en16155764

Mytton, D. (2021). 'Data Centre Water Consumption'. *NPJ Clean Water*, *4*(1), 11. DOI: 10.1038/s41545-021-00101-w

Nadjahi, C., Louahlia, H., & Lemasson, S. (2018). A review of thermal management and innovative cooling strategies for data center. *Sustainable Computing : Informatics and Systems*, *19*, 14–28. DOI: 10.1016/j.suscom.2018.05.002

Nahid, F. A., & Islam, M. M. Md. Kamruzzaman, and Ahmed Al Mansur. 'The Rising Energy Footprint of Data Centers: A Review of Global Trends, Challenges, and Opportunities'. 2025 IEEE 7th International Conference on Sustainable Technologies For Industry 5.0 (STI), 11 December 2025, 1–6. https://doi.org/DOI: 10.1109/STI69347.2025.11367578

Nahid, F. A., Islam, M. M., Kamruzzaman, M., & Al Mansur, A. (2025, December). The Rising Energy Footprint of Data Centers: A Review of Global Trends, Challenges, and Opportunities. In *2025 IEEE 7th International Conference on Sustainable Technologies For Industry 5.0 (STI)* (pp. 1-6). IEEE.

National Aeronautics and Space Administration. (2023). *NASA POWER data access viewer*. https://power.larc.nasa.gov

National Oceanic and Atmospheric Administration. (2023). *Climate and weather data services*. https://www.noaa.gov

Naug, A., Guillen, A., Kumar, V., Greenwood, S., Brewer, W., Ghorbanpour, S., . . . Sarkar, S. (2025). Lc-opt: Benchmarking reinforcement learning and agentic ai for end-to-end liquid cooling optimization in data centers. *arXiv preprint arXiv:2511.00116*.

Niemeyer, S. (2020, January). Deliberation and Ecological Democracy: From Citizen to Global System. [DOI.org]. *Journal of Environmental Policy and Planning*, *22*(1), 16–29. DOI: 10.1080/1523908X.2019.1661232

NVIDIA Corporation. (2024). *NVIDIA H100 Tensor Core GPU architecture: Technical overview*. NVIDIA Developer Documentation. https://developer.nvidia.com/blog/nvidia-hopper-architecture-in-depth/

NVIDIA. (2023). *Data center cooling solutions for AI workloads*. https://www.nvidia.com

OECD. (2015). *Stakeholder Engagement for Inclusive Water Governance*. OECD. OECD Studies on Water. *DOI.org (Crossref)*, .

OECD. (2023). *Water governance and sustainable infrastructure policies*. https://www.oecd.org

Ohadi, M. M., Dessiatoun, S. V., Choo, K., Pecht, M., & Lawler, J. V. (2012, March). A comparison analysis of air, liquid, and two-phase cooling of data centers. In *2012 28th Annual IEEE Semiconductor Thermal Measurement and Management Symposium (SEMI-THERM)* (pp. 58-63). IEEE. DOI: 10.1109/STHERM.2012.6188826

Ojadi, J. O., Owulade, O. A., Odionu, C. S., & Onukwulu, E. C. (2025). AI-driven optimization of water usage and waste management in smart cities for environmental sustainability. *Engineering and Technology Journal*, *10*(3), 4284–4306. DOI: 10.47191/etj/v10i03.36

Oró, E., Depoorter, V., Garcia, A., & Salom, J. (2015). Energy efficiency and renewable energy integration in data centres: Strategies and modelling review. *Renewable & Sustainable Energy Reviews*, *42*, 429–445. DOI: 10.1016/j.rser.2014.10.035

Patterson, D., Gonzalez, J., Le, Q., Liang, C., Munguia, L.-M., Rothchild, D., So, D., Texier, M., & Dean, J. (2021). Carbon considerations for large language model training. *arXiv preprint arXiv:2104.10350*. https://arxiv.org/abs/2104.10350

Patterson, D., Gonzalez, J., Le, Q., Liang, C., Munguia, L.-M., Rothchild, D., So, D., Texier, M., & Dean, J. (2021). *Carbon Emissions and Large Neural Network Training* (Version 3). arXiv. DOI: 10.48550/ARXIV.2104.10350

Pazienza, A., Baselli, G., Vinci, D. C., & Trussoni, M. V. (2024). A holistic approach to environmentally sustainable computing. *Innovations in Systems and Software Engineering*, *20*(3), 347–371. DOI: 10.1007/s11334-023-00548-9

Pham, A. D., Ngo, N. T., Truong, T. T. H., Huynh, N. T., & Truong, N. S. (2020). Predicting energy consumption in multiple buildings using machine learning for improving energy efficiency and sustainability. *Journal of Cleaner Production*, *260*, 121082. DOI: 10.1016/j.jclepro.2020.121082

Pimenow, S., Pimenowa, O., Prus, P., & Niklas, A. (2025). The impact of artificial intelligence on the sustainability of regional ecosystems: Current challenges and future prospects. *Sustainability (Basel)*, *17*(11), 4795. Advance online publication. DOI: 10.3390/su17114795

Pomnar, A., Rajawat, A. S., Tatkar, N. S., & Bhaladhare, P. (2023). Sustainable Power Prediction and Demand for Hyperscale Datacenters in India. Engineering Proceedings, 59(1), 124.

Prentice, C. M., Vergunst, F., Minor, K., & Berry, H. L. (2024). Education outcomes in the era of global climate change. *Nature Climate Change*, *14*(3), 214–224. DOI: 10.1038/s41558-024-01945-z

Privette, A. P., Barros, A., & Cai, X. (2026). Data centers water footprint: The need for more transparency. *AGU Advances*, *7*(2), e2025AV002140.

Provost, F., & Fawcett, T. (2013). *Data science for business*. O'Reilly Media.

Public Utilities Board. (2023). *Singapore's water story: Sustaining our water supply*. PUB, Singapore's National Water Agency. https://www.pub.gov.sg/watersupply/singaporeswaterstory

PwC. (2023). *AI and predictive analytics in resource management*. https://www.pwc.com

PwC. (2025). A new era of load growth: How data centers are becoming part of our communities and energy system.

Raihan, A., Paul, A., Rahman, M. S., Islam, S., Paul, P., & Karmakar, S. (2024). Artificial intelligence (AI) for environmental sustainability: A concise review of technology innovations in energy, transportation, biodiversity, and water management. *Journal of Technology Innovations and Energy*, *3*(2), 64–73. DOI: 10.56556/jtie.v3i2.953

Richards, C. E., Tzachor, A., Avin, S., & Fenner, R. (2023). Rewards, risks and responsible deployment of artificial intelligence in water systems. *Nature Water*, *1*(5), 422–432. DOI: 10.1038/s44221-023-00069-6

Ristic, B., Madani, K., & Makuch, Z. (2015). The water footprint of data centers. *Sustainability (Basel)*, *7*(8), 11260–11284. DOI: 10.3390/su70811260

Rogoway, M. (2023). Google's water use is soaring in the Dalles, records show, with two more data centers to come. Oregon Live.

Ruiz Carrillo, J. A., Huamaní Jordan, O., Mendoza Loor, E. G., & Espín Beltrán, C. X. (2026). Impact of artificial intelligence on the sustainable use of water resources. *Sustainability (Basel)*, *18*(8), 3864. Advance online publication. DOI: 10.3390/su18083864

Saboori, B., et al. (2006). ICT sector emissions and carbon intensity dependency. Environmental economics literature.

Sachs, J. (2024). AI-driven optimization for energy systems and data centers. Sustainability and energy systems research.

Sarkar, S., Naug, A., Guillen, A., Gundecha, V., Gutiérrez, R. L., Ghorbanpour, S., & Bash, C. (2025, April). Hierarchical multi-agent framework for carbon-efficient liquid-cooled data center clusters. *Proceedings of the AAAI Conference on Artificial Intelligence*, *39*(28), 29694–29696. DOI: 10.1609/aaai.v39i28.35370

Schneider Electric. (2023). *Data center cooling optimization strategies*. https://www.se.com

Semiconductor Industry Association. (2023). *Water usage in semiconductor manufacturing*.

Seyedzadeh, S., Rahimian, F. P., Glesk, I., & Roper, M. (2018). Machine learning for estimation of building energy consumption and performance: A review. *Visualization in Engineering*, *6*(1), 5. DOI: 10.1186/s40327-018-0064-7

Sharma, S. K., Seetharaman, A., & Maddulety, K. (2021). Framework for sustainable urban water management in context of governance, infrastructure, technology and economics. *Water Resources Management, 35*(12), 3903–3913. DOI: 10.1007/s11269-021-02916-1

Shehabi, A., Smith, S. J., Sartor, D., ... (2016). United States data center energy usage report. *Lawrence Berkeley National Laboratory*. https://eta.lbl.gov/publications/united-states-data-center-energy

Shehabi, A., Smith, S., Sartor, D., Brown, R., Herrlin, M., Koomey, J., ... & Lintner, W. (2016). United states data center energy usage report.

Shehabi, A.. (2016). *Data center energy and cooling systems analysis. Lawrence Berkeley National Laboratory*. LBNL.

Shehabi, A., Smith, S. J., Masanet, E., & Koomey, J. (2018). Data center growth in the United States: Decoupling the demand for services from electricity use. *Environmental Research Letters, 13*(12), 124030. DOI: 10.1088/1748-9326/aaec9c

Shi, W., Cao, J., Zhang, Q., Li, Y., & Xu, L. (2016). Edge computing: Vision and challenges. *IEEE Internet of Things Journal, 3*(5), 637–646. DOI: 10.1109/JIOT.2016.2579198

Shumba, N., Tshekiso, O., Li, P., Fanti, G., & Ren, S. (2025, July). A water efficiency dataset for African data centers. In *Proceedings of the ACM SIGCAS/SIGCHI Conference on Computing and Sustainable Societies* (pp. 453-460). DOI: 10.1145/3715335.3735483

Siddik, M. A. B., Shehabi, A., & Marston, L. (2021). 'The Environmental Footprint of Data Centers in the United States'. *Environmental Research Letters, 16*(6), 064017. DOI: 10.1088/1748-9326/abfba1

Siddiqui, A., Siddiqui, M., & Kulkarni, N. (2022). Artificial intelligence in water conservation: A meta-analysis study. *Prabandhan: Indian Journal of Management, 15*(3), 24–41. DOI: 10.17010/pijom/2022/v15i3/160407

Silva-Llanca, L., Ponce, C., Bermúdez, E., Martínez, D., Díaz, A. J., & Aguirre, F. (2023, September). 'Improving Energy and Water Consumption of a Data Center via Air Free-Cooling Economization: The Effect Weather on Its Performance'. *Energy Conversion and Management, 292*, 117344. DOI: 10.1016/j.enconman.2023.117344

Singh, J., Goyal, S., Kumar Kaushal, R., Kumar, N., & Singh Sehra, S. (Eds.). (2024). *Applied data science and smart systems* (1st ed.). CRC Press., DOI: 10.1201/9781003471059

Sit, M., Demiray, B. Z., Xiang, Z., Ewing, G. J., Sermet, Y., & Demir, I. (2020). A comprehensive review of deep learning applications in hydrology and water resources. *Water Science and Technology*, *82*(12), 2635–2670. DOI: 10.2166/wst.2020.369 PMID: 33341760

Sovacool, B. K., Bergman, N., Hopkins, D., Jenkins, K. E. H., Hielscher, S., Goldthau, A., & Brossmann, B. (2020, August). Imagining Sustainable Energy and Mobility Transitions: Valence, Temporality, and Radicalism in 38 Visions of a Low-Carbon Future. [DOI.org]. *Social Studies of Science*, *50*(4), 642–679. DOI: 10.1177/0306312720915283 PMID: 32375583

Springer, A., & Hammerla, N. (2025). Knowledge distillation and dataset distillation of large language models: Emerging trends, challenges, and future directions. *Artificial Intelligence Review*, *58*(4), 1–42. DOI: 10.1007/s10462-025-11423-3

Sunkara, K. C., & Konakanchi, R. (2025). Smart IoT-Based Leak Forecasting and Detection for Energy-Efficient Liquid Cooling in AI Data Centers. arXiv preprint arXiv:2512.21801.

Sunkara, K. C., & Narukulla, K. (2025). Power consumption and heat dissipation in AI data centers: A quantitative thermodynamic framework. *International Journal of Innovative Research in Science, Engineering and Technology*, *14*(3), 1–12. DOI: 10.15680/IJIRSET.2025.1403001

Sutton, R. S., & Barto, A. G. (2018). *Reinforcement learning: An introduction* (2nd ed.). MIT Press.

Synergy Research Group. (2024). *Data center market growth and analysis report*. https://www.srgresearch.com

Tang, Yu., Bai, X., Jin, C., Fu, Y., Mao, W., & Xu, X. (2022, March). 'Experimental and Numerical Study on the Thermal Environment of a Data Center'. *Building and Environment*, *212*, 108741. DOI: 10.1016/j.buildenv.2021.108741

Taylor Wiseman & Taylor. (2025). Your AI Assistant is Thirsty – Examining the Water Footprint of Data Centers.

The Green Grid. (2023). *Power usage effectiveness (PUE): A comprehensive examination*. https://www.thegreengrid.org

Tibshirani, R. (1996). Regression shrinkage and selection via the lasso. *Journal of the Royal Statistical Society. Series B, Statistical Methodology*, *58*(1), 267–288. DOI: 10.1111/j.2517-6161.1996.tb02080.x

Tranfield, D., Denyer, D., & Smart, P. (2003). Towards a methodology for developing evidence-informed management knowledge by means of systematic review. *British Journal of Management*, *14*(3), 207–222. DOI: 10.1111/1467-8551.00375

U.S. Department of Energy. (2022). *Energy efficiency in data centers: Best practices guide*. https://www.energy.gov

U.S. Energy Information Administration. (2024). *Annual energy outlook 2024 with projections to 2050*. U.S. Department of Energy. https://www.eia.gov/outlooks/aeo/

U.S. Geological Survey. (2023). *National water information system: Data from web interface*. U.S. Department of the Interior. DOI: 10.5066/F7P55KJN

UNESCO. (2023). *United Nations world water development report 2023*. https://www.unesco.org

United Nations. (2025). *Sustainable Development Goal 6: Clean Water and Sanitation*. https://sdgs.un.org/goals/goal6

United States Environmental Protection Agency (EPA). (2025). *WaterSense: Water Efficiency Management*. https://www.epa.gov/watersense

UN-Water. (2023). *Water scarcity and efficiency strategies*. https://www.unwater.org

UN-Water. (2024). *Progress on level of water stress: Mid-term status of SDG indicator 6.4.2 and acceleration needs*. United Nations Water. https://www.unwater.org/publications/progress-level-water-stress-2024-update

Uptime Institute. (2023). *Global data center survey 2023*. https://www.uptimeinstitute.com

Uptime Institute. (2023). *Global data center survey results 2023*. Uptime Institute LLC. https://uptimeinstitute.com/resources/research-and-reports

Uptime Institute. (2025). *Water Usage Effectiveness (WUE)*. https://uptimeinstitute.com/resources/glossary/water-usage-effectiveness-wue

Van Heddeghem, W., Lambert, S., Lannoo, B., Colle, D., Pickavet, M., & Demeester, P. (2014, September). 'Trends in Worldwide ICT Electricity Consumption from 2007 to 2012'. *Computer Communications*, *50*, 64–76. DOI: 10.1016/j.comcom.2014.02.008

Van Zetten, J. D., Cholette, M. E., & Bamdad, K. (2025, April). 'Data Centre Building Energy Optimization with Differential Temperature Control Utilizing I.T. Power Measurement'. *Journal of Building Engineering*, *100*, 111722. DOI: 10.1016/j.jobe.2024.111722

Vanham, D., Comero, S., Gawlik, B. M., & Bidoglio, G. (2018). 'The Water Footprint of Different Diets within European Sub-National Geographical Entities'. *Nature Sustainability, 1*(9), 518–525. DOI: 10.1038/s41893-018-0133-x

Vanitha, S., Radhika, K., & Boopathi, S. (2023). Artificial intelligence techniques in water purification and utilization. In Vasant, P., Litvinchev, I., & Rodríguez-Aguilar, R. (Eds.), *Human agro-energy optimization for business and industry* (pp. 202–218). IGI Global., DOI: 10.4018/978-1-6684-4118-3.ch010

Wang, E. H., Yan, W., & Huang, R. (2025). Thermal analysis of 3D GPU-memory architectures with boron nitride interposer. *MRS Communications, 15*, 1–8. DOI: 10.1557/s43579-025-00913-8

Wang, H., Yuan, X., Zhang, K., Lang, X., Chen, H., Yu, H., & Li, S. (2024). Performance evaluation and optimization of data center servers using single-phase immersion cooling. *International Journal of Heat and Mass Transfer, 221*, 125057. DOI: 10.1016/j.ijheatmasstransfer.2023.125057

Wang, Z., Zeng, C., Zhu, Z., Li, Y., Ma, X., & Zhao, X. (2025). 'Time-Series Machine Learning for Predictive Optimisation of a Highly Efficient Evaporative Cooling System'. *Building Services Engineering Research and Technology, 46*(2), 189–202. DOI: 10.1177/01436244251315047

Wang, Z., Zhang, Z., Lee, C.-Y., Zhang, H., Sun, R., Ren, X., Su, G., Perot, V., Dy, J., & Karypis, G. (2024). A comprehensive survey of continual learning: Theory, method, and application. *IEEE Transactions on Pattern Analysis and Machine Intelligence, 46*(8), 5362–5383. DOI: 10.1109/TPAMI.2024.3367329 PMID: 38407999

World Bank. (2023). *Water in circular economy and sustainable infrastructure.* https://www.worldbank.org

World Economic Forum. (2024). *Harnessing AI for sustainable resource management.* https://www.weforum.org

World Economic Forum. (2025). *Sustainable Data Centres and Infrastructure.* https://www.weforum.org/reports/sustainable-data-centres

World Health Organization. (2022). *Water quality guidelines and standards.* https://www.who.int

World Meteorological Organization. (2023). *Global climate data and monitoring systems.* https://www.wmo.int

World Resources Institute (WRI). (2025). *Aqueduct Water Risk Atlas.* https://www.wri.org/aqueduct

World Resources Institute (WRI). (2025). *Water Risk and Climate Data Integration*. https://www.wri.org/aqueduct

Wu, X., Kim, J. G., Hou, A., & Wang, S. (2023). Advanced AI computing enabled by 2D material-based neuromorphic devices. *npj Unconventional Computing*, 1, 1–9. DOI: 10.1038/s44335-025-00023-7

Yadav, K. D. (2024). Expansion of AI and ML Breakthroughs in HPC with Shift to Edge Computing in Remote Environments. In Ahmad, K. A., Hamid, N. A. W. A., Jawaid, M., Khan, T., & Singh, B. (Eds.), *High Performance Computing in Biomimetics* (pp. 129–145). Springer Nature Singapore., DOI: 10.1007/978-981-97-1017-1_5

Ye, Z., Yin, S., Cao, Y., Zhang, X., Wang, H., Li, J., & Chen, Q. (2024). AI-driven optimization of agricultural water management for enhanced sustainability. *Scientific Reports*, *14*(1), 25721. Advance online publication. DOI: 10.1038/s41598-024-76915-8 PMID: 39468244

Zhang, Q., Ng, K. K. W., Kazer, C., Yan, S., Sedoc, J., & Liu, V. 'MimicNet: Fast Performance Estimates for Data Center Networks with Machine Learning'. *Proceedings of the 2021 ACM SIGCOMM 2021 Conference*, 9 August 2021, 287–304. https://doi.org/DOI: 10.1145/3452296.3472926

Zhang, W., Fan, H., & Zhao, Q. (2023). Seeing green : How does digital infrastructure affect carbon emission intensity? *Energy Economics*, *127*, 107085. DOI: 10.1016/j.eneco.2023.107085

Zhao, R., Du, Y., Yang, X., Zhou, Z., Wang, W., & Yang, X. (2023, October). 'A Critical Review on the Thermal Management of Data Center for Local Hotspot Elimination'. *Energy and Building*, *297*, 113486. DOI: 10.1016/j.enbuild.2023.113486

Zhao, T., Song, C., Yu, J., Xing, L., Xu, F., Li, W., & Wang, Z. (2025). Leveraging immersive digital twins and AI-driven decision support systems for sustainable water reserves management: A conceptual framework. *Sustainability (Basel)*, *17*(8), 3754. Advance online publication. DOI: 10.3390/su17083754

Zhu, X., Jiang, W., Liu, F., Zhang, Q., Pan, L., Chen, Q., & Jia, Z. (2020, May). Heat to power: Thermal energy harvesting and recycling for warm water-cooled datacenters. In *2020 ACM/IEEE 47th Annual International Symposium on Computer Architecture (ISCA)* (pp. 405-418). IEEE.

Zimmermann, S., Meijer, I., Tiwari, M. K., Paredes, S., Michel, B., & Poulikakos, D. (2012). Aquasar: A hot water cooled data center with direct energy reuse. *Energy*, *43*(1), 237–245. DOI: 10.1016/j.energy.2012.04.037

About the Contributors

Hitesh Mohapatra received the B.E. degree in Information Technology from Biju Patnaik University of Technology (BPUT), Odisha in 2006, and the MTech. Degree in CSE from Odisha University of Technology and Research (OUTR) University, Odisha in 2009. He received his Ph.D. in Computer Science & Engineering in 2021 from Veer Surendra Sai University of Technology (VSSUT), Burla, India. He has contributed 150+ SCI and Scopus indexed research articles. He has authored two books on Software Engineering and C Programming respectively. He has 15 years of teaching experience both in industry and academia. He has served the research community with various capacities like session chair, technical chair, keynote speaker etc. His research interests include wireless sensor network, smart city, smart grid and smart water. Currently he is working as Associate Professor at School of Computer Engineering, KIIT (Deemed to be) University, Bhubaneswar, Odisha, India.

Amiya Kumar Rath received the B.E. degree in computer from Dr. Babasaheb Ambedkar Marathwada University, Aurangabad, in 1990, the M.B.A. degree in systems management from Shivaji University in 1993, the MTech. degree in computer science from Utkal University in 2001, and the Ph.D. degree in computer science from Utkal University, in 2005, with a focus on embedded systems. He is currently a Professor with the Department of Computer Science and Engineering, Veer Surendra Sai University of Technology, Burla, India. He has contributed over 80 research level papers to many national and international journals and conferences. He has published seven books by reputed publishers. His research interests include embedded systems, ad hoc networks, sensor network, power minimization, evolutionary computation, and data mining. He has been deputed as an adviser to National Assessment and Accreditation Council (NAAC), Bangalore, India from 2019 to 2023. Currently he is working Vice Chancellor for Biju Patnaik University of Technology (BPUT), Rourkela, Odisha since 2023 up to 2028.

Fadi Al-Turjman received the Ph.D. degree in computer science from Queen's University, Canada, in 2011. He serves as the "advisor to the chairman of the board of trustees on AI and informatics" with Near East University (NEU), Nicosia, where he is the Founding Dean of the AI and Informatics Faculty. He is the Head of the Software Engineering Department and the Founding Director of the AI and Robotics Institute and the International Research Center for AI and IoT, NEU. His publication history spans over 650 SCI/E publications, in addition to numerous keynotes and plenary talks at flagship venues. He has authored and edited over 100 books on AI, cognition, security, and the deployment of wireless sensor networks in smart IoT environments, which have been published by reputable publishers, such as IEEE, Taylor & Francis, Elsevier, IET, and Springer. He is a leading authority in the areas of smart and intelligent IoT systems, wireless and mobile networks' architectures, protocols, deployments, and performance evaluation in Artificial Intelligence of Things. He has been awarded the Lifetime Golden-Award of Dr. Suat Gunsel from Near East University in 2022, in addition to several other recognitions and best research awards at key international venues. He is leading several international conferences and workshops in prominent AI and IoT societies. He currently serves as a book series editor with distinguished publishers and as the lead guest or an associate editor for several top-tier journals, including the IEEE Communications Surveys and Tutorials (IF 23.9) and Sustainable Cities and Society (Elsevier, IF 10.8).

Manjur Kolhar is an Associate Professor of Health Informatics at King Faisal University, Saudi Arabia. His research focuses on artificial intelligence, digital health, machine learning, biomedical signal processing, predictive analytics, and healthcare cybersecurity. He has published more than 75 scientific articles and book chapters and has contributed extensively to health informatics education, quality assurance, and curriculum development. His work includes AI-based cardiovascular risk prediction, ECG analytics, telemedicine security, and intelligent healthcare systems. Dr. Kolhar's research has received recognition from the Saudi National Institute of Health Research for its contribution to impactful health informatics and data-driven healthcare innovation.

***

D.Rajeswari, currently working as a professor in the department of Data Science and Business Systems, School of computing, College of Engineering and Technology, SRM Institute of Science and Technology, Kattankulathur, India. Received Ph.D. degree in 2017 from College of Engineering, Guindy, Anna University under the guidance of Dr.V. Jawahar Senthilkumar, Professor, College of

Engineering, Guindy, Anna University. Completed M.Tech., Information Technology in 2010 from PSG College of Technology, Coimbatore, Anna University. Received GATE stipend from 2008 to 2010 for completing M.Tech. Completed B.Tech., Information Technology in 2008 from Annai Mathammal sheela Engineering College, Anna University. Received International Travel Support (ITS) from SERB to attend 2023 POMS Annual Conference from 21 May to 25 May in USA. Received 40 lakhs project grant from DST

Riadh Hanachi holds a PhD in Chemistry (2019) from the University of Tunis El Manar and is completing a Research Master's in Computer Science (2024–2026). He is a Researcher at LaDVEN, CERTE, Tunisia. His expertise spans computational chemistry, molecular modeling, machine learning, and AI, focusing on predicting physicochemical and biochemical properties and developing IoT-based applications for smart connected systems. He has conducted research stays at Hochschule Bonn-Rhein-Sieg, Ruhr University Bochum, University of Ulm (Germany), and Gustave Eiffel University (France). He contributed to European projects including TECHONEY, MareNostrum, Internisa, and WATERFRUIT-PRINT. His interests include AI-driven predictive modeling, data science, and digital tools for sustainable water and energy systems. He has authored peer-reviewed articles and presented at international conferences.

J. Dhanalakshmi was born on 6 th September 1992 in Tamilnadu, India. She received the B.E. degree in Computer Science and Engineering and M.E. degree in Computer Science and Engineering from Anna University, Tamil Nadu, India. She was ranked 31st at Anna University for her Masters degree in Computer Science and Engineering. She was a software developer at Fermtronics Technology Solutions until 2018. She completed her Ph.D. degree at the School of Computer Science and Engineering in B.S. Abdur Rahman Crescent Institute of Science and Technology, located in Chennai, India. She completed her postdoctoral fellowship at the National Kaohsiung University of Science and Technology, which is located in Kaohsiung, Taiwan. She is currently working as an Assistant Professor in Data Science and Business Systems, SRM Institute of Science and Technology, Kattankulathur, Chennai, India. She published more than 20 research papers in International, National Journals including Conferences and Book Chapter in SCI and Scopus Indexed. Her research interests include Energy Analytics, Artificial Intelligence, Deep Learning, Data Science, Internet of Things (IOT) and Machine Learning.

M. Robinson Joel received the Doctorate Ph.D degree in Computer and Information Technology from the Manonmaniam Sundaranar University, Tirunelveli in March 2019. He completed his M.Tech and MCA from the same university year

2008 and 2002 respectively. He is currently working as an Associate Professor in the Department of Computer Science and Engineering at KCG College of Technology, Chennai. He has 16 years of teaching experience.

M.G.Sumithra is currently serving as Professor in the department Electronics and communication engineering, Sri Krishna College of Technology, Coimbatore, India. She has completed her B.E., M.E., Ph.D. from Madras University and Anna University respectively. She is a university Rank Holder during her PG Studies. He has 30+ years of academic and research experience. She has guided many scholars towards the award of PhD under Anna University. She has authored many national and international research publications and authored many books. She is a renowned academic leader and a strategic planner.

Rampelli Manojkumar received his Ph.D. in Power Engineering from the Indian Institute of Technology Guwahati in 2023. He obtained his Master of Technology in Power and Energy Systems from the National Institute of Technology Karnataka, Surathkal, India, in 2015. He is currently an Assistant Professor in the Department of Electrical and Electronics Engineering at BVRIT Hyderabad College of Engineering for Women. He is currently the Editorial Board Member of Scientific Reports, Early Career Editor for the International Journal of Ambient Energy and the Guest Editor for PLOS ONE. Additionally, he is an active reviewer for around 30 SCI journals. He has been a speaker at the IEEE Smart Grid webinar series and has contributed to multiple funded research projects on smart transformer-based microgrid systems and distributed generation management. He has published 10 Q1 journal papers, including articles in IEEE Transactions on Power Systems, IEEE Transactions on Industrial Informatics, and Sustainable Cities and Society. His research expertise spans energy management, hybrid AC/DC distribution systems, energy storage, and power electronics applications in modern distribution systems.

J.Shanthini, currently serving as associate Professor in SRM institute of Science and Technology, Tamilnadu, India. She has completed B.E., M.E., Ph.D. from Bharathiyar University and Anna University respectively. He has guided 10 research scholars towards the award of Ph.D. degree under Anna University. She is a part of Doctoral committee members for many scholars guiding them in their research. She has rich set of academic and research experience. She is members of professional bodies. She is reviewer of many journals and conference.

Manisha Singh, Assistant Professor at VIT Bhopal University's Business School, holds a Doctorate in Management from Maulana Azad National Institute of Technology, Bhopal, specializing in Human Resource Management. With 18 years of combined experience in industry and academia, her research focuses on the impact of Employer Brand on Employee Retention. Dr. Singh has numerous publications in esteemed international and national journals, along with contributions to book chapters with reputable publishers & in SCOPUS. Her research areas span Organizational Behaviour, Human Resource Management, and Finance. She has presented research papers in various international and national conferences, seminars, and has organized several short-term courses. Additionally, Dr. Singh serves as an external paper setter in multiple universities and holds design patents with grants from the Department for Promotion of Industry and Internal Trade, Government of India. She is actively involved as a reviewer of board of several journals.

Sonali Srivastava is presently working as an Assistant Professor (Senior Grade) at Jaypee Business School, JIIT, Noi1da. She is a doctorate in Finance from Dayalbagh Educational Institute, Agra. She has an academic experience of more than 10 years with several reputed institutions i.e. Symbiosis University, VIT Bhopal University, MNNIT Allahabad, IIM Ahmedabad, Ministry of Finance and Amity University. She has published a number of research papers in peer reviewed international and national journals indexed in Scopus and WOS. She has also presented research papers in international and national conferences, seminars and conventions. Dr Sonali has organized a number of Conferences and Short-term courses. She has authored books with reputed publishers. She has also published Indian and UK design patents with grants. She is a member of various professional bodies and Member of the Reviewer Board of Various Journals.

Krishna Chaitanya Sunkara is an engineering leader specializing in AI-driven data center design, cloud security, and high-scale distributed systems. At Oracle, he spearheads automation and modeling of AI data centers, optimizing networking, HPC, and cloud security frameworks. With over 20 years of experience in large-scale infrastructure, he focuses on advancing research in data center efficiency, AI workload optimization, and next-gen cloud architectures.

Index

A

AI Data Centers 4, 11, 17, 18, 19, 33, 36, 43, 44, 45, 46, 47, 48, 49, 55, 56, 59, 65, 66, 101, 105, 114, 117, 119, 120, 147, 148, 149, 152, 156, 163, 175, 219, 231, 236, 247, 273, 276
AI workloads 1, 3, 4, 6, 7, 8, 9, 10, 11, 12, 16, 17, 19, 20, 27, 36, 37, 39, 40, 45, 61, 66, 67, 76, 78, 79, 86, 90, 108, 109, 114, 116, 142, 148, 151, 156, 164, 179, 249

C

carbon intensity 49, 50, 51, 53, 62, 63, 65, 79, 83, 86, 88, 90, 101, 105, 217
Cooling 1, 2, 3, 4, 5, 6, 7, 8, 9, 10, 11, 12, 13, 14, 15, 16, 17, 18, 19, 21, 23, 24, 25, 26, 27, 28, 29, 30, 31, 32, 33, 34, 35, 36, 37, 39, 40, 41, 42, 43, 44, 45, 46, 47, 48, 49, 50, 51, 52, 53, 54, 55, 56, 58, 59, 60, 61, 62, 63, 64, 65, 67, 68, 69, 70, 71, 72, 73, 74, 75, 76, 77, 78, 79, 80, 81, 82, 83, 84, 85, 86, 87, 88, 89, 90, 91, 92, 93, 94, 95, 96, 97, 98, 99, 101, 103, 105, 108, 109, 110, 111, 113, 114, 115, 116, 117, 118, 119, 120, 121, 122, 123, 124, 125, 126, 127, 129, 130, 131, 132, 133, 134, 135, 136, 137, 139, 140, 142, 145, 146, 147, 148, 149, 150, 151, 152, 153, 154, 155, 156, 157, 158, 159, 162, 163, 164, 165, 166, 167, 168, 169, 170, 171, 172, 173, 174, 175, 176, 177, 178, 179, 181, 182, 183, 184, 185, 187, 188, 190, 191, 192, 193, 194, 195, 196, 201, 205, 206, 207, 208, 212, 214, 216, 217, 218, 220, 221, 223, 225, 226, 227, 228, 229, 230, 231, 232, 233, 234, 237, 238, 239, 240, 242, 243, 244, 246, 247, 248, 249, 250, 252, 253, 254, 255, 257, 258, 260, 261, 263, 264, 265, 266, 269, 272, 273, 274, 275, 276, 278, 279, 280
Cooling Mechanisms 2, 11, 33
Cooling Technologies 1, 4, 5, 6, 9, 10, 11, 12, 13, 14, 16, 21, 23, 24, 29, 33, 36, 40, 43, 45, 46, 47, 56, 61, 63, 68, 78, 79, 81, 82, 83, 85, 87, 90, 94, 95, 108, 109, 111, 117, 120, 136, 181, 190, 193, 194, 205, 214, 223, 225, 231, 232, 234, 237

D

Data Center Cooling 11, 13, 14, 15, 16, 27, 29, 35, 40, 41, 42, 43, 47, 59, 60, 68, 78, 90, 113, 115, 120, 122, 125, 127, 129, 130, 135, 136, 139, 142, 151, 178, 214, 216, 226, 228, 234, 249, 275, 280
data centers 1, 2, 3, 4, 5, 6, 7, 8, 9, 10, 11, 12, 14, 15, 16, 17, 18, 19, 20, 21, 23, 25, 27, 28, 29, 30, 31, 32, 33, 34, 35, 36, 37, 38, 43, 44, 45, 46, 47, 48, 49, 52, 53, 55, 56, 58, 59, 61, 62, 63, 64, 65, 66, 68, 69, 75, 78, 83, 84, 91, 92, 93, 94, 101, 104, 105, 106, 111, 113, 114, 115, 116, 117, 118, 119, 120, 121, 122, 123, 124, 125, 126, 127, 135, 136, 137, 138, 139, 141, 142, 143, 145, 146, 147, 148, 149, 150, 152, 156, 157, 163, 175, 176, 177, 178, 179, 181, 182, 183, 186, 187, 188, 190, 191, 192, 193, 194, 195, 196, 199, 200, 201, 203, 204, 205, 206, 207, 209, 210, 212, 214, 215, 216, 217, 219, 220, 221, 223, 224, 226, 229, 230, 231, 233, 236, 237, 238, 239, 240, 241, 242, 243, 246, 247, 250, 265, 273, 275, 276

E

energy 1, 2, 3, 4, 5, 6, 7, 8, 9, 10, 11, 12, 13, 14, 15, 16, 18, 19, 20, 21, 22, 23, 24, 25, 26, 27, 28, 29, 30, 31, 32, 34, 35, 36, 37, 39, 40, 42, 44, 45, 46, 48,

49, 50, 51, 52, 53, 55, 56, 58, 59, 60, 61, 62, 63, 64, 65, 66, 67, 68, 69, 71, 72, 73, 74, 76, 77, 78, 79, 80, 82, 83, 84, 85, 86, 87, 88, 89, 90, 91, 92, 94, 95, 96, 97, 98, 100, 101, 103, 104, 107, 108, 109, 110, 111, 114, 116, 117, 118, 119, 123, 125, 126, 127, 128, 131, 132, 133, 134, 135, 136, 137, 139, 140, 141, 142, 143, 147, 148, 149, 150, 156, 161, 162, 169, 174, 176, 177, 178, 179, 183, 184, 186, 187, 188, 193, 194, 196, 199, 200, 201, 203, 204, 205, 206, 207, 208, 209, 210, 212, 214, 215, 216, 217, 218, 220, 221, 223, 224, 225, 226, 227, 228, 229, 230, 231, 232, 233, 234, 235, 238, 239, 240, 242, 243, 244, 245, 246, 247, 248, 249, 250, 252, 254, 256, 258, 259, 261, 262, 264, 265, 269, 270, 272, 273, 274, 275, 276, 277, 278, 279, 280
energy–water nexus 61, 79, 90, 104
Environmental Impact Assessment 239
Evaporative Cooling 2, 3, 6, 7, 8, 9, 10, 11, 12, 13, 14, 15, 23, 26, 27, 28, 29, 32, 35, 46, 53, 55, 56, 63, 74, 75, 78, 79, 88, 89, 91, 92, 116, 119, 146, 148, 158, 159, 163, 169, 193, 196, 201, 205, 218, 220, 221, 223, 226, 228, 229, 231, 233, 242, 260, 266, 278

F

Freshwater Consumption 33, 35, 46, 54, 55, 266

G

Green Data Centers 137

H

Heat Recovery 4, 13, 23, 25, 27, 42, 68, 82, 88, 92, 97, 105, 109, 110, 178, 248, 250, 264, 265, 269, 270, 271, 276, 279
Hyperscale Infrastructure 249

L

Liquid Cooling 6, 9, 12, 13, 14, 15, 23, 24, 26, 27, 29, 30, 32, 33, 35, 40, 41, 42, 44, 46, 47, 49, 53, 55, 58, 59, 65, 67, 68, 76, 77, 78, 81, 82, 83, 84, 88, 89, 96, 105, 109, 110, 116, 137, 193, 194, 196, 201, 205, 208, 232, 247, 248, 250, 254, 255, 265, 266, 272, 273, 275, 276, 278
Liquid Cooling Systems 6, 13, 14, 24, 116, 194, 205, 232, 254, 276

M

Machine Learning 9, 30, 31, 59, 99, 113, 114, 115, 121, 122, 124, 126, 127, 137, 138, 139, 140, 142, 145, 147, 151, 153, 156, 163, 173, 175, 177, 179, 212, 214, 218, 221, 223, 227, 233, 238

P

Power Usage Effectiveness 2, 21, 32, 35, 59, 67, 78, 79, 104, 125, 137, 143, 170, 187, 216, 223, 226, 231, 257, 280
Predictive Analytics 113, 116, 120, 121, 122, 123, 124, 125, 135, 136, 137, 138, 139, 140, 142, 145, 147, 150, 151, 152, 154, 155, 156, 157, 158, 159, 164, 169, 170, 171, 172, 175, 176, 179, 220, 233, 238

R

Rack Cooling 257
Resource Management 3, 4, 113, 121, 123, 125, 126, 127, 137, 139, 142, 143, 170, 183, 204, 212

S

sustainability 1, 2, 3, 4, 5, 7, 8, 9, 11, 14, 16, 17, 19, 20, 21, 22, 25, 26, 27, 28, 29, 30, 31, 32, 35, 36, 37, 40, 42, 44, 45, 52, 53, 55, 56, 58, 61, 62, 63, 68, 78, 80, 86, 91, 103, 107, 109, 110,

116, 119, 120, 121, 124, 135, 136, 141, 146, 147, 148, 155, 156, 157, 160, 164, 170, 171, 172, 174, 175, 183, 190, 193, 197, 203, 204, 206, 212, 213, 214, 215, 217, 218, 219, 220, 221, 222, 223, 224, 226, 227, 228, 229, 231, 233, 235, 236, 237, 239, 240, 241, 242, 244, 245, 246, 264, 280
Sustainable AI Infrastructure 5, 16, 29, 34, 44, 56, 146, 175, 215, 243
Sustainable Computing 31, 177, 244
Sustainable Digital Infrastructure 4

T

Thermal Management 2, 6, 16, 31, 32, 46, 47, 62, 64, 68, 76, 98, 115, 146, 194, 218, 231, 238, 243, 244, 272, 277, 278
Thermodynamics 30, 247, 252
Time-Series Analysis 113
Time-to-Inference-Ready 258, 280

W

Water 1, 2, 3, 4, 5, 6, 7, 8, 9, 10, 11, 12, 13, 14, 15, 16, 17, 18, 19, 20, 21, 22, 23, 24, 25, 26, 27, 28, 29, 30, 31, 32, 33, 34, 35, 36, 37, 38, 39, 40, 41, 42, 43, 44, 45, 46, 47, 48, 49, 50, 51, 52, 53, 54, 55, 56, 58, 59, 60, 61, 62, 63, 64, 68, 69, 70, 73, 74, 75, 78, 79, 80, 81, 83, 84, 85, 86, 87, 88, 89, 90, 91, 92, 93, 94, 95, 96, 100, 101, 103, 104, 105, 106, 107, 108, 109, 110, 111, 113, 116, 117, 118, 119, 120, 121, 122, 123, 124, 125, 126, 127, 128, 129, 130, 132, 134, 135, 136, 137, 138, 139, 141, 142, 143, 145, 146, 147, 148, 149, 150, 151, 152, 153, 154, 155, 156, 157, 158, 159, 160, 161, 162, 163, 164, 165, 166, 167, 168, 169, 170, 171, 172, 173, 174, 175, 176, 177, 178, 179, 181, 182, 183, 185, 186, 187, 188, 190, 191, 192, 193, 196, 197, 199, 200, 201, 203, 204, 205, 206, 207, 208, 209, 210, 212, 213, 214, 215, 216, 217, 218, 219, 220, 221, 222, 223, 224, 225, 226, 227, 228, 229, 230, 231, 232, 233, 234, 235, 236, 237, 238, 239, 240, 241, 242, 243, 244, 245, 246, 247, 248, 249, 250, 254, 255, 256, 257, 258, 259, 260, 261, 262, 263, 264, 265, 266, 267, 268, 269, 270, 271, 272, 273, 274, 275, 276, 277, 278, 279, 280
Water-Carbon Trade-off 53, 60
Water-Energy Nexus 135, 220, 224, 225, 226, 246
Water–Energy Nexus 1, 2, 3, 5, 6, 32, 135
Water Footprint 2, 3, 4, 5, 8, 12, 19, 20, 22, 23, 27, 29, 31, 32, 33, 38, 39, 44, 45, 47, 55, 56, 59, 79, 80, 81, 85, 89, 90, 100, 105, 108, 117, 145, 186, 187, 190, 205, 206, 209, 215, 217, 218, 219, 220, 221, 222, 223, 225, 226, 228, 230, 231, 233, 234, 235, 237, 238, 241, 243, 246, 265, 276
Water Management 22, 23, 24, 28, 31, 37, 113, 118, 120, 121, 123, 124, 128, 129, 130, 137, 138, 139, 145, 147, 150, 151, 152, 153, 165, 171, 172, 173, 175, 207, 219, 236, 237, 241, 242, 243, 244, 245
Water Resource Management 113, 121, 123, 125, 127, 139, 204
Water Sustainability 8, 22, 27, 36, 40, 56, 155, 170, 175, 206, 219, 236, 239, 246
Water Usage Effectiveness 2, 4, 7, 20, 21, 32, 35, 60, 63, 79, 119, 137, 141, 155, 156, 170, 171, 178, 179, 187, 188, 226, 231, 237, 262, 276, 280
Water Usage Effectiveness (WUE) 63, 79

www.ingramcontent.com/pod-product-compliance
Lightning Source LLC
LaVergne TN
LVHW080900300926
849162LV00003B/489